THE HISTORY OF THE FARADAY SOCIETY

THE HISTORY OF THE FARADAY SOCIETY

PART I

1903–1945

LESLIE SUTTON

PART II

1945–1971

MANSEL DAVIES

The Faraday Division of The Royal Society of Chemistry

ISBN 0-85404-863-4

A catalogue record for this book is available from the British Library.

© The Royal Society of Chemistry 1996

All rights reserved.

Apart from any fair dealing for the purposes of research or private study, or criticism or review as permitted under the terms of the UK Copyright, Designs and Patents Act, 1988, this publication may not be reproduced, stored or transmitted, in any form or by any means, without the prior permission in writing of The Royal Society of Chemistry, or in the case of reprographic reproduction only in accordance with the terms of the licences issued by the Copyright Licensing Agency in the UK, or in accordance with the terms of the licences issued by the appropriate Reproduction Rights Organization outside the UK. Enquiries concerning reproduction outside the terms stated here should be sent to The Royal Society of Chemistry at the address printed on this page.

Published by The Royal Society of Chemistry,
Thomas Graham House, Science Park, Milton Road,
Cambridge CB4 4WF, UK

Typeset by Vision Typesetting, Manchester
Printed in the UK by Athenaeum Press, Gateshead

Contents

Foreword ... ix
Preface ... xi

PART I *1903–1945* ... 1
LESLIE SUTTON

Chapter 1 *The Early Years, 1902–1920* ... 3
1.1 The First Five Years ... 3
1.2 People, 1907–1918 ... 25
1.3 General Running ... 31
1.4 Membership and Finance ... 37
1.5 Domestic and International Cooperation ... 42
1.6 War-oriented Research, 1915–1918 ... 47
1.7 More about People ... 51

Chapter 2 *The Inter-war Years and the Second World War: Council and the General Disscussions* ... 54
2.1 1918–1922 ... 54
2.2 1922–1924 ... 63
2.3 1924–1926 ... 66
2.4 1926–1928 ... 71
2.5 1928–1930 ... 77
2.6 1930–1932 ... 82
2.7 1932–1934 ... 93
2.8 1934–1936 ... 102
2.9 1936–1938 ... 111
2.10 1938–1940 ... 121

2.11	1940–1945	127
2.12	Retrospect	128

Chapter 3	Publication, Finance and General Administration 1918–1945	145
3.1	The General State of Affairs in 1918	145
3.2	The Finance Committee and the Publications Committee	146
3.3	The Immediate Causes of the Post-war Crisis	149
3.4	Attempts to Increase Income	150
3.5	Economies, Minor Savings	155
3.6	The Exchange with the American Electrochemical Society	157
3.7	Controlling the Cost of *Transactions*, 1915–1924	159
3.8	Controlling the Cost of *Transactions*, 1925–1933	162
3.9	Renewed Efforts to Control Costs, and Eventual Success, 1934–1939	166
3.10	The General Financial Picture	170
3.11	Membership and Changes in the Character of the Society	172
3.12	A Pot-pourri of Administration	174
	(a) Income Tax	174
	(b) Britain and the Gold Standard	175
	(c) The Ordinary Meetings	176
3.13	Accommodation	178
3.14	External Activities, especially Cooperation with other Scientific Bodies:	183
	(a) General	183
	(b) The Journal of Physical Chemistry	188
	(c) Cooperation with the Chemical Society	192
	(d) The Society and Politics	194
3.15	*Transactions* and Change	201
3.16	People	205
	(a) The Spiers Memorial Lectures	206
	(b) Sir Robert Hadfield, F.R.S.	207
	(c) Sir Robert Mond, F.R.S.	208

(d) Lieutenant-Colonel J. J. Bourke, I.M.S., C.I.E.	208
(e) Michael Faraday	210
(f) Honorary Membership	211
(g) The Twenty-fifth Anniversary	212
(h) Dr. Henry Borns	213
(i) Dr. Roland Edgar Slade	214
(j) George Stanley Withers Marlow	216
(k) The Assistant Secretaries	217

PART II 1945–1971
MANSEL DAVIES

		221
1	Accommodation	223
2	Early Post-war Years	226
	The Loss of Marlow	231
	The Appointment of Tompkins	234
3	Publications and Finance	236
	The Publications Committee 1947–1951	241
	The Financial Situation 1951–1959	243
	A New Treasurer	247
	Publication and Finances after 1960	248
4	The Colloid Committee, the Colloid and Biophysics Committee and the Biophysical Society	253
	Developments after 1954	258
5	The Legal Status of the Faraday Society	262
6	The Fiftieth Year Jubilee	266
7	The Faraday Society General Discussions	272
	General Character of the Discussions	272
	Consideration of the Discussions in Council	274
	Informal Discussions and the Formation of a Standing Committee on Conferences	276
	Listing the Discussions and the Groups of their Topics	281
	The Record of the General Discussions	284
	The Symposia	322
	The Assessment of the General Discussions	324

8	The Council of the Society	330
	Biographical Notes on Members of the Faraday Society Council 1945–1954	337
	1955–1964	340
	1965–1971	342
	Comments on the Presidents and Membership of the Faraday Society Council	344
9	Honorary Life Membership	346
10	The Bourke Lectureship, the Marlow Medal and the Spiers Memorial Lectureship	350
11	The Chemical Council	356
12	The Annual General Meetings	358
13	Miscellaneous Items of Faraday Society Council Business	360
14	Amalgamation	365
	The Background Situation	365
	The Initial Stages and the Proposals Accepted	367
	The Structure of the New Society and the Changes Involved	374
	The Terminal State of the Society	375
	Comments on the Amalgamation	377
	The Consequences of the Amalgamation	380
	Envoi	384
Appendices		385
Index of Names		400

Foreword

THE FARADAY SOCIETY was a child of its time, founded at the turn of the 20th century by a group of imaginative and enthusiastic 'scientific inventors'. Their early aim was for a Society that would be a bridge between science and technology, and its main interests were initially closer to 'electro-metallurgy' and 'electro-chemistry' than to the broad span of physical chemistry and the science lying at the 'interface between chemistry, physics and biology'. The Society lived an extraordinarily vigorous and distinguished life of three score years and ten during which it evolved into a unique scientific forum for the presentation, discussion and dissemination of new scientific knowledge. It achieved this, despite chronically straitened circumstances, by sustaining the imperatives of scientific quality and imagination; through the dedication and humanity of its servants; and through the unique vehicle of its General Discussions which provided an unparalleled record of free scientific debate – and which still do. The most important traditions of the Society continue to be nurtured in its 'afterlife' within The Royal Society of Chemistry.

The History of the Faraday Society takes the reader on a voyage of rediscovery. The log book charts the very human story of its passage from the early pioneering days; through to the intense, scientifically creative period of the 1930s; its survival during the catastrophe of the 1939–45 war; and its arrival as the leading European physical chemistry society in the years that followed. The Society inspired great affection among those it touched, most commonly through the medium of its General Discussions. The origins of this affection were best expressed by its legendary Assistant Secretary, Miss Beatrice Kornitzer, who said:

'If I were to write a History of the Faraday Society . . . I would say that what impressed me most was that when any major decision had to be made Council almost unanimously considered the scientific reputation and integrity of the Society before its financial profit. It seemed to me that they had a right sense of values and I was proud to work for them.

Two other things impressed me: the warm, informal atmosphere of the General Discussions which encouraged scientists, whether members or not, to meet together and discuss their problems with a great lack of regimentation and protocol . . . (and) the truly international flavour of the Society and its meetings.'

The History has been written by two authors who shared these feelings, but who did not allow them to cloud their objectivity. They wrote from the informed perspective of scientists who were important participants in the evolution of the Society, were personally involved with Faraday matters, and who had great insight into the factors which affected the policies, decisions, and affairs of the Society and the personalities involved. Sadly neither lived to see the publication of the History on which they had worked so hard. Leslie Sutton died in 1992 and Mansel Davies lived only for a few weeks after completing the manuscript of Part II. If you seek their memorial, read their History.

J.P. SIMONS
Oxford
June 1996

Preface

A HISTORY OF the Faraday Society? What is the point of that? What possible significance can it have?

Well, firstly, one should be clear what the Faraday Society was. For seventy years it was the only independent organized body in the English-speaking world devoted to the advancement of physical chemistry, that sizeable area in science where the two old-established continents of physics and chemistry overlap. The Society was principally concerned with the physical aspects of molecular science and, since its disappearance in 1972, no other society has taken its place.

Its significance? In 1946, when war-torn Europe was licking its wounds and actively considering how to re-establish the normalities of life, representative continental mainland scientists approached the Faraday Society with the suggestion that it should accept the responsibility for organising the overall need of integrated chemical journals for Western Europe. Of all European chemical societies it stood in the highest esteem.

The extraordinary basis for this appreciation is seen when it is revealed that up to 1946 the Society, in its executive function, consisted of one part-time qualified professional and one full-time secretary-typist. An analogy comes to mind: the sub-continent of India was efficiently administered by the four hundred civil servants of the India Office. The English genius for empirical methodology can claim some remarkable achievements.

Within the science community, the Faraday Society was recognized as being responsible for a high quality, but in no way extraordinary, research journal, the *Transactions of the Faraday Society*. However, for more than fifty years it organized two Faraday Discussions per annum for detailed appraisal of developments in topics of major current research interest: these international gatherings were generally acclaimed as unsurpassed in their quality and effectiveness: an early one was attended by seven actual or future Nobel prizewinners.

Preface

The reader will have seen that this volume consists of two parts, each written independently by two authors. Dr. L. E. Sutton (L.E.S.) prepared the account for the years 1903–45: Dr. Mansel Davies (M.M.D.) that for 1945–78. The way in which this came about is itself part of the Society's history. In what follows it is explained and the character of the two parts is clarified.

The Faraday Society ceased to exist when it merged with the (London) Chemical Society and others in January 1972, to become the Faraday Division in a much larger whole. At the first meeting of the Faraday Council under the new dispensation, the President (Professor J. W. Linnett) raised the question whether a history of the Society should be prepared. In M.M.D.'s memory this was immediately accepted as desirable, a response not conforming to the minute of the item (7th March 1972). So clear was the intention that Linnett quickly suggested (giving the impression it had already been considered—a characteristic feature of Faraday Council proceedings) that Dr. Leslie Sutton of Oxford be approached to undertake the task. It is a fact that M.M.D. offered the comment that in Professor R. P. Bell the Society had a Past President who had been intimately involved at many levels with Faraday affairs and he should also be considered. Another Council member mentioned the name of Dr. E. J. Bowen of Oxford. These details are minuted. However, the chair took or accepted the view that Dr. Sutton be asked if he would write the history.

That was in March 1972. When Dr. Sutton died in November 1992, it transpired that he had written three chapters, bringing the story up to 1945. M.M.D. had several times enquired of the Faraday Secretary how the history was progressing. Never did he ask L.E.S. as, sadly, he knew of the considerable family trials his friend was experiencing. That the two authors were close friends is documented within the first three chapters. There L.E.S. mentions more than a hundred names of those involved in Faraday Society affairs. Of only two does he use the familiar first name, and the first of those is 'Mansel'. (The other is 'Bea' Kornitzer.)

These circumstances led to M.M.D. being asked in May 1993 to complete the history. The invitation was accepted on the anticipation that senior colleagues could be called on for information and help. That has been generously given. The appearance of this History confutes a suggestion made in the 1991 volume, *The Royal Society of Chemistry: The First 150 Years*, p. 36.

Now it must be explained that Part I is not exactly as L.E.S. completed it. He had offered much detail on items which were only marginally relevant. For instance, in Chapter Two he presented biographical accounts of each member of the Faraday Council of which a few have been curtailed. One of these ran to 180 words for an elected member who never attended a single meeting before resigning; another of 1150 words on the Treasurer R. E. Slade, was mostly concerned with details of activities within Messrs. I.C.I. Ltd. taken from Volume 2 of its history by W. J. Reader. About half-a-dozen such items have been rewritten more briefly. Like at least ninety percent of L.E.S.'s accounts, a 500-plus word biography of James Kendall remains unaltered.

There are other changes. Several L.E.S. paragraphs ran to about a thousand words: they have, without alteration of content, been broken up. A few small corrections have been made: where new (brief) material has been inserted, it is identified as being by M.M.D.

To indulge in such, even minor changes, is both a highly sensitive operation and, perhaps, not permissible. They were reckoned necessary to promote acceptable significance throughout the published account. The whole of the original and all such changes have been independently assessed: Lord Dainton has been so good as to check and then to approve of the emendations.

L.E.S. fortunately made a pattern of his account clear in its opening paragraphs. In Part II a different approach was taken. The total activities over 1945–71 have been reported under some fifteen headings, in each of which, where appropriate, the details come chronologically. A great proportion of the facts came from the minutes of the Faraday Society Council, a total of over 600 close-filled typed foolscap sheets. Less complete sets of minutes for the Finance and Publication Committees were found. Unless otherwise stated the frequently inserted dates throughout Part II are those of the Council meeting for which the item appears in the minutes.

From 1945 to 1971 there were a total of 55 Faraday Society General

Discussions. In their published report volumes they generated 14670 text pages of scientific material. The Discussions (and later the Symposia), their organization, location, participants, and content have each been briefly presented. The Discussion pattern and again the significance of the amalgamation carry comments by M.M.D. which cannot be other than one person's evaluations.

A sad circumstance relevant in the writing of Part II must be recorded. In July 1993 M.M.D. wrote to Emeritus Professor F. C. Tompkins, the third and last Secretary, asking for his guidance and comments on Faraday Society affairs. It transpired that illness precluded his response to this request. This has been a major loss and the interest of other senior past officers of the Society is accordingly all the more appreciated.

Many colleagues have given M.M.D. the benefit of their observations on Faraday Society affairs. A number are quoted in the text and identified there. To all these grateful thanks are extended. From Professor David King of Cambridge, who had been a research student of Tompkins, came valued comments. On the circumstances attending the amalgamation a significant letter came from Mr. J. Ruck-Keene. Three further colleagues deserve special thanks. Professor J. S. Rowlinson of Oxford had several times helped L.E.S. with comments but now, even more importantly, he arranged and effected the transfer of L.E.S.'s files on the Faraday Society and his copies of the Discussion volumes from Oxford to Criccieth. Lord Dainton has contributed more generously than many others comments on Faraday Society affairs, and also given his time to assessing Part I and the changes proposed in it: he has read Part II and offered corrections and comments. Mention must also be made of Dr. David Young's interest: he wrote many detailed insights but as most of them related to the years after 1971, that is, to the Faraday Division, they have not been used here. Thanks are also tendered to Professor J. P. Simons, Faraday Division President in 1993–94, to Mrs. Angela Fish, Secretary of the Faraday Division,* and to Dr. John Gibson, Secretary, Scientific

---

* Until her illness took her from us in August 1995.

Affairs, in the Royal Society of Chemistry, for their help in handling the material of this History.

One final comment. It has been a somewhat eerie experience writing for several months of close friends and colleagues so many of whom are no longer with us.†

Ave atque vale egregii amici.

MANSEL DAVIES
Criccieth
Caernarfonshire
October 1994

Acknowledgements

Thanks are due to Nicola Best, Library and Information Centre of The Royal Society of Chemistry, for archive and picture research and to Janet Williams, a former Staff Editor of the *Transactions* and *Discussions*, for further research and for editing the text. Unless stated otherwise, the copyright of all photographs resides with The Library and Information Centre of The Royal Society of Chemistry.

---

† Alas this is now true of both Professor Davies and Professor Tompkins.

PART I

1903–1945

LESLIE SUTTON

CHAPTER ONE

The Early Years, 1902–1920

1.1. THE FIRST FIVE YEARS

MANY OF THE details of the founding of the Faraday Society are now lost. None of the founding members is still living. But enough early documents and personal statements are available to make possible a reasonably clear reconstruction. The Society was founded by a group of friends who met in 1902 'in a back room which opened on to somebody's roof, which was cool in summer' (Sir James Swinburne, Bt., *Trans. Faraday Soc.*, 1953, **49**, 471).

There is a clear but rather bald statement in the first Proceedings of the Society (*Proc. Faraday Soc.*, 1905, **1**, 171):

'The idea of forming an English Society of Electrochemists on the lines of that existing in Germany, and of the one then about to be formed in America, first took practical shape in the mind of Mr. Sherard Cowper-Coles, and it was in great part owing to his initiative and enthusiasm that the subsequent formation of the Society was due. An informal meeting to discuss the subject was held in his office, at the invitation of Mr. Cowper-Coles, on March 4, 1902, Mr. J. Swinburne presiding on that occasion. As a result of that meeting a provisional Committee was appointed to consider the desirability of creating a separate Electrochemical Society, and to report on the best method of procedure should it be decided to adopt that course. The Committee consisted of the following gentlemen: Mr. J. Swinburne (chairman), Mr. W. R. Cooper, Mr. Sherard Cowper-Coles, Dr. F. G. Donnan, Dr. F. Mollwo Perkin, and Mr. H. V. Simpson, who at first acted as honorary secretary, but who afterwards resigned, his place being taken by Mr. F. S. Spiers.'

It will prove important for the understanding of later events to know more about these persons – who they were, what kind of men they were, and what their aims were.

One of the friends was Dr. J. W. (later Sir Joseph) Swan, F.R.S.[1] He was one of those magnificent Victorian figures who, despite having had very little formal education – he left school at 12 – managed to educate himself effectively and make an outstanding career in technology. He had a great thirst for knowledge and understanding combined with great vigour. He had the equivalent of green fingers. He invented a carbon filament lamp in 1860, 20 years before Edison did, and years also before it was of practical use because the magnetodynamic generation of electricity on a large scale had still to come. It was he who discovered that the sensitivity of a silver bromide–gelatine emulsion could be immensely increased by heating it during its preparation, and so 'inaugurated a new era in photography'. He developed the so-called carbon process for making photographic prints (1864), and bromide paper (1879). What is less well known is that he made an important contribution to electrochemistry by establishing the conditions necessary for 'the satisfactory electrodeposition of sound copper' (*Proc. R. Soc. London, Ser. A*, 1919–20, **96**, ix; *The Electrochemist and Metallurgist*, 1903, **3**, 335). The written accounts show that he was widely regarded with respect and affection. He had the generosity of mind sometimes found in the successful, in that he wanted the next generation to have educational opportunities that he had not had. He expressed two of his major beliefs in his Presidential address to the Society of Chemical Industry in 1901: 'A scientific training of university standard for our manufacturers and our technical chiefs is an absolute necessity' and 'One of the most pressing requirements of the moment . . . is *adequate endowment and encouragement of research.*' (*The Electrochemist and Metallurgist*, 1901, **1**, 223). Swan was 74 in 1902 and had been a Fellow of the Royal Society since 1894.

Another of the friends was James Swinburne, who later succeeded to the baronetcy. He was a polymath. As an electrical engineer, he made substantial contributions to dynamo design and alternating current

---

[1] From a letter written 18th February 1963 by W. C. Prebble, of whom more later, to Professors E. Bishop, W. F. K. Wynne Jones and G. S. Nancollas.

measurement. In addition he was very successful as a consultant, an expert witness, and an industrial chemist. He had greatly improved the production of phenolformaldehyde polymers, but found himself anticipated, by one day, by L. H. Baekeland when he came to file his key patent. There was a business connection between him and Swan, for in 1880, when he was only 22, he had set up a lamp factory in France for Swan and had run it very successfully. He was the original editor of the Physical Society's scientific abstracts. In private life he was a very accomplished musician. F. A. Freeth (*Biographical Memoirs of Fellows of the Royal Society*, 1959, **5**, 253) describes Swinburne as 'the Victorian gentleman at his very best and a most lovable genius'. He had an office on the first floor of Grosvenor Mansions, 82 Victoria Street, London (*The Electrochemist and Metallurgist*, 1904, **3**, 740).

A third friend was Sherard Cowper-Coles, who was 36 in 1902. He was an entrepreneur with considerable practical flair. By accident he discovered the process for zinc-plating iron or steel articles by heating them with zinc dust in a closed vessel at relatively low temperatures, which he called 'Sherardizing'. He was much concerned with electrodeposition; for example, he devised a process for making parabolic reflectors for searchlights. He was the moving spirit of several firms, one being the Cowper-Coles Galvanizing Syndicate Ltd., which were mainly concerned with processes and plant for industrial applications of electrochemistry. In 1901, through another of his companies, he started a monthly journal, *The Electrochemist and Metallurgist*, of which he was the editor. It contained articles, abstracts from contemporary foreign journals, abstracts of patents, and book reviews in the field indicated by the title. He had an office and laboratory on the sixth floor of 82 Victoria Street (*Electrochemist and Metallurgist*, 1904, **3**, 740), and this was where the friends met in 1902.[2] W. C. Prebble, whose letter about this meeting has already been mentioned in connection with Swan, was articled to Cowper-Coles in 1900 as an assistant and was the first student member of the Society.

A fourth friend was F. G. Donnan, who had joined Sir William Ramsay as an assistant lecturer at University College, London, in 1901

---

[2] The old no. 82 has since been demolished and replaced by a post-second-world-war building. In 1980 the ground floor was a steak house.

(at a salary of £100 per annum) and in 1902, when he was 32 years old, became Assistant Professor there. He had worked with Wilhelm Ostwald in Leipzig and with J. H. van 't Hoff in Berlin on various aspects of solutions, including electrolyte solutions; so he brought first-hand knowledge of German research in the new discipline of physical chemistry.

According to Swinburne's description of that first meeting (*loc. cit.*) another person present was Vernon Boys, later Sir Charles Vernon Boys, F.R.S., who was 47 in 1902. He was at that time already a rather grand figure, being a Metropolitan Gas Referee (*Dictionary of National Biography*, 1941–50, p. 96) in which capacity he had devised an automatic recording calorimeter. He was 'an experimenter and above all an inventor of instruments'. He had successfully developed a quartz fibre suspension which he used in his radiomicrometer and in his apparatus for re-measuring Newton's gravitational constant. Swinburne said that he distracted the meeting by making paper butterflies. But Boys's connection with the Faraday Society seems to have been as ephemeral as his butterflies, for his name does not appear as a holder of office or as a member of Council.

Another of the founding friends was F. Mollwo Perkin. He was the youngest son of Sir William Perkin, F.R.S., and half-brother to W. H. Perkin, Jr., F.R.S. and A. G. Perkin, F.R.S. At that time he was Head of the Chemistry Department at the Borough Polytechnic Institute, where he was doing pioneering work on electrochemistry. He was particularly concerned with the use of electrolytic methods for the preparation of organic compounds. He too had an interest in chemical industry, for he later (1909) resigned his academic post to go into practice as a consultant. Like Donnan he was at that time a young man, being 33 in 1902. Prebble later became one of Mollwo Perkin's research pupils and had a high regard for him.

Prebble said in his letter that F. S. Spiers (born 1875), who became the first Secretary of the Society, was the editor of *The Electrochemist and Metallurgist*. This cannot be confirmed from the journal itself, for the early issues do not give the editor's name and the later ones name Sherard Cowper-Coles with W. Pollard Digby as assistant; but it is possible that Spiers's association with the early meeting and the Society

The Early Years, 1902–1920

was through Cowper-Coles. He published a few papers on electrochemical topics.

Swinburne (*loc. cit.*) mentioned 'my own assistant' as being present. This must have been W. R. Cooper (see later). The writer has discovered nothing about H. V. Simpson, who was the first honorary secretary of the provisional Committee. These nine people – Swan, Swinburne, Cowper-Coles, Donnan, Boys, Mollwo Perkin, Cooper, Simpson and Spiers – are the known *dramatis personae*: there may have been others but the evidence is inadequate. According to Donnan's account *Trans. Faraday Soc.*, 1953, **49**, 511, of that first meeting 'a letter from Sir William Ramsay was read, in which he objected to the appearance of a new scientific journal, saying that he had not time enough to read the existing ones!'. It seems probable that he was invited but declined his support.

What were the aims of these friends and what kind of Society did they want to found? The thoughts of one of them, Cowper-Coles, can presumably be read in the editorials of his journal. In the August issue of the *Electrochemist and Metallurgist* for 1901 (Vol. 1, p. 187), for example, the editor writes:

'Much good might be done, and much direct stimulus given to electrochemical activity, if the Society of Chemical Industry and the Institution of Electrical Engineers were to combine together in some way, either by means of a special committee on electrochemistry common to both, or in other ways which might be suggested, for the consideration of some of the problems (many of which are indicated in Mr. Swan's luminous address) that our borderland science will one day be imperatively called upon to solve, to the mutual advantage of the two great industries that these institutions serve, and to the welfare of the country at large.'

In Swan's Presidential address to the Society of Chemical Industry, in 1901, which is referred to, there is no urging of any such joint action; but the passages already quoted (*v.s.*) show how strongly he felt about the encouragement of scientific education and research. The writer has not found any contemporaneous expressions of view by any other of the friends; but because Cowper-Coles was able to crystallize their thoughts into action we may take it that they broadly shared his views.

We will consider later the magnitude of the task which faced these gallant innocents.

In *The Electrochemist and Metallurgist* for 1901 (Vol. 1, p. 226) is a note giving the composition of a Sectional Committee on electrochemistry and electrometallurgy set up by the Institution of Electrical Engineers which names Cowper-Coles, Spiers and Swinburne. However, the friends had evidently concluded by 1902 that evolutionary reform of existing institutions would not proceed fast enough; so they decided to take the revolutionary path and found a new Society. In the Council Report for 1903–04 we read further:

'The Committee went very carefully into the question of the possibility of working in conjunction with one of the existing Societies, and several Societies were approached on the matter, but the view was finally adopted that such an arrangement was not likely to have the desired effect, namely the bringing together of those interested both theoretically and practically in the subject of electrochemistry, and thus encouraging a science that was daily becoming of increasing theoretical, technical, and commercial importance. It was therefore decided that the Society should be independent.'

We do not know which of the established Societies were approached. It is noteworthy that there is no mention of the Chemical Society either in the editorial of August 1901 or in an article along similar lines by E. G. P. Bousfield, who was both an Associate of the Institute of Electrical Engineers and a Fellow (as members were then styled) of the Chemical Society (*The Electrochemist and Metallurgist*, 1901, 1, 192). Swan and Cowper-Coles never were Fellows of the Chemical Society and they were two of the most influential of the founding friends. For them the founding of a new Society was therefore not a secession from the Chemical Society. Swinburne, Mollwo Perkin, and Donnan were Fellows; but the presumption is that for them too the Chemical Society of that time seemed irrelevant. For its part the Chemical Society seems to have known nothing about the Faraday Society, at an official level, until the Spring of 1903. There is a revealing Minute, almost worthy of Lady Bracknell (*Chemical Society Minutes*, April 22nd, 1903, pp. 180–181): 'With regard to an application from the Faraday Society for the use of the Society's rooms for its meetings it was resolved that the Secretary should write and ask for further particulars about the Faraday Society'. We may note that permission was later given (*Chem. Soc. Minute* of May 7th, 1903, p. 190); but then there was a correspondence between the two Secretaries in the course of which Spiers was told firmly that smoking could not be allowed (*Chem. Soc. Minute* of November 5th, 1903, p. 220). We shall see the consequence. There are no other references to the Faraday Society in the Chemical Society Minutes from 1902 to 1906 inclusive.

The friends certainly could not have felt that they were seceding from the Society of Chemical Industry or the Institution of Electrical Engineers, for Swan was President of the former in 1901 (and had been of the latter in 1898–99), while Swinburne was President of the latter in the crucial year 1902–03. What one of them felt can presumably be gathered from a characteristically exuberant editorial describing the first ordinary meeting of the Society: 'It has been felt by the founders of the Society that, for the present, anyhow, nothing less than the individuality of the separate organization, judiciously holding the balance between the theoretical and practical aspects of the subject, will be able to pierce our apathy and indifference to the claims of the new Sciences and thus stimulate that interest in them without which further advance will be impossible' (*The Electrochemist and Metallurgist*, 1903, **3**, part 1). Swan put it more succinctly in an address read to that first meeting: 'We are not in competition with any other society. There is more than enough of useful and necessary work to be done on unoccupied ground'.

From the Council Report for 1903–04 we have already seen that the friends were strongly influenced by the foundation of the Deutsche Elektrochemische Gesellschaft (in Kassel in 1894), which in 1901, after a deal of argument, changed its name to the Deutsche Bunsen-Gesellschaft für angewandte physikalische Chemie in order to show a broadening of interests. Mollwo Perkin and Swinburne were members of the D.B.G. The friends were further influenced by a well advanced plan to form an American counterpart. Presumably these Societies also had originated from a feeling of need for independence. It seems unlikely that the friends had a clear aim of founding a large society even by the much more modest standards of 1903. They certainly wanted it to be influential, but they probably wanted it to have the form of a cosy discussion group, and this is what it was for a long time.

After the 1902 meeting the friends pressed on, making approaches and soundings of opinion. In *Proceedings* (1905, *loc. cit.*) we read 'During the summer of 1902 some 630 circular letters were sent out by the provisional Committee, and these elicited 106 definite promises of membership, and 21 replies unfavourable to the movement.' Apparently there was (*Proceedings* for 1914, July, p. 1) a meeting at St. Ermin's Hotel on January 6th 1903 at which Swan moved the resolution which created the Society, and then another meeting, this time a public one, on Wednesday 4th February (*Proc. Faraday Soc.*, 1905, *loc. cit.*) at which 'on the motion of Dr. J. W. Swan, seconded by Mr. Alexander Siemens, the Society was formally inaugurated and the present Officers and Council elected'. A week later, on 11th February 1903, the first meeting of Council took place at 82 Victoria Street, probably in Swinburne's office where later meetings were certainly held. By then there were 160 promises of membership.

It is clear that Swan had been very active, for of the friends only he is likely to have been able to persuade so many grandees to lend their names to the Society as Vice-Presidents. The balance between science and industry is evident. Crum Brown was an academic figure, a philosopher and a chemical theorist: he is remembered for his rule about further substitution into a substituted benzene compound, and he had devised a model of sodium chloride with coordination 6 round sodium and chlorine. Lord Kelvin had had a profound influence on academic science but also had strong views about the need for science to be useful. In a lecture to the Institution of Civil Engineers in 1883 he said: 'There cannot be a greater mistake than that of looking superciliously upon practical applications of science. The life and soul of science is its practical application; and just as the great advances in mathematics have been through the desire of discovering the solution of problems which were of a highly practical kind in mathematical science, so in physical science many of the greatest advances that have been made from the beginnings of the world to the present time have been made in the earnest desire to turn the knowledge of the properties to some purpose useful to mankind.' (*Dictionary of National Biography*

1901-11, p. 510). Sir Oliver Lodge was both an experimenter and a philosopher. His discoveries were of great practical importance in the very early spark 'wireless telegraphy' and he developed the moving boundary method of measuring ionic velocities. Ludwig Mond was co-founder of Brunner, Mond Ltd., which was to be one of the founding firms of Imperial Chemical Industries Ltd. He was also one of the co-discoverers of nickel carbonyl. Lord Rayleigh had immense standing in the scientific establishment of the time although 'of striking discoveries or inventions practically none stand to his credit with the single exception of the discovery of argon' with Sir William Ramsay (Sir James Jeans, *Dictionary of National Biography*, 1912-21, p. 514). He had 'a massive, precise, and perfectly balanced mind' and 'the capacity for understanding everything just a little more deeply than anyone else. . .'. Alexander Siemens was a distant relative, a third cousin once removed, of Sir William Siemens, who was the founder of the Siemens concern in England. Born in 1847 in Hannover he came to England first in 1867 and, after going back to fight in the Franco-Prussian war, he returned to settle here. In 1888, after the death in 1883 of Sir William, who had no child to succeed, he became head of the English firm.

The members of Council were likewise largely experimentalists and practical men. Some have already been mentioned. George Beilby must be George Thomas Beilby (later Sir George) who was a distinguished industrial chemist and, as a pioneering metallurgist, is remembered for his hypothesis of the 'Beilby layer' of amorphous metal produced at a bearing surface. Bertram Blount was consulting chemist to the Crown Agents for the Colonies. He was an expert on the chemistry of cement and had also written a book on 'Practical Electrochemistry'. Charleton was a mining engineer with very wide, international experience who was President of the Institution of Mining and Metallurgy in 1902-03. W. R. Cooper was the author of a book on primary batteries and had been a member of Swinburne's staff in his consulting practice since 1895. A. K. Huntington was another delightful, heroic figure. 'The subject of flight engaged Professor Huntington's attention at an early date, and he made frequent balloon ascents, a form of sport for which his steady nerves, trained by long practice in the hunting field and polo ground, rendered him well fitted. He built and owned his balloons...' and he was 'inclined to enjoy a technical controversy' (*Journal of the Institute of Metals*, 1920, **23**, 554). Before he became Professor of Metallurgy at King's College, London, he had 'assisted William Siemens in his early experiments with the electric arc furnace'. In 1913–14 he was President of the Institute of Metals. R. A. Lehfeldt was Professor of Physics at the East London Technical College and the author of a book on 'Electro-Chemistry': in 1906 he emigrated to South Africa and became interested in economics. W. S. Squire was an academic electrochemist and O. J. Steinhart an industrial one.

At that first Council meeting Mollwo Perkin was elected Treasurer, an office which he was to hold until 1917. Spiers appears to have been present as the 'Honorary Secretary' but he was not named in the Minutes until later when he was appointed Secretary at a salary of £100 per annum. The name of the Society was agreed, and perhaps we may quote more of Cowper-Coles' stirring prose (*The Electrochemist and Metallurgist*, 1903, **3**, 2): 'The wise choice of the grand name of Faraday – as the Germans chose that of Bunsen – as the designation of the Society, will meet with the approval of all. There can be no English chemist or physicist or electrical engineer through whom the mention of the name of Faraday does not send a thrill of pride and reverence, and of its suggestiveness in this connection there can be no question. Any possible criticism, on that score, however, from unimaginative members, has been met by the inclusion of an explanatory sub-title, 'To promote the study of Electrochemistry, Electrometallurgy, Chemical Physics, Metallography, and kindred subjects.'' the broadening of the aims was clearly intended to attract more members, for E. G. P. Bousfield (*loc. cit.*) had opined that, because there was no large electrochemical industry at that time in Britain, there might be only fifty or so people interested in electrochemistry itself. The Council Report for 1903–04 adds: 'the sciences specifically included ... are those branches of pure and applied physical chemistry which do not come precisely within the scope of existing scientific and technical Societies. It is worthy of note that the Faraday Society is unique in being the only scientific body in England which specifically aims at

The Early Years, 1902–1920

encouraging and combining both the theoretical and practical sides of the subjects that come within its scope.' The general aims of the founders must by now be clear; and it will prove important to remember them.

Council met five times in 1903 and, besides doing such obvious jobs as drawing up rules and setting up an Editing Committee 'to consider the question of Publication of Proceedings', made two especially important decisions. The first related to the evident intention to achieve international standing. Connections were sought, by the interchange of publications, with the American Electrochemical Society, although the Treasurer expressed concern about the cost. The Faraday Society had already received some recognition abroad for it was invited to nominate representatives on the Committee for the forthcoming Congress of Applied Chemistry in Berlin. The second decision related to the character of its ordinary meetings. These were to occur several times a year and were to be for the discussion of papers. We may quote Cowper-Coles again: 'The Council seems determined to run the Society on unconventional lines, where conventionality stands for what is antiquated and meaningless, so we find that, unlike the ordinary practice, papers will be printed – as a rule published in these columns, but in any case in proof form – and circulated among the members some time before being read, and that consequently they will be read, as a rule, in abstract only; a most salutary reform. By this means it is hoped greatly to increase the value of the discussions, and at the same time absent members, who will thus be led to take a keen personal interest in the affairs of the Society, will be given an opportunity of expressing their views to the meeting by correspondence.' (*The Electrochemist and Metallurgist*, 1903, **3**, 2). Notes on the discussions of the papers so presented were to be obtained or taken by Spiers and printed subsequently. Papers for a meeting were not required to be on any one, closely defined topic. The medium chosen for publication was, singularly enough, Cowper-Coles's journal, which then acquired the sub-title of *The Organ of the Faraday Society*.

The first ordinary meeting was arranged for Tuesday evening, June 30th 1903, in the rooms of the Chemical Society, when four papers on various aspects of electrochemistry were to be discussed. 'Out of a present membership of 238, of whom at least one-half lives away from London, there were present close on 50 members'. Swan, unfortunately, could not come. At a Council meeting only eight days before, when the adamant refusal of the Chemical Society to allow smoking was reported, the Secretary had been asked to try to arrange for the meeting to be held in the library of the Institution of Electrical Engineers; but it was too late. However, refreshments were served; and at subsequent meetings members of the Society smoked themselves and each other *ad libitum* at the I.E.E.

At this meeting a paper by W. C. Dampier Whetham, F.R.S., was discussed in which he set out very lucidly the physical arguments for the existence in solution of ions with some degree of independent mobility. Dampier Whetham, later Sir William Dampier, was a Fellow of Trinity College, Cambridge, who had worked with J. J. Thomson and who later became eminent in British agricultural economics. Perhaps most notably he was the first secretary of the Agricultural Research Council. H. E. Armstrong, who at that time was Professor of Chemistry at the Central Technical College in Exhibition Road, attacked the theory of free ions because it had no chemical explanation. Unfortunately he had no clear alternatives to offer. It is difficult to realize now that in 1903 the concept of the orbital atom with energy levels would not be enunciated for another ten years, and that although G. N. Lewis (1902) and Abegg (1904) are known to have appreciated that the Mendeléeff classification of elements offered a hint of a shell structure for atoms and ions even J. J. Thomson's 'plum pudding' atom would not be proposed for another year. There was therefore no physical basis for a theory of ionization, but nevertheless the experimental case was very well established and this was appreciated by the physically minded. To the best of the present writer's recollection, at a lecture given in the late 1920s or early 1930s Armstrong was still refusing to accept the case for free ions.

There was only one more meeting in 1903, on Thursday 8th December, when the members dined together beforehand. This was announced in the usual place (*Electrochemist and Metallurgist*, 1903–04, **3**, 239) – 'The Society is going to try the experiment of arranging a dinner before the meeting for the convenience of those members who, on account of the exigencies of distance, might otherwise be tempted to stay away altogether from evening meetings rather than endure the silent discomforts of the solitary diner.' There were, however, eight meetings in 1904, all in London at the I.E.E. and all of an evening. At the first of these the members and their friends dined together; but it is not clear that this became a regular custom.

The papers read at these early meetings were heavily weighted to applications of electricity to chemistry and particularly to industrial aspects, viz. mostly to primary and secondary batteries, to processes based on electrolysis e.g. Cowper-Coles on, *inter alia*, the coppering of ships bottoms (*The Electrochemist and Metallurgist*, 1903–04, **3**, 244), or on the very high temperatures that could be generated electrically. Sometimes, however, they were on metal technology or theory, e.g. Cowper-Coles's 'Notes on the Welding of Aluminium', or G. T. Beilby's paper on 'The Hard and Soft States in Metals' (*Loc. cit.*, p. 240, p. 806). In the industrial papers what we now call 'hard science' often formed a minor part; but there were full descriptions of practical procedures often illustrated by beautiful line drawings and photographs which now seem very nostalgic, and there were long discussions of costing and profitability, e.g. James Swinburne, *The Electrochemist and Metallurgist*, 1903–04, **3**, 75; Kr. Birkeland, *Trans. Faraday Soc.*, 1906, **2**, 98. Some of the papers were more academic, e.g. that of Dampier Whetham's on the theory of electrolytes (*v.s.*) or on the preparation of compounds using electrode processes. Pertinent to all this were some remarks by J. B. C. Kershaw in an illuminating, and in places very funny, report on the eleventh annual meeting of the Deutsche Bunsen-Gesellschaft at Bonn in May 1903 (*The Electrochemist and Metallurgist*, 1903–04, **3**, 736)

'The German is by nature and training more interested in the discovery of laws and principles than in their practical applications, and hence the general character of the papers read at the annual meeting of the Bunsen Gesellschaft always tends towards theory rather than towards practice.

Since the tendency of English and American scientists in this branch of study is towards the other extreme, I can recommend attendance at the annual meetings of the Bunsen Gesellschaft as a useful corrective influence, ...'.

There was at that time undoubtedly a very great difference in the vigour and effectiveness with which physical ideas and methods were being applied to chemical problems in Germany and in Britain, whether to academic work or to industry. As we have seen, it was a major concern of the founders of the Faraday Society to improve the understanding, by those working in British industry, of the principles underlying practice; and so necessarily to remedy this disbalance. In part the difference may have arisen by accident for it seems that in Germany a number of very able individuals with a mathematico-physical training had taken an interest in chemical problems and had set a trend, whereas in Britain this had happened much less by the late nineteenth century, despite the earlier examples set by Faraday and Graham and despite the flourishing state of physics in this country. (Science and Technology did not face the same bias in Germany as in the U.K., where the upper classes disdained industry as a dirty business. The dominance of Public School classical education led to science being associated with craftsmen and their working-class Institutes. In Germany the respect for science, in part traceable to Goethe, meant that scientists and engineers themselves developed industries, as the names Siemens and Linde remind us. [M.M.D.]) Barring a major structural difference, such as the degree of support of research generally, there was a hope that such a trend could be established in Britain by a suitable stimulation of interest, so this is what the founders set out to do. Progress at first was slow; but eventually the Society played a major part in effecting change. Ironically, however, as it did so it became a very different society from that which the founders envisaged.

Two reasons why that task took so long were that it did indeed prove to depend upon more generous support of research from private and, especially, from State funds, and on a change in the training of chemists. Up to about the 1920s few British physicists made much direct impact on chemistry, although there were outstanding exceptions such as the Braggs, father and son. Even after the 1920s, however, physical chemists were not recruited mainly from among physicists, as in Germany, but from chemists. There were exceptions again, such as Sir John Lennard-Jones and Charles Coulson; but by and large the chemists had to 'lift themselves up by their own boot straps'. This meant that the physical chemistry departments in Britain, when they did eventually develop, were closer to the other chemistry departments than was the case in Germany, and that had marked advantages.

To resume our account of more bread-and-butter matters, membership and finance soon became matters for concern. At the first ordinary meeting the number of members was 238, as has already been noted. By the end of 1903 this had risen to 254 of whom 30 lived abroad and 6 were student members; but some of them did not pay their subscriptions promptly or even at all. The subscriptions were £2 per annum for a member and £1 for a student, with an entrance fee of £1 for members. When we remember that there has been roughly a twenty-fold price inflation between 1903 and 1977 we realize that these subscription rates were high, e.g. the annual subscription was 2% of Donnan's stipend as an assistant lecturer. This makes the reluctance to pay less surprising. On 13th April 1904 concern was expressed in Council at the deficit of £93 10s $4\frac{1}{2}$d on the first year's working, and an estimate of a further deficit of £52 by the end of the year. The cost of postage for a monthly publication containing extraneous matter that added to the weight, particularly the cost of sending some 600 copies to the members of the American Electrochemical Society, and the charge for the physico-chemical sections of *Science Abstracts* (from the Institution of Electrical Engineers) which were incorporated, were all anxiously considered. Concern about membership and finance was, indeed, to become part of the enduring pattern of Council meetings.

This led to a very important development. At the Council meeting on 9th June 1904 the Treasurer was asked to examine the possibility of the Society publishing its own Proceedings, and on 27th July he reported that the annual cost of publishing quarterly Transactions and monthly Proceedings, with provision of advance proofs to members and subsidy to *Science Abstracts*, was estimated to be £293 7s 0d, while the present cost of the arrangement with *The Electrochemist and Metallurgist* was £405. In view of the great saving and the desirability of independent publication, Council resolved to give notice to the Proprietors of *The Electrochemist and Metallurgist* terminating joint arrangements, and that this was to be confirmed at the next Council meeting on 19th September. At that meeting, which Cowper-Coles, representing the Proprietors, said he preferred not to attend but at which Swan made one of his rare appearances, the resolution was confirmed unanimously. At the same time Spiers became Editor as well as Secretary and the Editing Committee became more formally organized, presumably in order to meet its new responsibilities. It started to keep Minutes of its meetings from 22nd July 1904 and these have all been preserved. The Minutes show that its main business was appointing referees, accepting papers for publication, arranging for them to be read at (Ordinary) Meetings, sometimes recommending candidates for election as members, and recommending advertising agents or printers as occasion required. Because so few papers are actually mentioned in these Minutes it seems likely that many were dealt with by the Chairman or the Secretary and Editor, or perhaps were passed unspecified in groups as issues of Transactions. The Minutes give an impression that procedure was often informal. The members of the original (1903) Editing Committee were Messrs. B. Blount, W. R. Cooper, O. J. Steinhart and F. S. Spiers. Those mentioned in the Minutes of 22nd July 1904 were F. Mollwo Perkin, W. R. Cooper and the Secretary (Spiers): 'Mr. Digby was present on behalf of the Editor of *The Electrochemist and Metallurgist*'. Blount's name reappears in Minutes of later meetings. In April 1905 the Committee was reconstituted as the 'Editing and Publishing Committee' and Huntington began to act as Chairman.

The new *Transactions* first appeared in January 1905 and killed Cowper-Coles's journal, which did not appear after 1904. According to an announcement it was to reappear under the title *Metals*; but from the catalogues there is no sign that it ever did. So Cowper-Coles saw one of his enterprises fail and lost his pulpit. He must have felt that the Society had shown him scant gratitude; and he may have been even more annoyed by the fact that *Transactions* was printed by the same printer and that many of the type faces were carried over from *The Electrochemist and Metallurgist*. In June 1906, when he was due to come off Council, he resigned from the Society but, more happily, he eventually rejoined in 1911 although he seems never again to have been closely involved with its affairs. Despite his enthusiasm, fertility of mind and practical ingenuity, illustrated by his contributions in the

The Early Years, 1902–1920

Abstracts of Patents in *Proceedings*, he never had much commercial success. He died in 1936, when he was 70, with his affairs in a muddle. In his citation in *Who Was Who* he describes himself as 'founder of the Faraday Society'. If this is meant to imply that he was the sole founder it is an overstatement; but he certainly played a central and vital part. He deserves our recognition and our gratitude. A vivid and sympathetic biography by William H. Tait is to be found in the *Metallurgist* (1962–63, **2**, p. 219). For the Society the publication of its own journal was to be essential, although it was painful to one who had been so influential and central. Evidently the Society was, like a growing child, beginning to develop an independent corporate life and to move away from some of the early formative influences. We shall see how this process continued.

It is worth noting here that most of the great names played little or no part in the day-to-day running of the Society. By 22nd June 1906 there had been 18 Council meetings. Swan, the President 1903–05, had attended only two; but it is evident from the Minutes that he had an effective influence through discussions with individuals. Of the Vice-Presidents Crum Brown, Kelvin, Lodge, Mond and Rayleigh had not attended one. Siemens had attended only three but helped with advice on finance; he was a member of the first Finance Committee. Swinburne was much more active: he attended 12 and often took the chair. Mollwo Perkin and Spiers had attended all 18. Of the members of Council, Cooper had attended 16, Blount had attended eight as had that intrepid Balloonist Huntington who several times took the chair, and so too had Steinhart. Cowper-Coles had attended six. Donnan was unable to attend at all. He went to Dublin in 1903 and had then tendered his resignation: this was eventually accepted in 1904. Little over half-a-dozen people, therefore, actually ran the Society in those very early days.

In 1904 there occurred an interesting contretemps. It was reported to Council in June that Professor Svante Arrhenius, Sir William Ramsay, Sir Oliver Lodge, Sir William Crookes and Professor J. J. Thomson had been approached to give a special lecture during the summer, but all had declined. However, Arrhenius promised to address the Society in the Autumn on his way back from the St. Louis meeting of the American Electrochemical Society. Although the details of the visit had not been settled, Swan, in high hopes, invited about 40 people – Vice-Presidents and Members of Council of the Faraday Society and 'Presidents of other kindred scientific institutions with the addition of a few other sympathetic friends' – to a private dinner at the Trocadero Restaurant on 10th October. Over 20 accepted. He also cabled to Arrhenius and sent a new-fangled 'Marconigram' to him at sea. By the 8th it was clear, alas, that Arrhenius was not coming; but the dinner went ahead and Swan had to toast the guest of honour in his absence, explaining that Arrhenius had had an urgent call so he had left St. Louis two days early and gone straight to Hamburg, bound for Berlin and Copenhagen. Thorpe (this must have been T. E. Thorpe) who was one of the guests is reported to have said 'Arrhenius is quite impossible'. The incident shows Swan's concern to gain international recognition for the Faraday Society and, perhaps, how far the Society still had to go.

There was in fact growing international recognition, for the St. Louis meeting was intended to be a joint one with the Deutsche Bunsengesellschaft and the Faraday Society, although in the end no members seem to have gone; and the Bunsengesellschaft invited the Faraday Society to send a representative to its annual meeting on Ascension Day. The urge for internationalism seems to have been limited in those days, however, for on 9th June 1904 Council unanimously agreed not to act with the American Electrochemical Society to develop a universal system of notation.

After 1904 the Society began to settle into a routine. In 1905 there were seven ordinary meetings, all at the I.E.E. library in London and of an evening. The problems of non-payment of membership subscriptions continued to vex Council, as did other causes of the chronic financial difficulties. That the Society was steadily gaining recognition was shown by the appointment of Mollwo Perkin to represent it on the organizing committee of the International Congress of Applied Chemistry which was to be held in Rome in 1906.

The main event of the year was the appearance of *Transactions* and *Proceedings*. The former contained the papers presented at Ordinary Meetings and from 1907 at General Discussions together with the full

The Early Years, 1902–1920

discussions thereon. The latter, intended to appear monthly, contained official notices, book reviews, précis of the papers presented at General Discussions and of the discussions thereon, précis of the discussions on papers presented at Ordinary Meetings and, once a year, the Council report, annual accounts and proceedings at the Annual General Meeting. It also contained the sections from Science Abstracts dealing particularly with physical chemistry and its applications, and Abstracts of English and American Patents relating to electrochemistry and metallurgy. Altogether, then, *Proceedings* was a substantial publication, with a certain amount of overlap with *Transactions*. Volume 1 contained 532 pages. The Society had embarked on a programme of publication that proved as a whole to be over-ambitious. Already, by the end of 1905, the publication of Science Abstracts was discontinued, which saved 297 pages, and the book reviews were transferred to *Transactions*. Later, as we shall see, *Proceedings* was pruned further.

In the summer of 1905 Kelvin succeeded Swan as President. He was then 81 years old and in poor health, so it is hardly surprising that he was not able to do much for the Society; but he did manage to preside over the ordinary meeting on October 31st 1905. He was succeeded as a Vice-President by Sir William Huggins, O.M., F.R.S., an eminent astrophysicist who pioneered the study of the spectra of the stars and comets. Sir William Ramsay had been the first choice of Council as the new Vice-President and Mollwo Perkin offered to see him personally; but Ramsay seems to have declined nomination. The new members of Council were W. Murray Morrison, M.I.C.E., who pioneered the development of the aluminium industry in Britain and of hydroelectric power in the Highlands of Scotland, and Ernest Wilson, M.I.E.E., who was Professor of Electrical Engineering at King's College, London. Thus, the mixture of pure and applied science, of chemistry, physics and engineering was maintained.

In 1906 the pattern of activity continued. There were six ordinary meetings and a special one at the rooms of the Royal Society of Arts, when Professor Kristian Birkeland read a paper on 'The Fixation of Atmospheric Nitrogen'. Council decided to strike off those who still had not paid their subscriptions, and its worries were increased by ten resignations, including that of Lehfeldt who was emigrating to South Africa.

Herr Geheimer Regierungsrat W. Hittorf (the title reminds us that the eminent German academics of those days were made Privy Councillors), although in retirement since 1890, became a Vice-President and so did Professor J. J. Thomson. One of the three new members of Council was R. S. Hutton, who had earlier pursued researches using the electric furnace with Professor Moissan in Paris. In 1906 he was a lecturer in electrometallurgy at Manchester University; two years later he joined the family silver business in Sheffield and eventually (1931) he became Professor of Metallurgy at Cambridge. Another was Herbert Jackson, who at the time was Professor of Organic Chemistry at King's College, London but who later became Director of the British Scientific Instruments Research Association and was knighted. The third was T. Martin Lowry, who was assistant to Professor H. E. Armstrong at the Central Technical College in Exhibition Road, and Lecturer in Chemistry at the Westminster Training College. He attended Council meetings faithfully and seems soon to have become very influential.

At a Council meeting on 27th November 1906, Swinburne (in the Chair), Huntington, Cooper, Lowry, Mollwo Perkin and the Secretary being present: 'It was decided to circularise persons who were particularly interested in pure physical chemistry informing them that it was among the objects of the Society to encourage chemical physics and inviting them to help strengthen that side of the Society by bringing their work before the Society for publication.[3] In order to initiate the movement it was agreed to ask the Early of Berkeley to show his osmotic pressure apparatus at one of the Society's meetings, and to initiate a discussion on that subject.' (*Council Minutes* of that date).

This proved to be a most significant step. It established the form of the Discussion Meeting since, for the first time, the papers presented at one meeting were all to be on a common theme and, following the custom established earlier for Ordinary Meetings, they were pre-

---

[3] The 'uniqueness' was qualified at least by the existence of the Royal Society [M.M.D.].

printed and circulated before the meeting. What ultimately proved even more important was that it pointed to the future path of development by the Society which was to be increasingly towards physical chemistry. The use of the adjective 'pure' indicated a change from the earlier preoccupation with applied science. This development set the Society more obviously in competition with the Chemical Society, which certainly regarded physical chemistry as being in its purview, e.g. from the start in 1904 the famous *Annual Reports on the Progress in Chemistry* included a section on general and physical chemistry. As we shall see, the resulting conflict of interests was between institutions rather than individuals. At first it mattered little because the discrepancy of power between the two Societies was so great; but as the Faraday Society grew in power so this conflict became more important. Eventually, as we know, it had to be faced and resolved.

The meeting to discuss 'Osmotic Pressure' was held on 29th January 1907, presumably in the usual place, the contributors being the Earl of Berkeley, W. C. Dampier Whetham, F.R.S., T. Martin Lowry and Dr. L. Kahlenberg of Wisconsin (*Trans. Faraday Soc.*, 1907, **3**, 12). The Earl of Berkeley was a striking person. He was another polymath, being later both F.R.S. and M.F.H. (i.e. Master of Fox Hounds, see *Obituary Notices of Fellows of the Royal Society*, 1942–44, **4**, 167; and *Dict. Nat. Biog.*, 1941–50, p. 72). Handsome, egocentric, impatient of authority, good at all games that he took up, 'wild as a hawk, ready to turn a back somersault off the mantelpiece for a bet and for any other physical feat of daring', he hardly fits the 1977 picture of a scientist. He became interested in chemistry while in the Royal Navy, when a chaplain in one of the ships lent him books on the subject, so he left the Service and started to study at Imperial College with T. E. Thorpe. A bout of double pneumonia compelled him to leave London, and he came to work in Oxford where he intended to establish a research laboratory for the University. However, he wished to be allowed to count term while living up in the healthier air of Boars Hill, but the University was inflexible and would not make an exception to the rule that residence must be within one-and-a-half miles from Carfax, so Berkeley would have no more dealings with it and set up a private laboratory in his large house. With E. G. J. Hartley as his assistant he built an apparatus for the direct measurement of osmotic pressure, displaying great engineering skill in so doing.

Louis Kahlenberg was German born (1870): he took his D.Phil. degree at Leipzig. He emigrated and settled in Wisconsin, U.S.A., and from 1900 had been professor of physical chemistry at the State University.

Another discussion meeting, on 'Hydrates in Solution' (*Trans. Faraday Soc.*, 1907, **3**, p. 123), was held in June of the same year, Lowry being one of the organizers and also an author. A major concern still was to explain the existence of ions in solution. The Minutes show that several attempts were made to organize discussion meetings on more applied topics viz. on 'Experimental Electric Furnaces', on 'Producer Gas (a) physico-chemical theory (b) the metallurgical standpoint', and on 'Photo-engraving and Colour Photography' (one of Swan's interests) but these all failed, partly because 'the unfortunate reticence displayed by Members on questions of industrial science makes the arrangement of such meetings much more difficult than where the discussions are concerned with unapplied science' (Council Report in *Proceedings* for 1907, p. 47). There was, indeed, no further discussion meeting until April 1910, when again the subject was an academic physico-chemical one viz. 'The Constitution of Water'. Professors P. Walden of Riga and Phillipe Guye of Geneva were there. This was made an occasion for a grand, formal dinner at the Trocadero Restaurant which was well attended by Members and their friends and was adorned by representatives of the Royal Society, the Chemical Society, the Physical Society and the Society of Chemical Industry.

A discussion on a rather more practical topic, viz. 'High Temperature Work', was held at last in May 1911. In the afternoon preceding the meeting the opening speaker, Dr. Arthur Day, who was Director of the Geophysical Laboratory of the Carnegie Institute at Washington, D.C., with a party of Members and their friends visited the National Physical Laboratory, by the kind invitation of the Director, who was to become President that year, to inspect the high temperature equipment there.

The drive towards physical chemistry evidently led to correspondence about the possibility of joint publication with the *Journal of Physical Chemistry*, for there is a Minute of 19th February 1907 recording that Professor [Wilder D.] Bancroft had replied saying that this could not for the present be discussed.

Kelvin survived his two years as President but only just, for he died in late 1907. As President for 1907–09 Council nominated Sir William Perkin, F.R.S., who then was 69 years old. Although he was younger than either of his predecessors he died in 1907, very soon after assuming office. The Society then elected Sir Oliver Lodge, F.R.S., to take office early in 1908 and hold it for the remainder of the normal term. He was 57 and had been the next choice to Perkin. He gave an address to the Society 'On certain aspects of the work of Lord Kelvin' (Sir Joseph Swan being in the Chair); but there is no record of his ever attending a Council meeting or taking the Chair at an ordinary meeting: indeed, before the end of his term of office he wrote expressing his desire to retire from the Society at the end of the year. He was dissuaded; but he resigned for a second time in 1911. J. J. Thomson's resignation was announced to Council in December 1909, soon after his term of office as Vice-President had ended. These particular busy physicists do not seem, therefore, to have given the Society a high priority.

By now it is clear that the interactions between people and events are central to this historical account, and the problem is how best to present this complex, continuing relationship. In the writer's view it is first to make the acquaintance of some of the people who were recruited as Officers or members of Council in the years from 1907 to the outbreak of war in 1914, and then to consider the course of events.

1.2. PEOPLE, 1907–1918

The recruits in 1907 maintained the mixture of interests which was the early characteristic of the Society. They included, as Vice-Presidents, R. A. Hadfield (later F.R.S and knighted), the steelmaster who developed manganese steel, and Professor Arthur Schuster, F.R.S., a mathematical physicist and a man of independent means who, although he lived in Manchester, was often in London – he was, too, father-in-law to R. S. Hutton, who had joined the Council in 1906.

S. Z. de Ferranti became a member of Council. He was a pioneer in the technology of alternating current and founder of the Ferranti firm. Other new members were F. W. Harbord, a metallurgist who was interested in electric smelting and who had just set up in private practice as a consulting metallurgist; A. C. Claudet, who was assayer to the Bank of England and to the Royal Mint Refinery, and was President of the Institution of Mining and Metallurgy in 1906–07; also N. T. M. Wilsmore, an Australian by birth who had worked at Göttingen and was an Assistant Professor of Physical Chemistry at University College, London; and Julius L. F. Vogel who was a consulting metallurgist and a partner of O. J. Steinhart, a member of the 1903 Council.

In 1908 there was only one new face on Council, that of H. F. K. Picard, another assayer and metallurgist who, with his partner H. L. Sulman, was a pioneer in introducing the very important processes for separating minerals by froth flotation. Later he was President of the Institution of Mining and Metallurgy (1919–20). It is interesting to note that Lord Rayleigh was willing to come back to serve as a Vice-President, although he then was President of the Royal Society; so he seems to have had a considerable regard for the Society, but the Minutes indicate that he later declined nomination as President for 1911–13, as did also J. J. Thomson.

In 1909 Hittorf was succeeded by Ernest Solvay, the Belgian industrialist who made the ammonia–soda process work. It was in this year, also, that Swinburne became President. He was relatively young (51); and for the first time the Society had a President who could and would find the time to take a really active part in its affairs. The one newcomer to Council was E. J. Bevan, F.I.C., who had been President of the Society of Public Analysts from 1905 to 1907. In about 1881 he and C. F. Cross (later F.R.S.) had founded the consulting firm of Cross and Bevan which still flourishes. The partners, together with Mr. Clayton Beadle, discovered and patented the Viscose process of regenerating cellulose. They also played a part in the development of the bisulphite process for making cellulose pulp from wood, which had been originated by Ekman. This process and the rapid development of paper-making technology and machinery were among the major factors that caused the price of paper to fall almost to a half between 1882 and 1902 (estimates by T. Y. Nuttall, supplied by Dr. P. H. Sykes). Subsequently they became widely involved in the paper industry and eventually they gained international recognition (*World's Paper Trade Review*, 21st January, 1921). Their cellulose fibres were of great interest to Swan as a source of carbon filaments; and they pioneered the industrial production of cellulose acetate.

In 1910 Dr. R. T. Glazebrook, F.R.S., the first Director of the National Physical Laboratory and a great one, became a Vice-President as did also Professor James Walker, F.R.S., who by then had succeeded Crum Brown at Edinburgh and was eminent in the affairs of the Chemical Society. There were three newcomers to Council. One was Dr. J. A. Harker, F.R.S., who was at that time Chief Assistant at the National Physical Laboratory. Another was Robert Mond, the elder son of Ludwig, who became an outstanding figure in two worlds, not only that of scientific administration but also that of exploration and archaeology in the Middle East. The third was Dr. George Senter, who at that time was a lecturer in St. Mary's Hospital Medical School and who eventually became Principal of Birkbeck College.

In 1911 Glazebrook succeeded Swinburne as President. One of the Vice-Presidents was Professor Bertram Hopkinson, F.R.S., who had been called to the Bar but had turned to engineering and physics in trying to complete the work of his father who, with three other of his children, was killed in the Alps in 1898. He was at the time Professor of Mechanisms and Applied Mechanics at Cambridge and he had married the eldest daughter of Alexander Siemens, who came back as a Vice-President in that same year. This, and the similar relation between Hutton and Schuster, perhaps indicates how small and close-knit was the scientific society of that time. The new members of Council were Dr. H. Borns and a lawyer-scientist W. R. Bousfield, K.C. and later F.R.S. Borns, born Heinrich in 1854 near Stettin, began his scientific life as an organic chemist at Greifswald. Some time later he settled in England, changed his first name to Henry, and became a scientific journalist and translator from Swedish and German. He seems to have developed wide interests in electricity, including electrochemistry and electrometallurgy. Bousfield was an outstanding patent lawyer, so probably was a friend of Swinburne's, who became interested in experimental research in physical chemistry. He did some research with Lowry; and he became especially interested in electrolysis. Both Borns and Bousfield were founder members of the Society and both were to play very useful parts in its affairs. The E. G. P. Bousfield whose views on the formation of the Society were mentioned earlier was one of W. R. B.'s sons, Edward George Paul, who started his career as an industrialist but later turned to medicine and became a specialist in nervous diseases. He did not become a member of the Society.

The general impression from these names is that the chemists were rather outweighed in eminence by the physicists and industrialists but, once again, an analysis of attendances at Council meetings shows little correlation between eminence and involvement. Some of the Vice-Presidents would occasionally attend to take the Chair; but the others appear to have been purely ornamental. Probably, also, they were very busy. It remained true, therefore, that the Society was actually run by a small group: the attendance at Council meetings averaged just over six. Those who attended most frequently from 1906 to 1911 were Swinburne, Mollwo Perkin, Lowry, Senter from 1910, Wilsmore until he returned to Australia in 1909, and of course Spiers the Secretary. Most of them were chemists.

In this context it is interesting that Glazebrook found himself very heavily involved in his work for the N.P.L., where he was preoccupied with the establishment of aeronautical research and with the necessary building of wind tunnels. In late 1912 he wrote regretting that he could give so little time to the affairs of the Society and wanting to resign, but he was persuaded to serve for the rest of his term of office.

In 1912 Hadfield came back as a Vice-President, possibly as a grooming for the Presidency, and the new eminent foreign Vice-President was Professor Kr. Birkeland of Kristiania (Oslo) who had lectured to the Society in 1906 on the Birkeland–Eyde process for fixing nitrogen. There were three newcomers to Council, viz. Reginald Belfield, M.I.E.E., Dr. Richard Seligman and Maurice Solomon. In 1908 Belfield, described by a kinsman as being an engineer of exceptional ability, was Director of the Westinghouse Metal Filament Lamp Company which made tungsten filaments by an alternative to the Osram process (*Engineering*, 1908, **65**, 714). He was also a dedicated amateur photographer. Of German ancestry, Seligman gained a Ph.D. at Heidelberg in organic chemistry. He had previously worked with Armstrong from whom, it is said, he had gained little physical chemistry but much stimulation and understanding of the role of scientific experiment. He became involved in aluminium welding, then a difficult process and one with no obvious industrial applications. When he saw an opportunity to apply it in the brewing industry, and later in the dairy industry, he launched out and with Dr. E. R. Moritz founded the Aluminium Plant and Vessel Company in 1910. By 1912 he was getting going: he proved to be a first-rate innovator and entrepreneur (G. A. Dummett, *Chem. Ind.*, 1978, p. 396) and earned wide recognition. Maurice Solomon was the Local Honorary Secretary for the Birmingham area. He was earlier concerned with carbon filament lamps and then helped develop the Osram process for making tungsten filaments (introduced ca. 1908). Later, in 1915, he became a Director of the General Electric Company.

In 1913 there were several changes that were to prove important. Three were translations from Council member to Vice-President for Bousfield, Huntington and Lowry. New to Council were Donnan, who had just returned to University College, London, where he had succeeded Ramsay as Professor of General Chemistry, Emil Hatschek, who was Lecturer in Colloid Chemistry at the Sir John Cass Technical Institute (later the City of London Polytechnic), Professor Alfred W. Porter, F.R.S., who was an Assistant Professor in Physics at U.C.L. with a special interest in thermodynamics, and E. H. Rayner of the Electricity Department at the National Physical Laboratory. Especially important was it that Hadfield became President. He had been first choice, with Alexander Siemens second choice. At first he had declined with great regret, but Huntington who was a valued friend volunteered to discuss the matter further with him and evidently prevailed. As things turned it was fortunate that he did, for Siemens had been born in Germany and an infantry private in the Franco-Prussian war; so in the rather hysterical anti-German atmosphere that developed soon after the 1914–18 war broke out, although he had been naturalized since 1878 and was 67 in 1914, he might have become a liability to the Society (see Sect. 2.1). Hadfield, on the other hand, proved to be a great asset as we shall see.

In 1914 one Vice-President was a newcomer: he was Eugen Haanel who was born in Breslau, emigrated to the U.S.A. and then to Canada where he became Director of Mines and was sent over to Europe by the Canadian Government in 1905 as a commissioner to report on electric furnaces. Bertram Blount, one of the original members of Council, became a Vice-President. The new members of Council were A. Gordon Salamon, F.I.C., and Cav. Magg. (Major) E. Stassano. Salamon was an analytical and consulting chemist who, although he had a wide range of technological interests, was particularly concerned with the brewing industry and became President of the Institute of Brewing in 1907. He was very articulate, a good linguist and he was very well informed about French business and legal customs. Stassano had designed and built a rotating electric arc furnace for smelting iron and steel, which was installed in Turin and was one of those inspected by Haanel. The continued power of the Society to attract the interest and support of very practical-minded men is striking.

In May of that year Swan died, 86 years old. His dreams for the Society had not yet been realized but he had at least seen the beginnings of fruition. Had he survived only a few more years he would have seen a much richer harvest.

In 1915 there were two newcomers on Council, namely Dr. C. H. Desch, who then was in the Metallurgical Laboratory in Glasgow, and Cosmo Johns, F.G.S., M.I.MECH.E., who first served in the shops and drawing office of the Landore Siemens steelworks in South Wales and eventually became manager of a steel mill in Sheffield. In that year members agreed, at a Special General Meeting, that the rule regarding changes in the Presidency and the Vice Presidents be suspended for the duration of the War; and in 1916 they agreed similarly that there should be no changes in Council. Consequently there were, with one major exception, no new faces until 1918 and these we will consider later. The exception was that in early 1917 Mollwo Perkin resigned as Treasurer, probably because of the pressure of his technical work, which was largely concerned with the production of fuel oils from coal and peat and with low-temperature carbonization. Until he gave up his

The Early Years, 1902–1920

academic post in 1909 his attendance at Council meetings had been exemplary; but then his appearances became less regular and between 1913 and 1915 he missed nine in succession. His resignation was received with great regret. Fortunately Robert Mond was persuaded to accept the Office.

Turning now to the affairs of the Society in the period 1907 to about 1918 we can distinguish several themes such as its general running, its finances, and its international relations although these are often interlinked. Let us consider some parts of the first of these.

1.3. GENERAL RUNNING

One of the Society's major activities was publication: it certainly was the main financial burden. In 1904 the estimated cost of printing *Transactions* and *Proceedings* and of distribution, including sending out advance proofs to members, sending copies of *Transactions* to members of the American Electrochemical Society, and paying a subsidy to *Science Abstracts*, totalled £293 7s 0d as has already been mentioned (Sect. 1.1). In 1906 the actual costs, given in the Report of Council (*Proc. Faraday Soc.*, 1907, **III**, Part 4, p. 34) are quoted below. The notes in brackets, [], are the writer's.

General printing	£ 62 18s 4d
Transactions [probably 1000 copies]	£100 18s 3d
Proceedings (including £26 paid to *Science Abstracts*)	£ 99 6s 1d
Expenses connected with the American Exchange [probably postage to the U.S.A.]	£ 36 18s 3d
Postage [probably mainly for distribution in the U.K.]	£ 28 14s 4d
Total	£328 15s 3d

The total income was £478 14s 2d, so publication took more than two-thirds of it. Because volumes of *Transactions* did not then coincide with years it is not certain how much printing was paid for, but it was probably about 240 pages of text liberally illustrated with line drawings and half-tone plates. *Proceedings* has already been described as a substantial publication (Section 1.1). Its cost was also substantial.

Heavy as this burden then was, it would be much heavier today (1978). To print 1000 copies of Volume 2 of *Transactions* would cost 40–50 times as much as the £101 that it seems to have cost in 1906, whereas the inflation ratio for prices generally is about 20-fold. This huge increase, due mainly to the cost of the labour intensive processes, can only be mitigated by changing printing techniques and by increasing the circulation. However, although library budgets and personal incomes have increased greatly since 1906 there is now a plethora of publications. Furthermore, what we may call 'Ramsay's problem' (Section 1.1) is aggravated by there being many more things for private persons to spend money on. So the competition for increased circulation is intense and the problem of choice for both libraries and individuals is great. There are now very few personal subscribers to primary journals. The Faraday Society was certainly brave; but it was also lucky to be born when opportunities were great and easier to seize, so that it could do things that now seem impossible.

According to Council Minutes the Editing Committee of 1903, which became the Editing and Publishing Committee of 1905 (Section 1.1), passed through a series of further metamorphoses. It became the Editing and Financial Committee in 1906, the Editing and Executive Committee in 1907, reverted to being the Editing Committee in 1909, becoming eventually the Publications Committee in 1915. The only definition of its powers was in 1903 but it was re-elected annually 'with the usual powers'. Several members became very long-serving such as Huntington who was Chairman, W. R. Cooper, Mollwo Perkin and, of course, Spiers. Spiers's job as Secretary and Editor was a spare-time one (Section 1.4) and his salary continued at £100 p.a. Considering his probable work load, with *Transactions* reaching 338 pages in 1913, this does not seem over generous today, even when multiplied by twenty. Publication was irregular, being apparently determined by when enough papers had accumulated. The original intention was that parts should appear quarterly; but in fact the intervals varied from two to nine months up to 1914, and during the war there was one of ten and another of fourteen months. Candidates for membership were sometimes elected by the Committee and sometimes by Council, presumably in order to speed election, but this system once produced a happy

The Early Years, 1902–1920

state of confusion when, in 1913, two candidates were elected by the Committee two months after they had already been elected by Council.

Despite the efforts to attract papers on 'pure physical chemistry' (Section 1.1) the character of the papers in *Transactions* only slowly became less applied. For example in 1909 there was a paper by Samuel Rideal (father of Eric Keightley, President 1938–45) on 'The Application of Electrolytic Chlorine to Sewage Purification and Deodorization' (*Trans. Faraday Soc.*, 1909, **4**, 179) and one by Mollwo Perkin on 'The Industrial Uses of Ozone, particularly for the Purification of Water' (*ibid.*, p. 81), the latter provoking a vigorous attack by a member of the hypochlorite school. Moreover, as we shall see, the interest in metals and especially in electrometallurgy was to continue for a long time to come.

Maintaining interest in the Ordinary Meetings proved difficult. On average attendance in 1903–04 was about 40, which was good. Already, however, in the Report for 1906 (*Proc. Faraday Soc.*, 1907, **3** Part 4, p. 31) 'Council regrets to note that attendances at meetings have often left much to be desired.' As an added inducement light refreshments were served in 1909 'at negligible cost to the Society' (3d per head, *Council Report, and Minute* of 7th June, 1909). In May 1913 an Ordinary Meeting was cancelled because of poor attendance, so a Committee which had been set up to consider the questions of the subscription and of ways of increasing membership and attendance advised that Ordinary Meetings should be reduced in number to perhaps two a year and also that there should be two Discussion Meetings annually plus occasional special lectures (*Council Minute* of 18th June, 1913). It was to be a long time before this pattern stabilized.

Up to the time of this recommendation there had been a total of six Discussion Meetings, the first four of which were mentioned earlier. We may infer that they were at least more successful in attracting members than were the Ordinary Meetings. They appear gradually to have become a form of reverse missionary activity i.e. attempts to import knowledge. In the first, on 'Osmotic Pressure', only one of the four papers was by a foreign author (U.S.A.), and in the second none of the four was; but at the third one, on 'The Constitution of Water'

(1910), three of the five papers were by Continental European authors. At the one on 'High Temperature Work' (1911) two of the five papers were by Continental authors and one by an American. At the 1912 Discussion on 'Magnetic Properties of Alloys', initially organized in conjunction with the Physical Society, out of the eleven papers five were by Continental authors, two were Anglo-German, and one was by an American; while at the March 1913 Discussion on 'Colloids and their Viscosity' only one of the five papers was by a British author – Emil Hatschek, who was born in Budapest.

On the last of these occasions the intellectual food was supplemented by a dinner between sessions, at the Trocadero Restaurant (in 1914 this cost 6s to 7s 6d per head). Moreover, for the first time the Society paid some of the travelling expenses of the foreign speakers. Council agreed to provide up to £20 from the Society's funds and also to have a whip-round among themselves. The total paid out was £35. Dr. W. Pauli of Vienna got £15, Dr. H. Freundlich of Brunswick got £10 as did Dr. Wolfgang Ostwald of Leipzig. These sums are roughly equivalent to current tourist return air fares. The whip-round produced £24 10s 0d, including £5 each from Hatschek and Beilby, so the Treasurer had only to find £10 10s 0d.

Later in 1913 there were two other General Discussions, one on 'The Corrosion of Iron and Steel' held at the Manchester School of Technology conjointly with the local Section of the Society of Chemical Industry. This was the first outside London: at it all four papers were by British authors, one of them being Bertram Lambert, who later was Tutor in Chemistry at Merton College, Oxford. The other was on 'The Passivity of Metals' when the dominance of foreign authors was again demonstrated, five of the eight papers being by Continental authors and one by an American. The picture was less unflattering to the home side at the Discussion on 'Optical Rotatory Power', with no less than fourteen papers, when Continental and British authors each contributed seven.

The outbreak of war in August 1914 disrupted plans. Several Discussions which had been considered were eventually abandoned, including one on 'The Theory of Solutions' mentioned in a Council Minute of April 1914 to which was added the note 'Invite Armstrong

The Early Years, 1902–1920

to summarize his views and ask Prof. Arrhenius or other to oppose them'. Others, like one held in November 1914 on 'The Hardening of Metals', were hurriedly organized. Nearly all those held from then on until the end of 1918 were of a practical nature and many of them were on topics related to metallurgy. It is worthwhile listing them: 'The Transformation of Pure Iron' (1915), 'Methods and Appliances for the Attainment of High Temperatures in a Laboratory' (1916), 'Refractory Materials' (1916), 'Training and Work of the Chemical Engineer' (1917), 'Osmotic Pressure' (1917), which was the one obvious exception to the rule, 'Pyrometers and Pyrometry' (1917), 'The Setting of Plasters and Cements' (1918), 'Electrical Furnaces' (1918) held in Manchester, 'The Co-ordination of Scientific Publications' (1918), and 'The Occlusion of Gases by Metals' (1918), which was suggested by Cosmo Johns. Despite the difficulties of the time some papers by foreign authors – American, French (including M. Henri Le Chatelier), Belgian, Dutch and Swiss – were communicated at these Discussions, although the authors were not able to attend.

Hadfield gave substantial introductory addresses to several of these Discussions, and there can be little doubt that he strongly influenced the choice of topics. In 1914 he gave a very long Presidential Address on 'Advances in the Metallurgy of Iron and Steel'. Huntington moved a vote of thanks in the course of which he said 'the Faraday Society, which stood for the union between theory and practice, could have chosen no more suitable President than Sir Robert Hadfield, who was recognized both here and abroad as one of the greatest living exponents of the application of science to industry and industry to science.' Hadfield was not a poor boy who had made good, as Swan was. He had had good opportunities and he had made full use of them. He was an extremely hard worker – probably an obsessive one. The reason why he hesitated to accept the offer of the Presidency was, in his own words, that 'not very robust health resulting from overwork made me very chary of undertaking what I feared might not be possible for me to fulfil.' (*Trans. Faraday Soc.*, 1914–15, **10**, 1). He was a great believer in research and he carried out many painstaking, systematic, mainly empirical investigations. It was in this way that he discovered the remarkable properties of steels containing manganese or silicon. He was a very good organizer both of himself and of others. He had liberal views on labour matters and was one of the first employers to introduce an eight-hour day into his works, in 1891. He was very hospitable and also generous: in 1914 he offered a very large Research Prize, in £200, for the best paper on subjects dear to his heart, viz. carbon in iron, steel, and alloys of iron (*Dict. Nat. Biog.*, 1931–40, p. 384; *R. Soc. Obits.*, 1939–41, **3**, 647). Small wonder that he made a marked impact on the Society.

In the Council Report for 1918 it was stated that the 14 symposia held in the period 1914–18 were attended by a total of about 3000 members and others, an average of over 200 at each, which in those days was remarkable. Indeed, the Discussion on 'Refractory Materials' was attended by 'some three hundred persons from all parts of the United Kingdom, representatives of the most varied scientific, technical, and industrial interests, and the discussion was continued until a later hour of the night. The 200 page Report was in great demand.' (*Proc. Faraday Soc.*, 1917, June issue, p. 2). Clearly they were very important in establishing the influence and status of the Society. The Council Report continues 'These Discussions will doubtless remain a characteristic feature of its work, but now that the war is over and research will be resumed on more normal lines, the Council hopes that the consideration and publication of original work in physical chemistry, pure and applied, will become an equally noteworthy feature of the Society's work. It will again be possible to hold meetings and to publish the *Transactions* at more regular intervals.' A quick inspection of the wartime and immediate post-war volumes of *Transactions* **12–17**, 1916–22 shows that the General Discussions provided roughly two-thirds of the titles and a similar proportion of the pages. They had, indeed, become the Society's major public activity. The procedure for organizing them had been regularized in 1915 by providing that Council should appoint a Special Committee for each one and that the Chairman of the Publication Committee, or a nominee thereof, should be an *ex officio* member. Thereafter, also, Council regularly reviewed progress by the Special Committees. Valuable as they had evidently proved for communication between scientists and industrialists the Discussions were not, however, enough in themselves. The concern for physical chemistry is repeated later in that Report: 'Emphasis may be laid on the growing importance of physical chemistry in the development and understanding of manufacturing processes, and members are asked to make known to those engaged in technical research the advantages which the Society offers by the publication of their investigations.' In fact there was no sudden switch to pure scientific topics in either the General Discussions or the papers.

In addition to the General Discussions there were a few special lectures. They proved difficult to arrange, partly because in the early days the Society could not afford to pay speakers their expenses. The abortive attempt to attract Arrhenius in 1904 has been described (Section 1.1) and there were other unsuccessful bids for eminences which, fortunately, were less public. Nevertheless there were four notable lectures, in 1906 by Kristian Birkeland (Section 1.1) and in 1911 by A. Scott-Hansen, also of Kristiania (Oslo), on 'Hydro-electric Plants in Norway and their Application to Electrochemical Industry', by Ernst Cohen of Utrecht, on 'Allotropic Forms of Metals', and by Edward G. Acheson of Niagara Falls, on 'Researches on Electric Furnace Products'. The topics further illustrate the interests of the Society of that time. Pursuing another of its interests, the Society took advantage of these occasions to hold informal dinners. In May 1914 Professor Paul Sabatier was in London lecturing to the University on 'Catalysis', so the Society gave him a formal dinner at the Trocadero Restaurant. Hadfield proposed his health and was supported by no less than five other speakers. Bousfield proposed the other guests, one of whom was no other than our errant friend Arrhenius: so honour was at last satisfied. In October 1913 Council invited Alfred Werner to lecture before Whitsun 1914; but the lecture was later postponed to October 1914 and in the end Werner did not come. In addition to these Special Lectures there were a few Presidential Addresses. Lodge gave one and so did Hadfield; but most of the Presidents were too old or too busy.

1.4. MEMBERSHIP AND FINANCE

An essential if unglamorous task behind the public life of the Society was that of building up membership and establishing sound finances. At the end of 1903 the nominal membership was 254 but, as has already been said (Section 1.1), a number of them never paid their subscription. In the Council Report for 1906 the Treasurer remarked bitterly 'It is difficult to understand how persons can join a Society, and while taking advantage of the privileges of membership, yet refuse to pay their subscriptions. If they find themselves unable to meet the amount of the subscriptions one would expect them to resign and not cause the Society the expense of sending them notices, papers, and publications which they never intend to pay for.' Gradually it was necessary to write off the names of those who did not pay, and the numbers fell to 210 in 1908. From an analysis of the accounts published for the years 1903–11 inclusive it seems that the total sum due for subscriptions (other than the life subscriptions) was £3601, while the total sum actually received was £3303, some of the subscriptions coming in very late, so the gross shortfall was £298, which is 8.3% of what was due. That is not too bad a figure. £201 had been written off by 1911. Later accounts do not show the subscriptions paid late or the actual sums written off, so it proves possible only to find an upper limit for these latter. There certainly was continued writing off of these bad debts, which were described (1916) as 'a blot on the accounts', and it seems highly likely that by 1919 the Society had lost several hundred pounds more. In 1908 the Treasurer ruefully compared the Society's membership of 210 with that of the American Electrochemical Society – nearly 800: 'a comparison anything but flattering to our national vanity'. By 1914 the American Society's membership had grown to 1600, and in negotiations over a possible exchange of publications the membership of the Deutsche Bunsen-Gesellschaft was stated to be ca. 800 (*Council Minutes* of 19th April 1914), but that of the Faraday Society had actually fallen to 200 so its disadvantageous position is painfully clear.

Because of its small membership and ambitious programme of publication (Sections 1.1 and 1.3) the Society's finances in the very early years were precarious. There were losses in the first two-and-a-half years' working of £119 8s 9d and £19 11s 3d. The cost of *Proceedings* proved to be greater than had been estimated, so an early saving of ca. £50 was made by not publishing *Science Abstracts* after 1905 (Section 1.1) although a £26 fee had still to be paid to the I.E.E. for 1906. A further saving of £60–80 p.a. was made by ceasing to publish *Abstracts of Patents* after 1909: Dr. H. Borns had been fighting for this economy since 1906. These savings and the steady sales of *Transactions* at about £50–70 p.a. gradually helped produce a modest working profit on the income and expenditure account; but because this included the outstanding subscriptions which proved so difficult or impossible to recover there was no true profit for several years. It was not until 1910 that earlier deficits had been paid off and that there was a credit balance in the balance sheet. By June 1912, also, the Society had received fourteen life subscriptions of £20 each, so the Treasurer was authorized to purchase £400–500 worth of Trustee Stock from the cash balance. At the same time the Society felt able to make a few modest grants, such as one of £5 in 1911 to the International Commission of Physical Constants, for which Wilsmore first asked in 1909. In 1912 two sums of up to £10 were authorized, one for the purchase of a 'lantern' (slide projector) in order to avoid the out-of-pocket expenses incurred in using such facilities in the I.E.E. large lecture theatre, and the other for a typewriter. In 1914 a sum of no less than £10 was provided for clerical assistance to Spiers. The first grants towards the travelling expenses of foreign speakers at a General Discussion in 1913 have already been noted; and in 1914 Council decided to offer second-class return fares within a limit which was to be £20 for that year. This was more generous than it now sounds, for in those days there were three classes of railway travel in this country and even more on the Continent.

The Society had no rooms of its own. Its official address was 82 Victoria Street, Swinburne perhaps providing an accommodation address. Council meetings were held in his office. Ordinary or Discussion Meetings were held in the I.E.E. rooms or later, sometimes in the Chemical Society rooms. Spiers used his works office as the office of the Society. In 1910 the Junior Institution of Engineers invited the Society to become tenants in a building it was proposing to erect but Council 'regretted not being in a position to accept the offer, as the Society could not afford at present to rent an office.' Overhead costs were thus kept very low, but only because of the generosity of older, larger, and better established Societies, and sometimes by the generosity of individuals, in subsidizing the fledgling.

By 1918 the inflation caused by the 1914–18 war had begun to bite. Spiers' salary was increased to £150 *per annum*, but probably his task was heavier, for £1 per week was allowed now for clerical assistance. In 1917 the allowance had been raised to 15s per week from the £10 *per annum* voted in 1914. In 1918 and again in 1919 there were deficits on the year's working of £202 15s 5d and of £173 15s 5d that were attributed largely to 'the present extravagant cost of printing and paper'. In the Council Report for 1918 it was stated that this had practically doubled. In the 1919 Report there is a table, which is worth reproducing, intended to show more clearly the effects of these increases. The writer has added the data in square brackets, using the published accounts.

Year	No. of pages in Transactions	Cost of printing		Total income (£)	Δ (£)	Cost per page run of Transactions (£)
		(a) Total (£)	(b) Transactions only (£)			
[1906]	[240[a]]	[237]	[101]	[479]	[242]	[0.42]
1914	300	212	[125]	439	[227]	[0.42]
1919	284	669	[425]	955	[286]	[1.46]
1920[a]	650	1200	–	1500	[300]	–

[a] Estimated.

The cost per page run of *Transactions* (calculated on the assumption that the runs were the same) appears to have been static from 1906 to 1914, but by 1919 it had increased 3.5-fold. In 1920 this was aggravated by the estimated doubling in size of *Transactions*. The total printing bills, i.e. for *Transactions*, *Proceedings* and 'advance proofs, reprints and sundries', took an ever increasing proportion of the total income; so the residue, Δ, which had to pay all the other costs, rose far too slowly. This was a recipe for disaster.

These difficulties were compounded in 1919 by the sudden need to pay for office accommodation. Spiers had given up his works office, and the room at 82 Victoria Street which had been shared with the British Scientific Products Exhibition had ceased to be available. He found an office at 10 Essex Street, Strand, at a rent of £80 p.a., which was shared with the newly founded Institute of Physics; but it had to be divided and equipped at a cost of about £88. The two working deficits produced a deficit in the balance sheet and therefore an emergency. Members were urged to pay a 'voluntary levy' of 10s per head as a suggested minimum. They responded well, and the threatened disaster was averted.

There was, at least, one good and hopeful development. After the long decline the membership suddenly started to rise in 1916, and by 1919 was 377. It is interesting to speculate why this came about. Over the years there had been a series of attempts to increase membership. In 1909 a meeting of members living in London was held at which they were exhorted to try to recruit more. Numbers fell. In 1910 local Honorary Secretaries were appointed in 'a few large centres' viz. Dr. C. H. Desch (Glasgow), Dr. F. G. Donnan (Liverpool), Dr. R. S. Hutton (Sheffield), Professor W. W. Haldane Gee (Manchester) and Mr. M. Solomon (Birmingham). Numbers rose from 207 to 216. In 1913 a further attempt was made, by allowing members of certain 'cognate societies' to apply for membership themselves and to be excused the entrance fee of £1. Hitherto the Society had requirements, rather like those of a private social club, not only that candidates for membership must be proposed and seconded by members but, for full as distinct from student membership, that they must be supported by not less than two other members, and that their applications must be displayed at the next Ordinary Meeting so that members could object within seven days if they wished. The first list of cognate societies was: the Chemical Society, the Physical Society, the Institution of Electrical Engineers, the Institute of Metals and the Iron and Steel Institute. It seems odd that the Society of Chemical Industry was not included. Later this privilege was extended to 'persons not residing in the United Kingdom' who might be accepted for election under the same conditions as members of the cognate societies. In 1916 the American Electrochemical Society was recognized as a cognate society. In the 1913 Council Report it was claimed hopefully that this relaxation of the Rules had already had a good effect; but it is possibly significant that no actual numbers of members are given in the Reports for 1912, 1913, or 1914. An actual count of the membership list for 1913 shows that there were only 197 full members and 4 student members. As a further resort two forms of corporate membership were introduced. One was that an officer of a 'provincial society having cognate objects' could be an *ex officio* member, so that the Society could receive publications and submit papers read before it for publication in *Transactions* (1915). The second was that a 'manufacturing company or firm may become a member of the Society, with power to send a properly accredited delegate to attend and vote at any meeting of the Society' (1918). Neither had an immediate effect upon the membership as judged by the 1919 membership list, although it is true that by 1926 there were nearly fifty corporate members viz. companies, research associations and libraries. What, then, had caused the increase? The writer's guess is that the relaxation of the Rules had some effect but that the main reason was the success of the General Discussions. Their form and content had met a real need. The pressures of War had developed this need and the consciousness of it: they emphasized the country's backwardness and lack of independence and so not only concentrated the minds of the scientists but convinced the politicians and industrialists, as we shall see, that much greater financial support for research was vital. To its credit the Faraday Society rose to the occasion and reaped success – though hardly material prosperity – as its reward.

1.5. DOMESTIC AND INTERNATIONAL COOPERATION

We have already noted (e.g. Section 1.1) the proper concern that the Society felt to create links with other scientific bodies, especially international ones. This theme merits some elaboration.

Very early on the Society started to cooperate with the International Congress on Applied Chemistry. In 1903 it was invited, through the Society of Chemical Industry, to nominate representatives to attend the 1903 Congress in Berlin, and subsequently to nominate members of the organizing committee. Several of the papers read at the London Congress in 1909 were published in *Transactions*. In 1909 also the Congress appointed a Commission to publish annually tables of chemical and physical constants and data, and Wilsmore became a Delegate. His first request for financial help was unsuccessful but in 1911, as has been noted (Section 1.4), the Society managed to make a grant of £5 and this was continued annually. In 1909 Wilsmore had also asked that the Society should cooperate with the Congress to consider amending electrochemical notation, so he was invited to represent it on the Symbols Committee of the Electrotechnical Commission. When he went to Australia in 1913 he was succeeded by Lowry in the first task and by Borns in the second.

In 1911 the Society became affiliated with the International Association of Chemical Societies and, once more, Wilsmore was asked to represent it on the Council. In 1914 this Association accepted recommendations of an International Commission for unifying physico-chemical symbols and passed them on. The Faraday Society, with true British independence, said that it would recommend them to its authors but that it would not seek to impose any fixed system.

The Society was not very successful, or lucky, in getting its members to go to foreign meetings. Nobody seems to have gone to the American Electrochemical Society meeting in St. Louis in 1904 (Section 1.1). It is true that J. B. C. Kershaw did represent the Society at the 1904 meeting of the Deutsche Bunsen-Gesellschaft; but after his report in which he commented very frankly on the German style of lecturing then current (*The Electrochemist and Metallurgist*, 1903–04, **3**, 736) he seems not to have been invited again. The Society tried twice (1912 and 1913) to organize a tour of Norway, to see the electrochemical works, but failed. In 1913 it was invited to take part in the International Engineering Congress to be held at San Francisco in 1915, but the 1914–18 War killed British participation. Similarly an invitation from the American Electrochemical Society to join with the Deutsche Bunsen-Gesellschaft and the French Société de Chimie Physique in a meeting 'at some industrial electrochemical centre' in 1916 or 1917 was accepted in 1913 but thwarted soon after.

The War also killed negotiations between the Society and the Deutsche Bunsen-Gesellschaft for an exchange of publications like that with the American Electrochemical Society, which had been started in 1913 and which, incidentally, showed that the German society was about four times as large as the British one. It is worth noting that although the war caused great bitterness towards Germany in the country as a whole, the Council decided in 1915 to leave the names of German members in the list soon to be published.

Without doubt, the Society's most ambitious and important international connection was that described in the first Council Report (*Proc. Faraday Soc.*, 1905, **1**, 72) viz. 'Another arrangement, which appears to be the first instance of its kind of such an interchange between scientific bodies, is the system of exchange with the American Electrochemical Society, by which members of each of the allied Societies receive the publications of the sister Society free of charge. The exchange came into force in August 1903, and it has given much satisfaction both here and in the United States, which is the home *par excellence* of electrochemistry.' As was said earlier the Treasurer had worries about this scheme from the start, and with good reason. Although the Faraday Society could charge 6s p.a. per member for any number of copies supplied over 600 it had, nevertheless, to supply two to three times as many copies as it received and pay a correspondingly large postal bill, made larger by the unwillingness of the American society to agree that each society should send copies in bulk for posting to individual members by the respective Secretaries. While the extra copies were at run-on cost, this would not have been negligible. The accounts are not explicit. There is an item 'Expenses connected with the American Exchange' that appears regularly from 1905 to 1918. For the first few years it was £30–40 p.a. and could have been the postage bill. At first there were no payments from the American society but from 1909 to 1917 there were and they roughly balanced the 'expenses' item.

In 1915 there was an embarrassment which arose from this very generous policy. A Dr. Turner discovered that it was possible to join the American Electrochemical Society for only 25s p.a. and so to get the Faraday Society *Transactions* much more cheaply than as a member of the Faraday Society (£2). He told Council which promptly asked the American society not to send the *Transactions* to its members in the United Kingdom or to accept transferences of membership therefrom. Later that year there was a crisis. Professor J. W. Richards, of the American society, gave notice to terminate the exchange 'in consequence of disturbed financial conditions.' It was suggested that an arrangement by which each society received the *Transactions* of the other on payment of a small sum might be arrived at. The President (Hadfield), with a grandly generous gesture, offered to make a contribution to the American society to help over the present difficulties. This was later transformed into an indemnity of £50 to the Faraday Society against loss. In 1918, after long negotiations by Hadfield and a temporary arrangement in 1916, eventual agreement was reached on the basis of a mutual payment of $2.25 per member receiving the *Transactions* of the other society. The American society generously made a retrospective payment back to 1917 on this basis: it totalled £56 3s 9d (*Council Minute* of 29th May 1919). From then on there is no further mention of the American society in the income and expenditure accounts.

From all this it is clear that the Society soon acquired an effective international voice and one that was not entirely muted by the War.

Turning to links within the United Kingdom, we find that as early as 1909 the Society was invited by the Ceramics Society to be represented on a Committee for grading and standardizing refractory materials. This cooperation presumably led to the outstandingly successful General Discussion in 1916 that was mentioned earlier, and later to participation in conferences as well as to its support for a National Association of Refractories Research, an early example of a Research Association. It became the custom of the Society to invite members of, to use its favourite adjective, 'cognate' societies to its special lectures, and we have seen how it eased conditions of membership for them. The 1914–18 war gradually brought a need for greater domestic cooperation and so led to the birth of new organizations. In 1915 Hadfield and Swinburne were appointed to serve on a General Committee formed by all the societies interested in chemistry 'to be thoroughly representative of chemical opinion'. In 1918 a Federal Council for Pure and Applied Chemistry was created and, at the invitation of the Chemical Society, the Faraday Society nominated a representative, namely W. R. Cooper. In 1915 the Royal Society sought to establish a Conjoint Board of Scientific Societies, and in 1917 Donnan and W. R. Cooper became the Faraday Society representatives. This Board urged 'closer coordination between cognate Societies' affiliated to it.

There were other joint bodies and projects to which the Society contributed experience or in which it expressed interest. One such was the British Scientific Products Exhibition, organized by the newly founded British Science Guild: Cooper spoke for the Society on electrical matters, while Hadfield, Mond and Mollwo Perkin were members of an Organizing Committee of which Spiers was Secretary. The Chemical Society also invited the Faraday Society to nominate a representative on a joint Committee which was being formed to consider publishing 'tables and other works of reference (in English) covering scientific and industrial chemistry'. Senter was appointed and with the help of Harker, Bousfield and other Faraday stalwarts managed to secure important modifications to the original proposals. The General Committee of Chemical and Allied Societies, which was the precursor of the Federal Council for Pure and Applied Chemistry already mentioned, resolved that 'it is expedient to publish an English work of reference covering scientific and industrial chemistry on lines similar to those of Beilstein.' These proposals proved much easier to make than to bring to fruition. They showed far too little appreciation of the scale of effort required and of the inadequacy of the resources of scientific manpower available in the United Kingdom.

Another example of greater cooperation was that the Chemical Society decided to make its library available to other societies and, with Senter and Spiers negotiating for the Faraday Society, a satisfactory arrangement was soon agreed which required the payment of 5s p.a. by those members who wished to use it. In August 1918, 27 did so wish. The Library Committee was enlarged and Senter was the Faraday Society representative.

One of the most important organizational tasks in which the Society was involved was the founding of the Institute of Physics, which it did in collaboration with the Optical Society and the Physical Society in 1918, joined by the Royal Microscopical Society and the Röntgen Society in 1919. Faraday Society connections were clearly evident, for Glazebrook was the first President, Hadfield the Treasurer, A. W. Porter the Honorary Secretary and Spiers the Secretary. In addition Cooper and Rayner were representatives on the first Board of the Institute. This was how the Institute and the Society came to share an office (Section 1.4).

In 1918 the British Non-Ferrous Metals Research Association turned to the Faraday Society for support in its attempt to obtain changes in the conditions of support of Research Associations by the recently formed Department of Scientific and Industrial Research.

All these examples show the Society acquiring a very definite status in the United Kingdom and being consulted and involved in all kinds of problems of scientific organization that wartime pressures helped to create. The Society, moreover, became very actively involved in some specific war activities which merit a brief review.

1.6. WAR-ORIENTED RESEARCH, 1915-1918

The first step taken by the Society, at the suggestion of Rayner who was at the National Physical Laboratory, was to place 'at the disposal of the Government such expert knowledge as in available among its members.' (*Council Minutes*, 29th June 1915). Appended to the letter conveying this offer was a list indicating some of the subjects on which the members had expertise. A reply was received expressing Mr. Lloyd George's appreciation (*Proc. Faraday Soc.*, 1916 [June], p. 3). Later in 1915, at the request of the Royal Society, a special war service register was compiled showing the special interests of each member, and those who were on active service. Many of the Officers and members were already on Government Boards and Panels that had been set up since the beginning of the War. Hadfield was prominent among them because his steel firm was an important manufacturer of munitions; and it is clear from the Minutes that he was a very useful channel for communication with Government.

At the invitation of the Chemical Society, again in 1915, the Faraday Society nominated members to a Special Committee 'to consider and advise the Council of the Chemical Society with regard to inventions and suggestions which might be submitted to it based on electrochemical and chemico-physical processes' (*Proc. Faraday Soc., loc. cit.*). In October 1915, 'In response to an invitation from the President of the Board of Education inviting the cooperation of the Society, it was decided to formulate some researches which could be carried out under the auspices of the Society and to apply to the Advisory Council for Scientific and Industrial Research for a grant to carry out the research or researches approved' (*Council Minutes* 4th October 1915). The

Advisory Council, precursor of the Department, had been created in August of that year. A Committee was formed and it later proposed research on 'The Setting of Plasters and Cements', a topic which does not seem of great relevance or urgency in the circumstances of the time. Lowry and Desch, who were to supervise the work, applied for a grant of £510 for 1916, but in March of that year the application was reduced to only £10, for documentation, and the research was not pursued. However, a new proposal that originated from Lowry was much more to the point. This was for research 'on the caking and disintegration of simple salts, in view of the immediate bearing of the subject on certain explosives problems', and obvious reference to the caking of ammonium nitrate. The Council Report for 1916 shows that a grant of £240 had been made, and in October 1917 Council agreed that an application for a further £190 should be made. This research was eventually described in a published report (T. M. Lowry and F. C. Hemmings, *J. Soc. Chem. Ind.*, 1920, **39**, 101T).

A much more important project started with the setting up of a small 'Nitrogen Products' Committee to study the question of the production of synthetic nitrogen compounds. According to the Council Minutes of 2nd March 1916 this was on the motion of Lowry. The members were Hadfield, Huntington, Cooper, Harker and Spiers, Harker (of the N.P.L.) being chairman. The Committee got to work with a feeling of urgency. It met four times in April 1916 and again in May. It co-opted Dr. H. C. Greenwood, Dr. Joseph Knox, Mollwo Perkin, Dr. R. Messel and Murray Morrison: and it started by gathering information.

That it started from scratch is indicated by the ordering of four copies each of Jobling's 'Catalysis' and of Knox's 'Fixation of Atmospheric Nitrogen' at 2s 6d and 2s 0d respectively. The Committee managed, through the efforts of Huntington and Harker, to interest the Explosives Department of the Ministry of Munitions. The Ministry formed a Nitrogen Products Committee of the Advisory Panel of the Munitions Inventions Department and invited the Society to appoint Huntington as its official representative. Council readily agreed. Through this channel the Society's Committee was informed that there was plenty of gasworks ammonia in the country but not enough ammonium nitrate or nitric acid, it was reported that cyanamide ammonia or gasworks ammonia was inferior to the pure, dry Haber ammonia for the Ostwald process of ammonia oxidation, because the impurities in the former tended to poison the catalyst, although there were claims that these difficulties had been overcome. They found that there was practically no information outside Germany about the details of a Haber process plant.

In May 1916 the Committee sent an interim report to the Ministry and in July, at the invitation of the Comptroller, it set up experimental research, mainly in Donnan's laboratories at University College, London.

This was financed by a direct Treasury grant, initially of £250 for six months. Harker, who had been seconded to the Munitions Inventions Department and was responsible for organizing the work on nitrogen products, became Director of this research group. The leader was Greenwood, and the group included E. K. Rideal who had been invalided out of the Army, E. B. Maxted, J. R. Partington and Hugh Stott Taylor: this interesting group had been assembled by Donnan. Greenwood and Rideal managed to build a prototype pressure reactor for the Haber process (see D. D. Eley, *Biog. Mem. Fellows R. Soc. London*, 1976, **22**, 383), presumably before October 1917 when the Explosives Department recommended to the Minister that a National Factory for the Haber and the Ostwald processes should be erected. Greenwood had worked with Haber, and Rideal had had experience with a 100 atm hydrogen furnace when working with Anschutz at Bonn; but even so they had been remarkably quick in achieving their purpose. Furthermore, Rideal and Taylor (both then 27) developed a catalyst for the selective oxidation of carbon monoxide in the presence of hydrogen (water gas), as part of the programme for developing a process for making pure, cheap hydrogen which was a vital requirement. The Society's Committee organized a report to the Ministry's Committee covering all aspects of the production of nitrogen compounds, and this formed the basis of a large, official report (357 pp.) by that Committee, published soon after the war (*Ministry of Munitions of War, Munitions Inventions Department*, 1919, *Nitrogen Products Committee, Command 482*, London, H.M.S.O.). It set out the economic merits and the current practicability of the several fixation and oxidation processes. It states that in February 1918 the final recommendation was made by the Explosives Department to develop the Haber process on an industrial scale and to give up attempts to develop the cyanamide process which, though less novel, required too much electrical energy. The decision was announced in the Commons in April 1918 by the Minister, Mr. Winston S. Churchill.

In February 1918 the Ministry's advisers expected to have a factory producing within a year (1919, *Command 482*, p. 142), but in reality there was a huge chemical engineering gap between the small-scale plant developed by Greenwood's team and a full-scale one. The Badische Anilin-und-Soda-Fabrik, which ran the Haber–Bosch process at Oppau and at Merseburg, had needed all the expertise of Krupps, the gun-makers of Essen, to develop a working plant over about 1908–11. One who seems not to have shared this optimism was Major F. A. Freeth (of whom more later). During the war he had been the moving spirit in the development by Brunner, Mond and Company of three double-decomposition processes for making ammonium nitrate (Sir Peter Allen, *Biog. Mem. Fellows R. Soc. London*, 1976, **22**, 105).

The day after the 1918 Armistice was declared he suggested to Lord Moulton, a mathematician turned lawyer who was Director General of Explosives Supplies, that a scientific mission should be sent to Germany 'to pinch everything they've got'. Moulton replied that a proposal for burglary should not be made to him, a Lord of Appeal. Nevertheless, in February 1919 a mission led by Brigadier Harold Hartley (of whom, also, more later) went to Germany. Their time at Oppau was too brief to be of much use, so in April 1919 a second mission went. This included Greenwood and two members of Brunner, Mond and Company (see W. J. Reader, *Imperial Chemical Industries Ltd., A History*, Vol. 1, pp. 354–355). The firm had been persuaded, half-reluctantly, to take an interest: Freeth said that Donnan played a considerable part in the persuasion (*Biog. Mem. Fellows R. Soc. London*, 1957, **3**, 26).

Its ambiguous attitude seems to have been due to feelings that the acquisition of this technology would be advantageous but that the task would be difficult and costly, and that the post-war market for nitrogen was likely to make the project of dubious commercial value. All three feelings proved correct. The second mission did their best to extract information, but the Germans knew that they were being burgled and were wholly uncooperative. Indeed, they burgled back. Somebody sawed through the floor of a locked and guarded railway van and took the notes which all the members of the mission, save one, had left in it at the end of their six-week task.

According to Reader (*op. cit.*) much the most valuable information was what the firm, after considerable hesitation, bought from two Alsatian engineers who had worked both at Oppau and at Merseburg and who offered it in the autumn of 1920. By mid-1921 ammonia was being produced at an experimental plant at Runcorn at the rate of about 200 tons p.a., and by the end of 1923 the first unit at Billingham came on stream. Thereafter production increased steadily and by 1925 was of the order of 1000 tons p.a., which was about one thousandth of that of B.A.S.F. in 1920. So, after a long and difficult start, a new industry was established in the United Kingdom, triggered by the Society's initiative of 1916 (Sir Robert Hadfield, *Trans. Faraday Soc.*, 1920–21, **16**, 467).

From their work on this project Rideal and Taylor became interested in solid catalysts (see D. D. Eley, *Biog. Mem. Fellows R. Soc. London*, 1976, **22**, 383) to the knowledge and understanding of which they later both made contributions of the greatest importance; and they formed a lifelong friendship.

1.7. MORE ABOUT PEOPLE

In a variety of ways the 1914–18 War presented the Society with a series of challenges which it met with great versatility and success. As an institution the Society had arrived. We have, however, consistently treated it not merely as an institution but also as a body of people; and it is appropriate to end this chapter as it was begun, by putting the emphasis on people, first mentioning again some whom we met earlier.

Glazebrook's development of aeronautical research at the National Physical Laboratory was largely responsible for the superiority of British aircraft in the early days of the war. Professor James Walker set up the manufacture of TNT in Edinburgh in 1915. In 1918 Sir Herbert Jackson became Director of another of the new research associations, namely that for the British scientific instruments industry. Beilby became Director of the Institute for Fuel Research in 1917 and so had a part in introducing the 'Therm' to the British public. Soon after the war began, Bertram Hopkinson was commissioned in the Royal Engineers, eventually becoming a Colonel. He made a number of important engineering contributions to the testing, use and effects of explosives; but he also had to do with other problems including enemy ciphers and the arming of aeroplanes. Because of the last duty he learned to fly, and in 1918 he was killed in a plane crash in bad weather. Huntington died in 1920, in his sixty-ninth year, having worked indefatigably for the Society's interests from its very beginning almost to his own life's end. He is one of its great benefactors. Capt. (Dr.) H. C. Greenwood, of the Haber process team, died that same year: his untimely end deprived the Society of one of the most promising of the younger members (*Council Report*, 1919, *Proc. Faraday Soc.*, July 1920).

Among the younger members of that time there are several who, for one reason or another, deserve mention. One was Dr. A. F. Joseph who, when he was Government Chemist in the Sudan, lived next door to a family named Crowfoot which included a daughter named Dorothy. Much, much later Dorothy Crowfoot Hodgkin, D.SC., O.M., F.R.S., told the writer that it was 'Uncle Joseph's' early encouragement that largely excited her interest in science and, later his introduction of her to Lowry, who advised her to work with J. D. Bernal, that indirectly turned her interests to the determination of the structures of biologically important molecules. Another was Dr. Alexander Fleck, later to become Chairman of the Main Board of Imperial Chemical Industries, Ltd. A third was T. R. Merton, whose first major interest was spectroscopy but who gradually turned from science to invention and who proved to be a most ingenious and prolific inventor of very useful devices based on physical principles. In the second World War, his invention of the long-persistence radar screen, the ultra-black paint for bombers and the diffraction range finder for fighter pilots chasing V1s were of the greatest value. In peacetime probably his most important technical contribution was the reinvention of the elastic nut

The Early Years, 1902–1920

engaging with many threads of a screw, the 'Merton nut', which greatly facilitated the ruling of diffraction gratings. The breadth of his interests and the simplicity and directness of his style are well shown by his final paper 'On a barrier against insect pests', including ants, earwigs and ladybirds. 'The latter were examined because they are closely related to the bed bug, which I have not had the opportunity of testing' (*Proc. R. Soc. London. Ser. A*, 1956, **234**, 218). Merton was a man of wealth, so he was able to indulge his love of Renaissance paintings of which he formed a fine and remarkably homogeneous collection (H. Hartley and D. Gabor, *Biog. Mem. Fellows R. Soc. London*, 1970, **16**, 421). A fourth was J. W. Nicholson, 'the brilliant but erratic and unfortunate genius' (Hartley and Gabor, *loc. cit.*), a mathematician who collaborated with Merton on spectroscopy, who proposed the model of an atom with a positive nucleus and planetary electrons before Rutherford did, and who in 1911–12 saw the essential quantizing condition for such a system, viz. that angular momentum must change by steps of $h/2\pi$, but who failed to propose the clear coherent theory that Niels Bohr did in 1913 and so received almost no recognition. A fifth young member was H. T. Tizard, who much later played a vital role in the application of science to military problems before and during the second World War. In the first World War he became a protégé of Bertram Hopkinson who was, he said, one of the great formative influences in his life. During the second World War perhaps his most important contribution was his mission to the U.S.A. in 1940 that really initiated work in that country on the proximity fuse, on certain aspects of radar, and on the atomic bomb which, however, he hardly could believe would work (Sir William Farren and R. V. Jones, *Biog. Mem. Fellows R. Soc. London*, 1961, **7**, 313). Finally we may mention a master at Eton College, Windsor, named John Christie, who is said to have taught physics unconventionally and whose home address was given as Glyndebourne, near Lewes, Sussex.

CHAPTER TWO

The Inter-war Years, and the Second World War: Council and the General Discussions

IN THE LATER part of Chapter 1 we saw how the Society was beginning to face peacetime conditions; but nothing was said about the men who were to lead it, except to mention the new Honorary Treasurer, Robert Mond.

This was deliberate, for it seems better that we take a careful look at the succession as our first task in this Chapter.

2.1. 1918–1922

In 1918 (*Minutes* for 29th August) Council decided that Professor A. W. Porter, F.R.S., should be nominated as a Vice-President. This nomination and Salamon's death in April 1918 created two vacancies on council for which G. S. Albright and Harold Moore were nominated. George Albright was the second son of one of the founders of the firm of Albright and Wilson, whose principal manufactures were phosphorus and potassium chlorate. He had been a founder member of the Faraday Society and during the war he was a chairman of the Nitrogen Products Committee of the Munitions Inventions Department (see section 1.6). Unfortunately he could not attend any of the Council meetings in 1918–19 and declined renomination in 1919. Harold Moore was chief metallurgist in the Chemical Research Department at Woolwich Arsenal. He had coped very successfully with an overwhelming load of work during the war and his group had increased from four (including himself) to forty. In 1918 he was 40 years old and in his prime. In 1920 he became Director of Metallurgical Research at Woolwich and in 1932 he became Director of the British

Non-Ferrous Metals Research Association. He attended Council meetings regularly.

In 1919 (*Minutes* of 29th May) Council unanimously asked Hadfield to be President for one more year: the war being technically not yet over this was within the terms of the 1915 special resolution. Also, in order to avoid the need of electing a complete new Council in 1920 the three senior Vice-Presidents viz. Bousfield, Huntington and Lowry offered to resign, as did members of Council. The upshot was that Cooper, Hatschek and Rayner were nominated as Vice-Presidents, and Professor A. J. Allmand, Dr. H. Borns, Dr. W. Rosenhain and Sir Thomas Kirke Rose were nominated for Council. Professor J. R. Partington was later added to replace Albright. Borns's return speaks for the continuing value of his services to the Society. Allmand had just become Professor of Chemistry at King's College, London. He had been one of Donnan's first research pupils at Liverpool and had then worked with Haber at Karlsruhe (1910–11) and with Luther at Dresden (1911–12). His first and main interest was electrochemistry, but he was becoming interested in photochemistry (*Obit. Not. Fellows R. Soc. London*, 1954, **9**, 3). Rosenhain, an Australian, was a metallurgist. At Cambridge he was co-discoverer with Professor Sir Alfred Ewing, F.R.S., of slip bands in metals. He then worked on optical glass with Chance Brothers but continued to work privately on metallurgical problems. In 1906 he joined the N.P.L. and in 1919 was Superintendent of the Department of Metallurgy and Metallurgical Chemistry. Kirke Rose was Chemist and Assayer of the Royal Mint. He had been President of the Institute of Mining and Metallurgy (1915–16) and a Vice-President of the Institute of Metals. He failed to attend any Council meeting, and in 1922 he resigned from the Society. Rosenhain resigned at the same time, but he had attended a few meetings. Partington, a Manchester graduate who had worked with Lapworth and then with Nernst in Berlin, had in 1919 become Professor of Chemistry at East London College (now Queen Mary Westfield College). He became well known as a writer with encyclopaedic knowledge, first about inorganic chemistry, then about physical chemistry, and later about the history of science.

Rather late in 1920 (*Minutes*, 28th October) Council decided on the remaining changes necessary to re-establish a peacetime mode of operation. Porter was nominated as President. Desch, Harker, Lowry and Senter were nominated as Vice-Presidents. Sir Robert Robertson, K.B.E., F.R.S., Mr. C. C. Paterson, O.B.E, Professor W. C. McC. Lewis and Professor A. O. Rankine, O.B.E., were nominated for Council.

Porter, who had been on Council 1913–18, was a kindly, solitary person, a bachelor and said not to be unduly ambitious. He was a Professor of Physics at University College, London. He was primarily a teacher. From his original contributions – 'numerous, notable and varied' – he appears as a critic of current theories rather than as a pioneer of wholly new ideas (*R. Soc. Obits.*, 1939–41, **3**, 87). He was the first academic the Society had had as President since Lodge (1908–09), and gave far more time and thought to its affairs than Lodge ever did. However, in the Officers and Council taken as a whole, the balance between pure and applied science remained much as it had been. Robertson, whose father had been acquainted with Faraday, was Director of Explosives Research in the Research Department at Woolwich Arsenal. He had a first-rate record for problem solving. He is described as being in his youth 'studious, persevering, pushful and highly self-confident. He was ambitious and determined to succeed in life.' He did. He was a dedicated man but he was described as a hard one, autocratic and ruthless with his juniors, a good man at the helm in an emergency, but not very loveable although he had some close friends (*Obit. Not. Fellows R. Soc. London*, 1948–49, **6**, 539). Peterson had been in the N.P.L., and by 1919 was Superintendent of the Light Division; but in that year he joined the General Electric Company to establish a research laboratory at the invitation of Lord Hirst, the founder of that company. He showed a flair for creating an environment where scientists and engineers could live in harmony. Lewis, a Belfastman, had been a research pupil of Donnan slightly later than Allmand and F. A. Freeth (of whom more later). He succeeded Donnan in the Chair at Liverpool in 1913. He was interested in the study of chemical reaction rates. Later, he became deeply interested in colloidal and physico-biological chemistry. He was a very retiring, sincere, kindly man who rarely attended scientific meetings but, at least at first, he did attend Council meetings (*Biog. Mem. Fellows R. Soc.*

London, 1956, **4**, 193). Rankine was an applied physicist, best remembered for his work on the viscosity of gases and for deriving sizes of molecules therefrom. He became increasingly a scientific administrator. He was a founding Fellow of the Institute of Physics, hence having close contact with the Faraday Society, and later was its Honorary Secretary (1926–31). In 1920 he was a Professor of Physics at Imperial College, London.

One feature of the new body of Officers and members of Council was that most of them were relatively young. Although some of the pre-war old hands were on the list, viz. Porter, Borns, Cooper, Hatschek and Lowry, only Borns was over 60 in 1920 (he was 66). Porter was the next oldest, at 57. Lewis, Allmand and Partington were young Turks in their mid-thirties, and the others were in their forties or early fifties. There was not one grand, ornamental old man like the early Presidents. Those who had served with Hadfield had become accustomed to the idea of a hard-working President. In effect, a new generation with new ideas had taken over. The problems that they faced certainly required of them all the vigour and flexibility of mind that they could muster. We may note that with Kirke Rose, Rosenhain and, especially, Desch and Moore on the list the Society maintained a strong interest in the science of metals.

Council decided in September 1921 to organise a dinner for Hadfield in recognition of the great services which he had rendered to the Society during his Presidency; but no details survive.

In the first Chapter we saw that by 1920 the Society was very lively and thriving scientifically. Membership, then 419, was showing a welcome rate of growth. However, the Society was still in an early stage of development. It had to promote greater activity in its chosen field. It had to attract more and better original papers and to proselytize industry. As time went on it had, moreover, to sort itself out: it had to discover what its central interest was. This continuing task proved more difficult than might be supposed. By 1920 the Society was sufficiently established for its influence to increase quasi-autocatalytically. It was consulted as a matter of course and it had to meet new responsibilities for representation. Commendable and hopeful as was all this scientific activity there lay beneath it, however, the constant, nagging problem of how to achieve financial stability. Thus, there are much the same themes to develop as before. Because the General Discussion Meetings had already proved so important in the life and growth of the Society we will look next at their development after 1918, and it will be necessary to consider many of them individually because they were very important occasions that had great influence not only in this country but internationally. A complete list of all that were ever held by the Society is given as Appendix C with standardized information for each. A series of reviews of the several fields in the purview of the Society, and of how the General Discussions had contributed, was published as part of the celebrations of the fiftieth anniversary of the founding of the Society (*Trans. Faraday Soc.*, 1953, **49**, 515–584). The Discussions were often closely linked with personalities in the Society; so we will note changes in Officers and on Council.

The first General Discussion in 1919 (in January) on 'The Present Position of the Theory of Ionization' reverted to the Society's very early interests. Papers were contributed by Svante Arrhenius (of Stockholm), S. F. Acree (of Syracuse University, New York,) and J. C. Ghosh (Calcutta) but none of them seems to have attended; so international communications had not yet been fully restored. With hindsight the brief note by S. R. Milner on 'The Effect of Interionic Forces in Electrolytes' (*Trans. Faraday Soc.*, 1919–20, **15**, 148) appears particularly significant, for he claimed to have shown that the effect of ion fields on the kinetic freedom of the ions can explain the properties of electrolytic solutions; but almost no attention was paid to him. Four years later P. J. W. Debye and E. Hückel, using a much more elegant approximate treatment, convinced everybody that the basic ideas were right. Lars Onsager nearly anticipated them. Thus there was a sort of spontaneous nucleation of these ideas within a short time (see e.g. *Biog. Mem. Fellows R. Soc. London*, 1970, **16**, 195–199). A postscript is added by Lord Dainton, F.R.S., who, while an Oxford undergraduate, was at home in Sheffield during a vacation and met Milner. They discussed the Debye and Hückel treatment: Milner obviously understood it fully but he was an old-style classical physicist who approved of rigour, disliked approximation, and was one of the world's worst expositors. Debye had a flair for making the fruitful approximation and, as the writer vividly remembers, was one of the world's best expositors.

The next five General Discussion were mainly, or largely, about practical matters. One, on 'The Examination of Materials by X-Rays' (April 1919), that had been suggested by Hadfield, was organized jointly with the former Röntgen Society which was mainly concerned with the physics of radiography, and which provided financial help. Alone among the radiographers W. H. Bragg pointed out that X-rays could also be used to investigate crystal structure and that this might prove important. He said 'Hardly anything has yet been done. The war is so recently over than no one, in this country at least, has yet been able to get on with the work, but it is to be hoped that it will soon be taken in hand very seriously.' His hopes were realized beyond his dreams. In Hadfield's introduction to the Discussion, he mentions only one foreign guest as being present viz. 'Dr. H. M. Howe, the well-known American Scientist and Metallurgist' who was not an author. There were several foreign authors, French or American, but there is no evidence that they actually came.

Another General Discussion, on 'The Microscope: Its Design, Construction and Applications' (January 1920) was also dear to Hadfield's heart and was organized jointly with the Royal Microscopical Society, the Optical Society and the Photomicrographical Society, which societies shared the costs of the meeting and of the Report, up to 2500 copies of which were to be printed. At that time the microscope was the key instrument in research on metals. Although there were certainly ten foreign authors, again French, Italian or American, it is not clear that any of them were present.

A Discussion on 'The Physics and Chemistry of Colloids and their bearing on Industrial Questions' (October 1920) was organized jointly with the Physical Society, and this time the cost of the Report was met by the Department of Scientific and Industrial Research, which arranged for it to be printed by His Majesty's Stationery Office. The moving spirit behind this Discussion, as behind the earlier one on colloids in 1913, must have been Emil Hatschek. The only foreign visitor certainly present was Dr. The Svedberg (Uppsala). His paper was a review of the subject in its still primitive state. He mentioned sedimentation rates and equilibria as means of determining the sizes or masses of colloidal particles, but said nothing about the ultracentrifuge, the construction of which was progressing but was not to be completed until 1924. Wolfgang Pauli (Wien), Wolfgang Ostwald (Leipzig) and H. Freundlich (Berlin-Dahlem) contributed papers but were not present although they had come to the 1913 meeting. This was the first occasion since the war when papers were invited from Germans or Austrians. Some scientists in this country had come to feel great bitterness toward them, as is described by Professor L. Badash (*Notes and Records, R. Soc. London*, 1979, **34**, 91–122). It directly affected two eminent members of the Faraday Society. Alexander Siemens, who had been a candidate for the Presidency in 1913 (Section 1.2), was forced to decline his re-nomination as Secretary of the Royal Institution, of which he had been a member for 33 years. Arthur Schuster, who had been a Vice-President of the Society in 1907–09 (Section 1.2), was under great pressure from some people of strong feelings to withdraw from re-election in 1918 as Secretary of the Royal Society but, at the urging of the Council of that Society, he did not and he was re-elected. The Faraday Society seems not to have been riven thus. As has already been noted (Section 1.5) Council decided in 1915 against expelling its German members. This was on the motion of the President, seconded by Lowry. From his writings, Hadfield appears as a single-minded, rather naive patriot; but evidently he was not a narrow-minded one. Borns, who was German-born, was re-appointed in 1916 to represent the Society on the International Electrochemical Commission; he also translated and abstracted German patents for the Nitrogen Products Committee (Section 1.6).

A more general reason for the manifest difficulties of restoring full international participation may have been the parlous state of the Society's finances immediately after the war. In 1914 (*Minutes* of 12th March) Council resolved that:

'(1) When foreigners were invited to take part in General Discussions second-class return fares might be offered.

(2) The Council should allocate a certain fixed sum for each General Discussion to the Sub-Committee having in hand the arrangements.'

The Inter-war Years, and the Second World War

For 1914 the Treasurer was authorised to spend not more than £20 for this purpose. There is no further Minute relating to travelling expenses until 6th March 1923 when it was agreed that Professors Goldberg and Luther could have £10 each, and on 17th July 1923 Council decided to offer not more than £5 each to any foreign contributor to a General Discussion approved by a Sub-Committee to enable him to attend: in 1925 Council voted £5 5s 0d each to Professors Bodenstein, Franck and Winther. Therefore it is possible that the 1914 resolution remained in abeyance after the war until 1923.

Also in 1920 there was held a Discussion on 'Electrodeposition and Electroplating', a topic very close to the Society's original interests. It was organized in consultation with the metallurgists at Sheffield and was the first o be held there. It was a purely domestic occasion. So, too, was one on 'Basic Slags: Their Production and Utilization in Agriculture' (1920). This was the first of two which arose from a proposal by Dr. W. H. Hatfield, Director of the Brown-Firth Research Laboratories at Sheffield, following the advice of Dr. E. J. Russell, F.R.S., who then was Director of the Rothamsted Experimental Station, to whom the proposal had been referred. The second one was on 'Physico-chemical Problems of the Soil' (1921). Russell gave the introductory addresses at both: and after the second of these meetings a party of members visited Rothamsted. This time there was one foreign visitor, namely Sven Odén from Uppsala. The Chair at the second meeting was taken by Sir A. Daniel Hall, K.C.B., F.R.S., who had preceded Russell at Rothamsted and then become Scientific Advisor to the Ministry of Agriculture and Fisheries. By a happy symbiosis the Treasury made a grant of £100 from the Development Fund to help pay for the cost of publishing the Report. Although crop yields were discussed and there were papers on organic constituents of the soil neither of these Discussions could be deemed biological. The soil was treated essentially as a physico-chemical system. They were, however, the nearest that the Society had yet approached to biology except for a paper on proteins in the 'Colloids' Discussion.

The pattern of joint organization and sharing of costs seen in a number of the post-war meetings helped the Society over a particularly difficult time and was a useful exercise in interdisciplinary discussion but, although it was occasionally repeated, it did not become usual.

We may note here that there were three recuits to Council in 1921. One was Cosmo Johns, the steelman, who returned. Next was Dr. J. N. Pring (born 1884), a Manchester graduate whose early research interests centred on high-temperature studies. After war service he rose to be Reader in Electrochemistry before leaving in 1921 to take charge of Physical Chemistry research at Woolwich Arsenal. The third man was Dr. E. K. Rideal, now 31 years old, back from a sojourn at the University of Illinois, and established in Cambridge as Humphrey Owen Jones Lecturer, who now made his first appearance on Council where he quickly became a power.

There were three other Discussions in 1921, so that in this year as in 1920 there were no less then four. They were a brief, domestic one on 'Capillarity', held in Manchester with the Literary and Philosophical Society, and two full-scale ones in London. The next, on 'The Failure of Metals under Internal and Prolonged Stress', was organized with six other societies concerned with engineering, ship-building or metallurgy. The last was on 'Catalysis with Special Reference to Newer Theories of Chemical Action': it was based on a suggestion by Rideal. This was indeed seminal. J. Perrin (Paris) and W. C. McC. Lewis presented their theories, evolved independently, that reacting molecules acquire the necessary energy of activation by absorbing ambient radiation, i.e. that all reactions are photochemical. To their surprise their views were criticised vigorously by F. A. Lindemann, newly come to Oxford as Dr. Lee's Professor of Experimental Philosophy (1919) from R.A.E. Farnborough, because these would mean that e.g. the inversion of sucrose in aqueous solution would go 5×10^{13} times as fast in sunlight as in the dark 'if poured in a fine spray from an ordinary watering can', which it certainly did not, as T. W. J. Taylor showed (*Nature (London)*, 1921, **108**, 210: but see W. C. McC. Lewis *ibid.*, 241). Lindemann showed also that the radiation theory was not necessary to explain 'unimolecular' gas reactions, the real existence of which had been shown only a few months earlier by F. Daniels and E. H. Johnston (*J. Am. Chem. Soc.*, 1921, **43**, 53), for this could be done by assuming that there was a time interval between activation by bimolecular collision and reaction, which was long compared with the time

The Inter-war Years, and the Second World War

required to establish the Maxwell–Boltzmann energy distribution. Irving Langmuir (General Electric, Schenectady) was at this meeting, elaborating his view that surface reactions should be treated by regarding the surface as a checkerboard on which molecules are held in a single layer. Svante Arrhenius was also there and took part in the discussion. This was the first meeting after the war at which there was a substantial foreign contingent.

2.2. 1922–1924

In the next five years about half of the General Discussions were on essentially practical or technological topics, several of them being on metallurgical ones. In March 1922 there was a Discussion on 'Some Properties of Powders with Special Reference to Grading by Elutriation' which arose from a suggestion by Lowry whose wartime concern with powdered materials has been noted (Section 1.6). In October 1922 there was one on 'The Generation and Utilisation of Cold', when a paper by H. Kammerlingh Onnes described how, by evaporating liquid helium, the lowest yet temperature of <0.9 K could be attained. Kammerlingh Onnes, with whom there had been correspondence as far back as 1912, had intended to come with his colleague Professor J. P. Kuenen, but Kuenen died suddenly and Kammerlingh Onnes, whose own health was failing, felt that he could not come. Dr. C. A. Crommelin came instead to present the Leiden work; and he also conveyed an invitation for about 20 members of the Society to visit the Leiden Laboratory, but we may note that this visit had not taken place by April 1923. Cosmo Johns speculated, with notable foresight, about the possibility of using oxygen-enriched air in metallurgical processes.

In 1922 Robertson succeeded Porter as President. There were two newcomers to Council. Dr. R. Lessing, German born (1878) and a former pupil of Willstätter, who had come to England as a young man, had become a consultant in chemistry and chemical engineering, with a special interest in coal and its products (D. T. A. Townend, *Chem. in Britain*, 1965, **1**, 321). Professor J. W. McBain, born in Canada in 1882, had worked in Leipzig and in Heidelberg (with Bredig) before the war:

he became the first Leverhulme Professor of Physical Chemistry at Bristol in 1919 and, up to 1922, he had specialized in studying the nature and the electrolytic conductivities of soap solutions. W. R. Pousfield, the K.C. and chemist (Sections 1.2 and 2.1, F.R.S. in 1916) was brought back on to Council in his late years (he then was 68) although he had warned that he would not be able to attend often. Harold Moore also returned, but after only a year off.

In 1923 no less than five General Discussions are listed (Appendix C). Two of these concerted the corrosion of metals: viz. one on 'Alloys Resistant to Corrosion' was the second Discussion to be held at Sheffield, while the other, which was hardly a full General Discussion, was the first Report by Dr. W. H. J. Vernon to the Atmospheric Corrosion Research Committee of the British Non-Ferrous Metals Research Association. Moore succeeded R. S. Hutton as Director of this Association in 1932, and it is likely that he was already interested in its problems. The other three were more academic in aim, but two of them, on 'The Physical Chemistry of the Photographic Process' (London) and on 'The Electronic Theory of Valency' (Cambridge) could not get very far because quantum theory was in a state of inadequacy, and even of confusion, that was only to be cleared up when matrix mechanics and wave mechanics were introduced in 1925–26. Wilder D. Bancroft (Cornell University), then Editor of the *Journal of Physical Chemistry*, was present at the former meeting which had originally been planned for 1914. The Eastman Kodak Research Laboratory at Rochester, N.Y. was represented by four papers at the London meeting. Of these S. E. Sheppard, a school friend and U.C.L. colleague of C. Kenneth Mees, was a co-author but seems not to have been present. Mees had set up and was the research director of the Rochester Laboratories. The meeting included a dinner at 7s 6d per head.

The second meeting was the first to be held in Cambridge: it had been suggested to Council by Lowry only four months before it was actually held. About 120 members and visitors attended, and amongst those contributing were seven elected or future Nobel scientists. J. J. Thomson presided over one session, when G. N. Lewis put the view that there could be gradual transition from electrovalency to covalency, or *vice versa*, whereas Sidgwick was inclined to regard them as different forms of bonding. It may be on this occasion that Lewis also visited Oxford and was introduced to C. N. Hinshelwood, then about 26, by his former tutor, Brigadier General (later Sir) Harold Hartley, who is reported to have said breezily 'Now, Hinshelwood, tell Professor Lewis about your book on thermodynamics'. This was the first General Discussion to last more than one day, and on the second day Robert Robinson presented his essentially empirical theory of polar effects in organic compounds. Members were charged 25s for meals and accommodation, which was about the actual cost to the Society. The third meeting, on 'Electrode Reactions and Equilibria', was less inhibited by the imperfections of current quantum theory. There were some useful semi-empirical discussions of the origin of electrode potentials from both the thermodynamic and kinetic standpoints.

In 1923 there were three newcomers to Council viz. Dr. W. H. Hatfield, Dr B. A. Keen and Mr. H. T. Tizard. Hatfield was one of the eminent Sheffield metallurgists whom we have already encountered (Section 2.1) and who, incidentally, commenced his industrial experience in the works of Sir Henry Bessemer and Company. Keen was another of the many graduates of University College, London, who became associated with the Society. After a brilliant career there he was invited by E. J. Russell in 1913, at the age of 23, to come to Rothamsted Experimental Station. By 1923 he was head of the soil physics department and Assistant Director. Tizard was a graduate of Magdalen College, Oxford, who, after spending a rather disappointing year with Nernst in Berlin (1908–09) returned to Oxford to work with N. V. Sidgwick and was elected to a Fellowship at Oriel College in 1911, at the age of 26. Already by 1913 he was feeling that life in Oxford was too comfortable, and although he returned to Oriel, after war service, in 1919 and was elected by the University to a Readership in thermodynamics in February 1920, he had concluded that he would 'never be outstanding as a pure scientist'. He seemed to need an external stimulus such as he had found during the 1914–18 war, which had shown him that he reacted strongly and effectively to the challenge of applying science to national affairs (see Section 1.7); so when he was invited to become an Assistant Secretary to the relatively new Department of Scientific and Industrial Research he accepted and left Oxford in September 1920. In 1922 he became Principal Assistant and in 1927 Permanent Secretary; thus, he joined Council soon after going to D.S.I.R., and when he was beginning to feel his feet on firm ground.

It may be mentioned that J. A. Harker died in 1923 at the early age of 53, soon after he came to the end of a period of office as a Vice-President. During the war, as Director of Research in the Inventions Department of the Ministry of Munitions, he had had to visit Canada and the United States. On the way back his ship was torpedoed off the Irish coast, in 1918; so for some hours he was in an open boat. He gave his greatcoat to a companion. This exposure undoubtedly weakened a constitution that was never very strong, and so it was that when he contracted pneumonia he died very quickly (see *Proc. R. Soc. London Ser. A*, 1924 **105**, xi). With his record of distinguished service to the Society it is clear that had he lived he would have had much still to give. His death was felt as a great loss.

2.3. 1924–1926

In 1924 there were again five Discussions. The first, initiated by the President, Robertson, described unsuccessful attempts to detonate the mixed salt ammonium sulphate–ammonium nitrate, about 4500 tons of which supposedly was the cause of a disastrous explosion at Oppau, in Germany, in September 1921. The second was a brief, further exploration of the Society's interest in soil problems, entitled 'Base Exchange in Soils', initiated by Dr. B. A. Keen. Dr. D. J. Hissink, Director of the Royal Dutch Agricultural Experimental Station at Groningen, gave the introductory address (see R. K. Schofield, *Trans. Faraday Soc.*, 1953, **49**, 522). There was another metallurgical one on 'Fluxes and Slags in Metal Melting and Working', which originated with R. S. Hutton, and two that explored new ground viz. 'The Physical Chemistry of Igneous Rocks', suggested by Dr. R. Lessing and organized with the Geological Society and the Mineralogical Society, and 'Physical and Physico-Chemical Problems relating to Textile Fibres'. Spiers proposed the latter topic in 1922: it was organized with the Textile Institute and held at the British Empire Exhibition at Wembley. There was a later, secondary discussion on textiles at Nottingham.

It may be remarked that this idea of a follow-on in 'a provincial centre' to a major Discussion held in or near London was one which Council for a while found attractive. Robertson, who attended the Nottingham meeting, reported that it was very successful; but in general the idea did not prosper. Council was anxious that only essentially new material presented at such meetings should be published. Obviously if there was much such material its omission at the main meeting would make that incomplete; while if there was not then the provincial meeting would fall rather flat. Eventually the problem of sharing the benefits of discussion with the provincial centres was solved by having the General Discussion meetings increasingly in those centres. By the end of 1924 only six of the 42 Discussions had been held outside London – in Manchester (3), Sheffield (2) and Cambridge. Between 1925 and 1940, however, 16 of the 28 (and one planned for the Autumn of 1939) were held outside. This development had to wait the growth of interest in the Society's aims in the provincial centres themselves, and this of course was what the Society was seeking to achieve; so there had to be positive feed-back, and gradually this was established. As we have seen, Liverpool pioneered in physical chemistry while, early on, Sheffield had a strong connection with the Society through the common interest in metal science and the personal link with R. S. Hutton. Manchester had a link through W. W. Haldane Gee, who was Professor of Pure and Applied Physics at the College of Technology. Cambridge established a school when Lowry went there in 1920 as the first Professor of Physical Chemistry. At Oxford there had been physico-chemical research from the days when A. G. Vernon Harcourt, F.R.S., who had a laboratory in Christ Church in the late 19th century, investigated reaction kinetics. He is remembered for the Harcourt–Esson reaction. Because the University Department was unhelpful the subject had to develop further in other College laboratories, viz. in one associated with Brig. Gen. (later Sir) Harold Hartley and later with C. N. Hinshelwood, E. J. Bowen and R. P. Bell, which was in the cellars of Balliol College and the converted bathrooms and lavatories of Trinity College; in the Leoline Jenkins Laboratory of Jesus College, run by D. L. Chapman; and in the Daubeny Laboratory of Magdalen College where N. V. Sidgwick, a pupil of Vernon Harcourt, worked before he moved to the new Dyson Perrins Laboratory for organic chemistry. None of these groups took much interest in the Faraday Society until after the 1914–18 war; but then things changed rapidly. Bristol we shall hear about later, as we shall about Leeds and Edinburgh.

In 1924 Donnan became President. To judge from the number of eminent members of the Faraday Society who had been his pupils, or had been influenced by him at Liverpool or at University College, London, he was then a key figure in British physical chemistry. His scientific papers were mainly on thermodynamics and he is remembered for his theory of membrane equilibria; but his influence rested more on his personal qualities. In Ramsay's time U.C.L. began to produce a large proportion of the small company of British physical chemists; so Donnan inherited a position of influence on which he built. He did this by his shrewdness in choosing people, his charm, his power to inspire enthusiasm, his willingness to give young people their head, and the ability to meet practical needs which he showed during the war (*Year Book R. Soc. Edin.* 1956–57; *Biog. Mem. Fellows R. Soc. London*, 1957, **3**, 23). His pupils remembered him with affection. One of them wrote to him just before his retirement dinner in 1937 'Memory is sometimes an impish thing and I am at this very moment recalling seeing you on Mafeking night gingering up our bonfire, mostly U.C.L. doors I fear, with about half a barrel of pitch, loaned from road-making operations in Euston Road;'. That was in May 1900, when Donnan was still a senior research student in Ramsay's laboratory.

There were three new faces on Council, namely those of Professor E. N. da C. Andrade, Dr. F. A. Freeth and Dr. F. C. Toy. At that time Andrade was Professor of Physics at the Ordnance College, Woolwich, the precursor of the Royal Military College of Science. Born in 1887 he had been a pupil of Trouton's at University College, London. He next worked with Lenard in Heidelberg and later with Rutherford in Manchester. He had a useful war record. He did sound experimental research in physics but is best remembered as a scientific historian. He was a superb writer. Freeth also was a colourful character. In speech, although he was 'often vehement to the point of extravagance', he was less caustic than Andrade. Both men had great ups and downs in their subsequent careers (see *Biog. Mem. Fellows R. Soc. London*, 1972, **18**, 1 and *ibid.*, 1976, **22**, 105). Freeth had been a pupil of Donnan's at Liverpool: in 1924 he was research manager of the Brunner Mond Company's Laboratory at Winnington, and by then he had initiated the high-pressure work which was to lead later to the discovery of polythene. Toy was another U.C.L. man. In 1924 he was a physicist with the British Photographic Research Association. Later (1930) he became Deputy Director of the Shirley Institute for cotton research and then Director. E. H. Rayner, of the Electrical Department of N.P.L., returned to Council after a short absence. He was one of those who served throughout the war.

Of the two General Discussions held in 1925 the first, on 'The Physical Chemistry of Steel-Making Processes' was a substantial contribution to the better understanding of the thermodynamic and kinetic aspects of steel-making. It was well in the Society's tradition. The Iron and Steel Institute gave generous support. The later one, on 'Photochemical Reactions in Liquids and Gases' was the first to be held in Oxford and the second to last two days. It is likely that Allmand proposed the topic and that the link with Oxford was through D. L. Chapman, since the meeting was held in Jesus College. Chapman and E. J. Bowen both read papers at the meeting, although neither of them was a member of the Society at that time. Following what may by then have become as established custom, the President publicly welcomed the foreign guests. Eleven are named, including Professors Bodenstein, Christiansen, Franck, W. A. Noyes, Jr., von Halban and Weigert, but others seem to have participated in the Discussion making about fifteen foreign guests in all, eight of them German. The ice had at last broken. There was strong emphasis on the Einstein law of photochemical equivalence. According to E. J. Bowen few photochemists appreciated how non-equivalence arose from secondary reactions, especially chain reactions; so those who did understand this, notably Bodenstein, performed a most useful service by clarifying ideas. In considering how absorption of a quantum could cause the dissociation of a molecule, Franck enunciated and used what later became known as the Franck–Condon principle: but Bowen says that because photochemists then had little knowledge or understanding of molecular spectroscopy – not that there was much to have – Franck's paper did not attract much immediate attention. It was a notable meeting.

In 1925 there were four recruits to Council. One was Professor J. E. Coates (then 42) who in 1920 had gone to Swansea to start a new chemistry department. After graduating from the University College of North Wales at Bangor he had worked with Haber on the ammonia project in 1908–09 and then gone to Birmingham. He became interested in the solvent properties of anhydrous hydrogen cyanide and according to his son Geoffrey 'There were litres and litres of the stuff about the labs'. By modern standards of 'Health and Safety at Work' this sounds shocking; but in those days chemistry departments consisted of small numbers of highly skilled people. In fact nobody was the worse for it; and Coates lived to be 90. The second was C. R. Darling (then 55), who was Assistant Professor of Science at the Royal Military Academy at Woolwich. He was a physicist, a graduate of the Royal College of Science in Dublin, who had been a Lecturer in Applied Physics at the Finsbury Technical College until that institution was closed. He was essentially a masterly teacher, and this included having a very lively interest in a variety of practical physical problems. One such, which he pursued in collaboration with Claude Freese-Green, was the development of cinematography when in its infancy (see *Proc. Phys. Soc.*, 1943, **55**, 251); and during the 1914–18 was he was chairman of the Admiralty Electric Welding Research Committee. Later he was a Vice-President of the Physical Society. The third was E. C. Williams, who had been with the British Dyestuffs Corporation and in 1923 had become the first Professor of Chemical Engineering at University College, London. His inaugural lecture still makes good reading. The fourth was Dr. R. S. Willows (then 50), who had been a pupil of J. J. Thomson's, had held a post in the physics department of the Sir John Cass Technical Institute in London, and in 1918 had been persuaded to become head of the research department of Tootal, Broadhurst, Lee and Company in Manchester. His main task there was to develop a cotton fabric which did not crease, and after ten years he succeeded (see Sir Lawrence Bragg, F.R.S., *Chem. in Britain*, 1970, **6**, 147).

2.4. 1926-1928

By 1926 the pattern of having two General Discussions annually, one in Spring and one in Autumn, had almost become established. What had happened was that over the years 1920-25 the large number of Discussion Meetings had produced a great amount of material for publication and, furthermore, this varied greatly from year to year with consequent effects on the size of *Transactions* and therefore on printing costs. Thus, volume **15** (1919-20) had 580 pp., **16** (1920-21) had 613 pp. plus the 'Colloids' offprint of 190 pp., **17** (1921-22) had 756 pp., **18** (1922-23) had 403 pp., **19** (1923-24) had 949 pp. and **20** (1924-25) had 626 pp. Such fluctuations could not possibly be met by any rational budgeting; so the Finance Committee began in 1922 to press for limitations on printing commitments. In 1924 (*Council Minutes*, 7th July) it urged that volume **20** should not be more than 600 pages, a target almost achieved, and that not more than two General Discussions be held in 1925. In 1925 the new Chairman of the Publications Committee, A. W. Porter, who in 1924 had succeeded W. R. Cooper whose health was failing, wanted a precise definition of the function of his Committee in relation to the material from General Discussions arranged by the *ad hoc* committees appointed by Council. A decision of Council on 29th June 1915 that the Chairman or a nominee of the Publication Committee should be an *ex officio* member of each such committee seems either to have been forgotten or to have become ineffective. Council decided that programmes of the General Discussions should be considered by the Publication Committee before definite arrangements were made and that all matter for publication should be passed by that Committee. Thus, some control was gradually established over the rather uninhibited enthusiasm of those who wanted Discussions *ad libitum* by those who were more concerned about paying for the consequences. Readers will doubtless have a feeling of *déjà vu*.

The first of the two General Discussions in 1926, both only one-day affairs, was on 'Explosive Reactions in Gaseous Systems'. Dr. W. E. Garner, then at U.C.L. in Donnan's laboratory, gave the introductory address. Although the theory of reaction kinetics was far too primitive for any deep understanding in terms of molecular processes to be achieved there was a substantial range of more empirical treatments in terms of macroscopic parameters such as ignition temperature, thermal conductivity, and heat of reaction, of reaction characteristics such as limits of inflammability and velocity of flame front or of detonation wave. There were also numerous good observations to report about explosions in engines, on which a paper was read by Tizard who may well have proposed this Discussion. At the second Discussion, on 'Physical Phenomena at Interfaces with special reference to Molecular Orientation', the introductory address was given by Rideal, who had about four years previously started to take an interest in surface chemistry. The meeting was largely concerned with studies of surface films of insoluble substances on water. Rideal and his near-contemporary N. K. Adam both contributed as did Gorter, who had produced protein films. Scherer showed how X-ray studies on crystals of long-chain compounds were beginning to throw light on the forces between such molecules. Electrical potential differences at interfaces also received attention, one contributor being R. Kenworthy Schofield, a former pupil of Rideal's, who later went to Rothamsted.

Professor C. H. Desch, F.R.S., became President in 1926. He had left Glasgow and become Professor of Metallurgy at Sheffield in 1918. By 1926 he was regarded as being in the top rank. He was then 52. He had come to metallurgy by an unusually roundabout route, from organic chemistry at U.C.L. via the chemical industry, where he learned photomicrography, then metallography and metal analysis applied to archeology, and metallographic microscopy in Huntington's department of King's College, London. He found Huntington 'quick-tempered and exacting'. One of his early tasks was to evaluate different resins for making his Professor's balloons less permeable (Sect. 1.1). His wide experience made him highly versatile and gave him very varied interests including French dialects, of which he had a remarkable knowledge, and sociology. His biographers describe him as somewhat austere and gentle in manner, not given to 'forceful rejoinder', and showing 'quiet kindness and diligence' in ensuring that his research pupils received full credit for their original work (*Biog. Not. Fellows R. Soc. London*, 1959, **5**, 49; *Proc. Chem. Soc.*, 1959, 134). There were only two new members of Council, namely Dr. N. V. Sidgwick, O.B.E., F.R.S., and Dr. R. E. Slade. Sidgwick was only the second man holding, or who had held, an Oxford teaching post to become a member of Council. The first was Tizard (in 1923), who had been his pupil at Magdalen College, Oxford. Sidgwick had already shown a remarkable flair for ordering known facts and thereby bringing new clarity and understanding into a field, and he was soon to do this with even greater effect in valency theory (*Obit. Not. Fellows R. Soc. London*, 1954, **9**, 237; *J. Chem. Soc.*, 1958, 310). Slade was a Manchester graduate who had held lectureships at Liverpool and at U.C.L., and then in 1918 had become Director of the British Photographic Research Association, but had very soon after (1920) joined Brunner, Mond and Company; so he may well have been proposed by Freeth. He was to play a considerable part in the life of the Society. Hatschek and Pring returned to Council after relatively short absences.

On May 21st, 1926, just before the Council met in U.C.L., 'a telephone message was received from the office of the Faraday Society stating that Mr. Spiers had suddenly died in his office' (*Minutes*). He was not yet 51 years old. For 23 years he had worked for the Society, as Hadfield put it, *con amore*. 'The ardour which he put into all his work was more than his slight body could sustain. On several occasions he had been obliged to take a prolonged rest. During the Spring of this year there were clear signs that he had been overworking.' 'It is not too much to say that it is due to his zeal and energy that the Faraday Society has developed as it has.' (A. W. Porter, *Trans. Faraday Soc.*, 1926, **22**, 207). There is a fine portrait of Spiers in *Transactions* (1927, **23**, opposite p. 56). 'The Council were deeply shocked and grieved' but the Society had to go on so a joint committee with the Institute of Physics was set up quickly to find a successor to be, like Spiers, Secretary of both institutions. Seven applications were considered. One of them 'was summoned to the meeting by telephone and had an interview with the members of the Council. It appeared from the discussion and interview that Mr. Marlow would be in many ways a very suitable person to appoint if possible to the permanent post' (*Minutes*, 3rd June 1926). He was appointed forthwith as Acting Secretary to both institutions at a salary of £400 p.a., later increased to £600 p.a. from the date upon which he could devote the full time required by his two employers. By the end of the year the joint arrangement was abrogated, so Marlow was offered the permanent post of Secretary and Editor to the Society at a salary of £450 p.a., and he accepted. According to Donnan [*Trans. Faraday Soc.*, 1948, **44**, 184 (i)] Marlow had been recommended to him by Sir Robert Robertson. Marlow had been holding a post within the Institute of Chemistry while he read for Bar Finals, and he had recently been called; so he combined his work for the Society with being an active and increasingly successful barrister. He was lucky to have only a one-day General Discussion as his first.

[It seems to M.M.D. that the debt of the Society to Spiers is inadequately expressed in the one paragraph which announces both his death and the appointment of Marlow. Even so the few sentences make it probable that overwork on behalf of the two societies Spiers served led to his premature death.

One had to search for statements from 70 years ago to provide supporting evidence of the Faraday Society's debt to Spiers. Whilst, as explained later (Chapter 4), minutes of the Annual General Meetings of the Society are generally not to be found in the archives, one old minute-book did contain records of those from November 1925 to July 1932. The bareness of these records makes the more noteworthy the entry for 2nd June 1926 where 'the death of Mr. Spiers, Secretary and Editor of the Faraday Society' is recorded, and 'high appreciation with regard to long years of highly efficient and devoted service which Mr. Spiers had rendered to the Society. It might well be said that had it not been for the enthusiastic work of Mr. Spiers, the Faraday Society would not have continued to exist.'

It must be remembered that this minute was written by Marlow, for whom it must have been an inspiration.

The remainder of the minute then evidences the slender significance attached to the A.G.M.s: 'An Annual Report of the Council was read in abstract by the President and accepted by the meeting.'

(signed) C. H. Desch.]

In 1927 there were nominally three General Discussions but the first was Dr. Vernon's second report on Atmospheric Corrosion which was only an evening meeting. The second was a major, two-day meeting again at Oxford, this time because of Brigadier General Harold Hartley's concern with electrochemistry. The topic was 'The Theory of Strong Electrolytes'. The famous papers of Debye and Hückel had been published only four years earlier (see Section 2.1) and Onsager's development of the conductivity theory only one year before; so it was the time of breakthrough in the understanding of the kinetic and thermodynamic properties of solutions of strong electrolytes. Debye unfortunately could not come, but Erich Hückel presented an introductory paper by him. The President welcomed also J. N. Brønsted, Niels Bjerrum, Kasimir Fajans, H. S. Harned, G. Hevesy, R. Remy, G. Scatchard, J. A. Christiansen, Lars Onsager and H. Ulich. In addition D. A. McInnes, R. H. Fowler, Charles Kraus and Merle Randall contributed papers. One could hardly have asked for more. Apart from the absence of Debye, the other assembled contributors formed almost the totality of the world's leaders in the then very active field of electrolyte studies. [The Society's status as an international institution was clearly established. M.M.D.] It is normal for any one country to import more scientific ideas than it exports; but at this meeting, as at the two photochemical meetings, it was evident that this country, although beginning to export, was still far behind Germany and the U.S.A. as an exporter.

It was on this occasion that R. P. Bell, later to be a President, had his first paper appear and met Brønsted, with whom he went to work in 1928. For the writer it was the beginning of an integrated memory of Marlow. Tall, thin, lantern-jawed, very striking in appearance, he would glide about the lecture room, stooped low and pretending to be invisible, giving papers to those who had contributed to the discussion, or conferring, constantly making sure that the meeting ran smoothly. He always beamed. The writer, too has a vague memory of an interminable argument which was resumed after every paper, and of Sidgwick telling him later that it was between two continental eminences who were at cross purposes because one of them meant by a 'strong solution' one which was about molar whereas the other meant one where the Debye–Hückel approximations were just breaking down, a confusion which Sidgwick claimed to have helped resolve after 24 hours. This may be exaggerated or even apocryphal, but it is true to the spirit of these Discussions. They had an informal, family atmosphere and private discussions were immensely valuable. An assessment of the meeting has been given by C. W. Davies (*Trans. Faraday Soc.*, 1953, **49**, 531: see also J. N. Agar, *idem.*, 533).

The third meeting, on 'Cohesion and Related Problems', at which Desch gave the introductory address, was concerned with the elasticity, flow, work-hardening and rupture of crystalline materials, especially metals, and with the great discrepancies between theoretical and observed tensile strengths. The old idea, derived from the Beilby layer, that there was amorphous metal in the slip planes, was no longer tenable in the light of X-ray crystallographic evidence. The modern idea of the lattice dislocation had not developed fully, but there was movement towards it. Michael Polanyi made his first appearance at a Faraday Society meeting. He was then at the Kaiser Wilhelm Institut for Physical Chemistry and Electrochemistry at Berlin-Dahlem, having previously been in R. O. Herzog's group at the Institute for Fibre Chemistry, where he was one of the pioneers in developing the X-ray diffraction study not only of organic fibres but of metal wires (*Fifty Years of X-Ray Diffraction*, Ed. P. P. Ewald, Int. Union for Crystallography 1962, p. 629). A colleague, Dr. G. Sachs, from the Institute for Metal Research also came, as did Professor A. F. Joffé (Leningrad) who described some ingenious work on the strength of rock salt. Professor J. E. Lennard-Jones and a pupil gave a paper on the force between an inert gas atom and a rock-salt type crystal. Although the meeting was only a one-day affair it was important because of its quality (see A. H. Cottrell, *Trans. Faraday Soc.*, 1953, **29**, 519).

There were only two newcomers to Council in 1927, namely Professor W. E. Garner and Mr. J. H. G. Monypenny. W. C. McC. Lewis returned. Garner, born in 1889, had graduated from Birmingham and after working with Professor P. F. Frankland, on organic chemistry, had spent a year with Tammann in Göttingen, where his interest in physical chemistry was aroused. In 1914 he had joined the staff of Woolwich Arsenal: he worked on the calorimetry of high explosives with the redoubtable Robert Robertson whose good opinion he seems to have earned. After the war he returned to Birmingham but soon moved to U.C.L. (Chapter 2, Section 2.4) where he became interested in reaction kinetics and heterogeneous catalysis, explosions, flames and spectroscopy, to say nothing of adsorption. In 1927 he succeeded McBain in the Leverhulme chair of physical chemistry at Bristol. He looked slightly like Mr. Punch and certainly was very shrewd and a doughty fighter. Monypenny was chief of the research laboratory of Brown Bayley's Steel Works Ltd. in Sheffield and an occasional contributor to *Transactions*.

2.5. 1928–1930

For some reason, possibly Spiers' sudden death, Council did not start considering General Discussions for 1928 until it was too late to arrange a Spring meeting, so only an Autumn one was held. This was in Cambridge, on 'Homogeneous Catalysis'. The topic had been proposed by Lowry who had made some notable contributions to the better understanding of acid–base catalysis and who gave the introductory address. He and J. N. Brønsted had arrived independently at a more general definition of acids and bases than Ostwald's; and, with H. M. Dawson, had shown that molecules as well as ions have catalytic power. They all three gave papers. Hinshelwood gave one about the general characteristics of homogeneous catalysis in gases and in solution, Boeseken and von Euler discussed the nature of the catalytic process. By 1928 the role of chains in photochemical reactions was fully accepted, and Semenov had recently shown that branching chains could explain explosive reactions; but J. A. Christiansen argued that linear chain reactions were much commoner in steady, thermal reactions than had been generally realized. M. Polanyi, now interested again in reaction kinetics, showed how chains could be initiated in the combination of chlorine with sodium vapour. It may have been this meeting that induced Hinshelwood to look more carefully for chains in non-explosive reactions. The general impression is that ideas then were neither clear nor settled, but that there was vigorous development (see C. E. H. Bawn, *Trans Faraday Soc.*, 1953, **49**, 558). It was an excellent meeting, with 21 papers being read and at least 10 eminent foreign scientists being present. The Honorary Treasurer, Robert Mond, very generously met the cost of entertaining the Society's guests.

1928 saw major changes in Council. Lowry became President, a natural sequel to his great influence on the Society's affairs from its early days (Section 1.1), arising from hard work and good ideas. There were four new members of Council viz. Professor W. E. Gibbs, Sir Harold Hartley, F.R.S., Professor M. W. Travers, F.R.S., and Professor F. G. Tryhorn. Gibbs had succeeded E. C. Williams in the Ramsay chair of chemical engineering at U.C.L. that very year. Born in 1889, and educated at Liverpool University, he must have been a pupil of Donnan's. He had been a metallurgist but he was also interested in clouds, smoke, dust and their relevance in industry. Hartley, who was in charge of the Balliol College laboratories in Oxford (Section 2.3) we have met before as Brigadier General: he was knighted in 1928. He is still a legend in Oxford. First an academic, though not in his own estimation a very good one, he was introduced to the greater world by his wartime experiences and found himself very effective therein. He became Controller of the Chemical Warfare Department of the Ministry of Munitions. Then, in words ascribed to Hinshelwood:

> 'When the war was over
> General Hartley
> Returned to civil life again,
> Partly.'

His outside interests became increasingly industrial and by 1928 he was a director of the Gaslight and Coke Company. His mannerisms could be and were much imitated; but it was easier to do this than to emulate his drive and ability. He became a notable industrialist, always with a scientific bent (*Biog. Mem. Fellows R. Soc. London*, 1973, **19**, 349). Travers had been co-discoverer with Ramsay of krypton, neon and xenon. He was a superb experimentalist and altogether had great practical ability. He became Professor of Chemistry at Bristol in 1904 where he played a major part in pressing for independent University status. During 1906–14 he was Director of the New Indian Institute of Science at Bangalore; then he went into industry; but from 1928 onwards he held an honorary post at Bristol and, among other activities, he investigated gas reactions. He was very independent-minded. He clashed with Hinshelwood because he did not believe that it was sufficient to follow a gas reaction by pressure change alone (*Biog. Mem. Fellows R. Soc. London*, 1963, **9**, 301). Tryhorn was a Liverpool graduate who also must have been a pupil of Donnan's, and who in 1928 had just become Professor of Physical Chemistry at University College Hull. Later he became an authority on forensic science.

The pattern of two General Discussions a year was re-established for 1929. Both were of major importance. The first originated from a suggestion by Lowry that Professor V. M. Goldschmidt (Oslo) should be invited to give a lecture and that a discussion on this and on other papers should follow. Goldschmidt talked on 'Crystal Structure and Chemical Structure', expounding his views on ionic radii and of the effect of their ratios on coordination number, and applying them to many examples. The other contributions, about crystal structures of inorganic compounds, organic compounds and metals, as well as on general matters, were from members of the international group of X-ray crystallographers which was forming. This included Sir William Bragg; W. L. Bragg, E. Schiebold, A. Müller, Kathleen Lonsdale, J. D. Bernal, A. F. Westgren, H. Mark, K. Weissenberg, W. T. Astbury and P. P. Ewald. C. Hermann, Ch. Mauguin, and W. Zachariasen were also present. There were no reports on large organic molecules or fibres despite Polanyi's pioneering work: Bernal talked about metals and Astbury about techniques. X-ray crystallography was then still in the very primitive days of trial structures without the benefit of Patterson functions; but a surprising number of structures had already been solved. Goldschmidt said that this was the most representative assembly of workers in the field that had yet taken place. Sir William Bragg took the opportunity to convene a meeting to discuss nomenclature and publication, at which it was decided to produce the *International Tables for the Determination of Crystal Structures*. The cooperation thus engendered led ultimately to the founding of the International Union of Crystallography (see E. G. Cox, *Trans. Faraday Soc.*, 1953, **49**, 540).

The second one was on 'Molecular Spectroscopy and Molecular Structure' and was the first to be held at Bristol. It had been urged by

Garner; but what made Bristol especially appropriate was that J. E. Lennard-Jones had come to the physics department there in 1925 and created a school of theoretical chemistry which quickly became very distinguished. The topic was particularly timely. Matrix mechanics had been presented in 1925, wave mechanics in 1926, the concepts of electron spin and nuclear spin in 1925 and 1927, the Pauli principle in 1925. A great outburst of activity had followed immediately, with the result that by 1929 many of the fundamental principles of molecular spectroscopy had been established. Electronic spectra were being rationalized: concepts of, and notation for electron configurations were being established; the *Aufbau* principle for molecules had been suggested; selection rules had been derived; Heitler and London's treatment of the covalent bond (1927), Hund's united atom (1927) and the beginning of the molecular orbital treatment had been put forward. Vibration and rotation spectra were being used to determine molecular structural characteristics, to 'finger-print' chemical bonds, to determine ground-state dissociation energies, isotope ratios and nuclear spins. The relation of Raman and absorption spectra was appreciated. All this had happened in three or four years, the greater part of it in Germany. No less than 39 papers were submitted for discussion: the list of authors included many of the foreign world authorities, e.g. R. M. Badger, E. F. Barker, R. T. Birge, G. B. Bonino, J. Cabannes, J. W. Ellis, Victor Henri, G. Herzberg, V. Kondratjew, J. Lecomte, R. Mecke, R. S. Mulliken, J. C. M. McLennan, Sir C. V. Raman, Clemens Schaefer, A. M. Taylor and R. W. Wood. There was a strong home team which included R. C. Johnson, J. E. Lennard-Jones, O. W. Richardson, Sir Robert Robertson and some bright young men such as C. F. Goodeve, F. I. G. Rawlins and C. P. Snow. To say this is not mere name-dropping: it is a simple way of showing that by 1929 the Faraday Society had acquired international respect.

There were four newcomers to Council in 1929. They were A. C. G. Egerton, F.R.S., Dr. R. O. Griffith, Dr. A. F. Joseph and Prof. J. F. Spencer. Although born to country house life, Egerton developed as a boy a passionate interest in science and in experiment; and with the support of Lord Rayleigh, who was a family friend, he persuaded his parents that he should read science rather than train for a naval career as they had intended. After leaving Eton he therefore went to U.C.L. where he worked with Ramsay, becoming a close friend of the Ramsay family. He taught for a while at the Royal Military Academy, Woolwich, and then went in 1913 to work with Nernst in Berlin where he met his contemporary F. A. Lindemann, later to become Lord Cherwell: both were born in 1886. Soon after the outbreak of war he was directed to the Ministry of Munitions where he was both adviser and man of action. His eminently sound judgement made him highly successful in a series of tasks. He was a member of the mission, suggested by Freeth and led by Hartley, which in 1919 went to investigate the war-time production of munitions in Germany: that mission's report led to Government support for the chemical industry in this country. Unlike Tizard and Hartley, Egerton was more attracted by doing research than by organizing it, so he was anxious to return to *academe*. Through Lindemann he was invited in 1921 to come to Oxford where he succeeded Tizard, also a friend of his, as Reader in Thermodynamics, and he stayed there until 1936: then he went to Imperial College, London, as Professor of Chemical Technology. He was an unusually nice man, of high ability yet most kindly and modest. It was an impressive sight to see him, in rainy weather, riding a bicycle and carrying an opened umbrella. Griffith was a lecturer at Liverpool University. His interests were in reaction kinetics. He was co-author of a book on *Photoprocesses in Gaseous and Liquid Systems* (1929) in a series edited initially by Ramsay and then by Donnan. He seems to have been the leading light of a small group which included A. McKeown and W. J. Shutt. Joseph was that same soil chemist from Khartoum who had encouraged Dr. Dorothy Crowfoot Hodgkin in her early days (Section 1.7). He had just returned to this country, after spending most of his working life abroad, to start the Imperial Institute of Soil Science under Sir John Russell. Soon afterwards he gave up science, and in 1937 was admitted to the Society of St. John the Evangelist (the 'Cowley Fathers') eventually holding high office in that order (*J. Chem. Soc.*, 1952, 4088). Spencer (born 1881) was a Manchester graduate who spent some time doing research in his native Liverpool, in Breslau with Abegg, and then in U.C.L. with Ramsay. Most of his working life was spent at Bedford College, in London, which had been founded only eight years when he joined the staff in 1906. In 1919 he became head of the Department of Inorganic and Physical Chemistry. With limited time and resources he carried out researches in organic chemistry, in electrochemistry and in magnetochemistry; but his energy went mainly into teaching and the affairs of the College which he served devotedly (*J. Chem. Soc.*, 1955, 3311).

2.6. 1930–1932

The two General Discussions in 1930 were on 'Optical Rotatory Power' and 'Colloid Science applied to Biology'. Both deserve comment. The former was proposed by Lowry who was a pioneer in the study of optical rotatory dispersion. The first useful theory had been Drude's (1896), which was based on an ingenious classical model in which an electron with a resonant frequency, but forced into vibration by an applied field of arbitrary frequency, moved in a spiral instead of a straight line. Unfortunately his mathematical treatment did not correspond to his model as Max Born and then Werner Kuhn later pointed out; but it gave results of the right form. One Drude term, or more in combination, could give the observed dependence of rotatory power on wavelength. Born (1915) put forward a theory based on four non-planar, coupled, isotropic electron vibrators, and by 1930 he was putting his theory on a proper quantum-mechanical basis; but for several reasons, some of them technological (see C. Djerassi, *Optical Rotatory Dispersion*, McGraw Hill Book Co., Inc., New York, 1960, Chapter 1), it was to be more than 20 years before the observations could be made easily and related quantitatively to molecular structure. Lowry, unfortunately, was not to see that day. In 1930 he and others were rationalizing phenomena – he showed, for example, how the absorption by a carbonyl group could be made to contribute to optical activity by the propinquity of an asymmetric carbon atom – but they were not able to derive much chemical information. Nevertheless his book on *Optical Rotatory Power* (1935) was of such lasting value that it had the rare distinction of being republished unaltered in 1964. It was not possible at this meeting to announce dramatic progress, but the occasion was a useful one. Among the ten foreign scientists present were Werner Kuhn from Heidelberg and P. P. Ewald from Stuttgart.

It is noteworthy that in his introductory address to the meeting Lowry mentioned that he had initiated the series of General Discussions in 1907. Although this was not clear from the Minutes of 27th November 1906 (see Section 1.1) it had seemed probable, from the names of those present, that the original suggestion was his. It is satisfying to have this surmise confirmed.

The second Discussion arose from a decision by the Society that was to prove very important. On 25th February 1929 (*Minutes*) the President (Lowry) reported to Council a proposal that a colloid symposium should be held annually in Britain. From the accounts given by J. A. V. Butler and E. K. Rideal (*Trans. Faraday Soc.*, 1953, **49**, 575 and 579) it is clear that the impetus came from Sir William Bate Hardy, F.R.S. He had become fascinated by cell division and he thought that physical chemistry and especially colloid science, to which he had himself made distinguished contributions, could help explain it. He had long wanted the Faraday Society to take part in bridging the gap between the physical sciences and biology. He was a formidable character. Tizard wrote that Hardy 'with his black beard and commanding air, looking rather like an Elizabethan pirate, was one of the great men of his time. I never met a man so exactly fitted by knowledge and intellect and force of character to be a leader of industrial research' (see *Biog. Mem. Fellows R. Soc. London*, 1961, **7**, 326). Council agreed that the Society should aim to take a large share in organizing such symposia, but it was concerned about the cost of extra publication and the possible need for separate publication which could lead to a loss to *Transactions* of the normal intake of papers about colloids. There were subsequent discussions, with Lowry, Andrade, Partington, Porter and Marlow representing the Society. Eventually it was decided that a new Society should not be formed but that the Faraday Society should form a Colloids Committee which would include representatives from the Royal Society, the Biochemical, Chemical, Physical and Physiological Societies and the Society of Chemical Industry, and that through this Committee the Society should organize meetings in 1930, 1932 and 1934 to discuss colloid problems following the pattern of its General Discussions (*Council Minutes*; 23rd May 1929, 7th Oct 1929). It was also agreed that the papers on colloids discussed at Ordinary Meetings might be collected and published together. The Society's representatives on this new Committee were Lowry, Porter, Marlow, Donnan, Hatschek, Rideal and Lewis. The September 1930 Discussion was the first fruit, the expressed aim being that biologists and physiologists should draw attention to problems where the assistance of physicists and physical chemists was essential.

It is a very interesting fact that about this time other, similar moves were being made spontaneously. In this country J. D. Bernal and W. T. Astbury became interested in applying X-ray crystallographic analysis to compounds of biological importance, while they were at the Royal Institution with W. H. Bragg. Bragg and his son, W. L. B., had shared out chemistry. Father took organic compounds and son took inorganic ones. So in 1926 W. H. Bragg, doubtless prompted by the work at Berlin-Dahlem, asked Astbury to take X-ray photographs of some fibres like wool and hair for a Royal Institution lecture that he was to give. This simple request determined Astbury's life work. He became fascinated by fibres and when in 1928 he went to Leeds as Lecturer in Textile Physics he began a vigorous attack on the structure of keratin. In 1927 Bernal went to the Mineralogy Department at Cambridge, where he had to divide his energies between examining amino-acid crystals and fighting for better facilities. These two men founded the British school of what is now called molecular biology. The brilliant group that had formed round Polanyi in R. O. Herzog's institute for fibre chemistry at Berlin-Dahlem had split up. Polanyi himself had changed back to reaction kinetics (see Section 2.5). In 1927 H. Mark, J. Hengstenberg and R. Brill joined Kurt H. Meyer at the I. G. Farbenindustrie laboratories at Ludwigshafen-am-Rhein. K. Weissenberg remained at Berlin-Dahlem. they were soon to be further dispersed by the rise to power of the Nazis, and this prevented the development of a school of molecular biology in Germany. In the U.S.A. Warren Weaver was soon (1932) to be appointed Director of the Rockefeller Foundation's programme in the natural sciences and to persuade it to shift its support from the physical sciences *per se* to their application to problems in biology: one of its grants was to enable Linus Pauling to start working on the structure of biologically important molecules.

The Inter-war Years, and the Second World War

The Cambridge Colloid Science meeting was a large one, more than 250 people being present which was so many that it had to be transferred forthwith to the Botany lecture theatre. The President (Lowry) 'introduced each of the overseas members and guests individually, calling on them (as is the custom of the Society) to rise in their seats, whereupon they were welcomed with acclamation by the Society'. Twenty visitors were so welcomed. Professor A. V. Hill, F.R.S., giving the first paper, emphasized that the characteristic of living systems was that overall they were not in thermodynamic equilibrium but were in more or less steady states of dynamic equilibrium. Sir Frederick Gowland Hopkins, F.R.S., in his introductory paper to the second part of the meeting, gave a remarkably foresighted and realistic appraisal of the relative importance of colloid phenomena – 'But those dynamic events and the diverse chemical reactions which, within the living cell, provide energy and specific chemical synthesis, are as much part of the essence of life as the behaviour of the colloidal apparatus in which they occur. Without them no colloidal system whatever can display, save in some accidental and unreal aspects (of which the importance is often exaggerated), the attributes of life. The organic chemistry of the cell with its special regard to molecular structure and the influence of structure is for biochemistry a study of co-equal importance with colloid chemistry. Each must prove essential for adequate description.' (*Trans. Faraday Soc.*, 1930, **26**, 770).

In 1930 it was very far from obvious how any hybrid of physics and chemistry could best be applied to biology. Biochemistry had been a molecular science from its start in Pasteur's time and had used chemical and physical techniques; but most biologists, while conceding that chemical molecules were the basis of living organisms, did not think in molecular terms. Nobody was sure up to what level the characteristics of such organisms could be explained in terms of chemical or physical concepts such as molecules and their structures, osmotic effects, solution equilibria, reaction rates and catalysis, and when some kind of vitalism took over. Certainly in those days chemistry and physics ceased to be useful at quite a low level. The real question was how far might that level be raised? Was there in principle a limit? Some physicist philosophers thought that there might be. For example the

Heisenberg uncertainty principle suggested to Niels Bohr that there might be an indeterminacy in living organisms 'that must be considered as an elementary fact that cannot be explained, but must be taken as a starting point in biology,' rather like the quantum of action or the fundamental particles. In 1930 the chemistry of biological systems was still in a relatively primitive state. No vitamin had yet been isolated in a pure form. The female sex hormone had been, but no chemical structure had been ascribed. No respiratory cycle had yet been established. The nature of proteins was still mysterious although some had long been known in crystalline form, and, as he reported to the meeting, Svedberg had recently shown that they are definite in molecular weight and much larger than had hitherto been supposed. The first isolation of an enzyme in pure, crystalline form had only recently been effected (1926). There was no idea that genetic material might be molecular.

A further difficulty was that a great division existed between the concepts of the colloid chemists who believed that complex materials were built from relatively small units held together as 'micelles' by rather mysterious forces – perhaps the secondary valency forces which Werner had postulated – and the ideas of a few like H. Staudinger who were convinced that there are macromolecules held together by genuine chemical bonds. The X-ray people were not yet agreed about the sizes of molecules in fibres from the evidence of their photographs. There was, then, great ignorance and great confusion. Dr. Robert Olby gives an excellent picture of this period in his book *The Path to the Double Helix* (Macmillan, London, 1974); and the writer is indebted also to Sir Hans Krebs, F.R.S., for his recollections.

To choose which line to take therefore meant following a hunch. The Faraday Society might have supported the structural attack following its very successful meetings in 1929 on X-ray crystallography and molecular spectroscopy. The physical methods for structure determination were, however, still relatively new, primitive and of uncertain promise. Only a few people anywhere, like Astbury, Bernal and Pauling, had the vision, the faith, the obstinacy or the burning enthusiasm that led them to persevere in trying to solve biological problems by these techniques. That there was no mention in the 1929 meeting of any examination of biological material, unless two very limited papers on long-chain fatty acids be counted, perhaps shows how little there was to say about such work. Structure was, moreover, not an aspect in which many physical chemists were interested at that time. They mostly regarded the scope of their subject as being the development and application of the principles of thermodynamics and of kinetics to macroscopic phenomena, e.g. states of matter, solutions, membranes, electrochemistry, reactions, catalysis and surfaces. One of the founding fathers of the subject, F. W. Ostwald, had for a time even questioned the existence of atoms and molecules, although his views attracted very little support and, indeed some of the topics mentioned above obviously require these concepts. Structure had been developed as a study by the organic chemists using their wonderfully ingenious and sophisticated methods based on organic reactions: at least it had been until the X-ray crystallographers established structures for inorganic compounds and minerals, and for a few organic compounds: but the crystallographers were aberrant physicists rather than chemists. A big switch of thought and of interest by the Society would therefore have been needed and there was little to encourage this at the time.

As Professor Jeremy Knowles, F.R.S., said 'taking a photograph of a horse does not necessarily tell you how fast it can run' (*Chem. in Britain*, 1974, **10**, 194), but if we know only that a horse is a something which runs and not even that it has four legs our vision is limited. In order to be able to make major predictions chemists and biologists need not merely a knowledge of the structure of a system nor merely to have a reaction scheme even with all the reaction parameters. They need to understand reaction processes. This requires a subtle interweaving of structure, kinetics and energetics. Indeed it requires more than static structures: it requires an understanding of the way in which these can be perturbed when reactants approach, a point which Robert Robinson and C. K. Ingold had already made in a limited context by 1930. Therefore structure is not the end but it is an essential beginning to the understanding of the very complex, interlinked reactions which occur in living material. Later in the 1930s some reaction schemes were established, e.g. the Krebs respiratory cycle (1936). The quantitative study of enzyme kinetics, which effectively began in 1913 with the work of Michaelis and Menten, was progressing steadily. In 1930 J. B. S. Haldane, whose formal training was in mathematics, published a book on *Enzymes* which was based on lectures that he had been giving since 1923 and in which he stressed the quantitative aspects of enzyme reactions especially the kinetic ones. 'It appeared at a time when enzyme research was moving out of the phase of 'natural history' and into the phase of detailed kinetic study' (N. W. Pirie, *Biog. Mem. Fellows R. Soc. London*, 1966, **12**, 235). Biological kinetics was therefore another approach which the Society might have chosen to support. Instead it chose to try to clarify biology by applying the rather clouded concepts of colloid chemistry, and this appears to have been partly because of personalities and partly from its long-standing interest in colloids which went back to 1913 (see D. C. Henry, *Trans. Faraday Soc.*, 1953, **49**, 571).

The study of colloids in this context must not be belittled. It led to some very useful results, e.g. Svedberg was led to develop his ultracentrifuge and Tiselius his electrophoretic separation methods, from their interest in colloids. As Olby points out (*loc. cit.*, p. 21) the concepts proved very useful to cytologists like Hardy and helped in the development of tissue culture. The distinctive emphasis on the importance of surface effects proved very stimulating. It was, however, the structural attack which eventually led to the breakthrough of ideas in the 1950s. This it was that raised beyond the hopes of anybody in the 1930s the level at which chemistry and physics became the essential stuff of biology, and showed that indeterminacy was not fixed but was rather an expression of current ignorance. Prediction from the current principles and knowledge of chemistry and physics always becomes more difficult the higher the level of biological organization that is being considered, but the absolute level at which it fails has not yet been fixed.

It may be re-emphasized that the Society had to make a choice. Although societies and their committees do not actually determine the directions in which science develops – for the vital ideas come from the genius of individuals – they can help create a milieu in which the ferment of ideas is encouraged, and they can affect the rates of development of particular ideas. But in deciding how they are going to do this they all have to accept that their resources are finite and that they cannot do all that they would wish. With hindsight we may feel that the Society backed the wrong horse in choosing to support colloid chemistry but, as we have seen, it was not clear in 1929 how many runners there were: one certainly was a shadowy creature. In the opinion of at least one eminent biochemist the Society deserves great credit and gratitude for having been willing to back any kind of biological horse at a time when other chemical societies, in this country and in Germany, were not so willing, although it should be said that the American Chemical Society was more helpful. Some organic chemists were suspicious, if not contemptuous, of any dealing with substances that could not be characterized and preferably crystallized. 'Tierchemie ist Schmierchemie' said one. In a sense they were right so to insist, for the isolation of pure chemical substances has proved crucial to the development of biology: but, although there were very honourable exceptions, many of them regarded living systems formally as a convenient source of materials for organic chemists to study, and they were not much interested in how those systems worked. In fact the Faraday Society did not take much further interest in biology until 1937, for the concern of the Colloids Committee was rather with colloids *per se*. It is very doubtful if the Society could have provided major support for the application of physical science to biology once this field of activity really developed.

Two general Discussions a year would nor have been enough to serve both old and new interests, so it would have been necessary either to increase the number or to give less attention to more traditional interests. Either way, it is a fair guess that *Transactions* could not have become the major journal for publication in this field unless it had grown even faster than it actually did. The difficulty of doing all this would have been partly of a structural nature, e.g. the capacity for organizing, and the need for a choice of interests, and partly due to the need for an increase of income, which would have required an increase of membership, greater than seems likely to have been possible. What the Society could hope to do, and what it had some success in doing, was to nudge some physical chemists into taking an interest in biological problems, i.e. in acting as a catalyst. This matter of trying to match possible tasks with actual or latent resources, given that neither tasks nor resources are fixed, is a complex one to which we shall have cause to return more than once.

Dr. Robert Ludwig Mond succeeded Lowry as President in 1930, when he was 63, taking over in the second part of the Cambridge meeting. We have already glimpsed his remarkable versatility and range of interests (Section 1.2). He was also rich and generous. He founded an Infants' Hospital in Vincent Square as a memorial to his wife who died in 1905, and he was a generous benefactor to the Society. He was knighted in 1932 before the end of his term of office. We can learn something about him from the second Spiers Memorial Lecture, entitled 'Michael Faraday', which he gave in 1931, a year of celebration of Faraday's discovery of the induction of electric currents (*Trans. Faraday Soc.*, 1931, **27**, 341). He tried to understand Faraday's genius and to suggest how one might produce or at least encourage future men of genius; but while he could describe the circumstances of Faraday's working life and give Faraday's own views on education he had no more success in explaining the origin of genius than anybody else has had. He urged the need of trying to bring 'the great pupil to the great teacher' and discussed how this might be done, a preoccupation that nowadays would be called 'élitist' and so by implication be condemned. Mond said 'Our modern civilization has painfully evolved an intellectual sausage machine, which, from heterogeneous raw materials, attempts to produce a uniform product, only distinguishable by the label attached by the manufacturer'. He then described with pride how his father, Ludwig Mond (Section 1.1), and he had created the Davy–Faraday Laboratory at the Royal Institution in the hope of helping the development of a few men of exceptional ability. His last act as Honorary Treasurer was to make up the Society's investments to a round £1000.

Newcomers to Council in 1930 were William Rintoul as a Vice-President and Emile S. Mond as Honorary Treasurer, with Professor J. E. Lennard-Jones and Professor J. C. Philip, F.R.S., as members. Rintoul, who was born in 1870, after a good training in chemical analysis in the laboratory of Mr. R. R. Tatlock, the Glasgow City Analyst, migrated southwards and first took a post with a London

The Inter-war Years, and the Second World War

firm manufacturing paint and varnish. He soon went to the Royal Gunpowder Factory at Waltham Abbey, where he worked under Colonel Sir Frederick Nathan, K.B.E., a martinet but an extremely able man. there he met Robert Robertson, with whom he collaborated: they became close friends. Then in 1914 he accompanied Nathan to the Nobel Explosives Company Ltd., later the Nobel Division of Imperial Chemical Industries Ltd., becoming Chief Chemist and staying there until 1929 when he became joint research manager of the newly formed I.C.I. Ltd. (*A History of Research in the Nobel Division of I.C.I.*, Nobel Division, 1955, pp. 49–50). Unlike his colleague Freeth (Section 2.3) he was very happy and successful in this post. Emile Mond, born in 1865, was a very civilized, international figure. He had been educated in Paris and in Zürich, where he specialized in chemistry. He joined his uncle Ludwig at Brunner, Mond and Co., Ltd., later becoming a Director of this firm and of the Mond Nickel Company Ltd. as well as being Chairman of the Boards of three other companies. He and his wife gave excellent concerts of chamber music over many years. One of his sons had been killed while flying in France in May 1918, so he founded the Francis Mond chair of aeronautical engineering at Cambridge (*Trans. Faraday Soc.*, 1939, **35**, 301). He was elected to membership of the Society on 28th November 1929. His cousin Robert was favoured as candidate for President at the Council meeting on 10th February 1930 and was formally proposed at the next meeting on 26th March when also Hadfield was proposed as Honorary Treasurer with Emile Mond as reserve candidate. Hadfield, who then was 72, evidently declined and Emile Mond was elected. He had great financial ability which he placed freely at the service of the Society. Lennard-Jones of Bristol, whom we have already met (Section 2.5), was then a bright young man of 36, a rapidly rising star. He won both great esteem and affection, founding a school of theoretical chemistry which became one of the world's best, and he proved to have exceptional power as an organizer and administrator, yet he was a most modest, kindly and warm-hearted person (*Biog. Mem. Fellows R. Soc. London*, 1955, **1**, 175).

Philip, born in 1873, after graduating from Aberdeen University went to work with Nernst in Göttingen and then worked for a year with H. E. Armstrong at the Chemistry Department of the City and Guilds College at South Kensington. In 1900 he joined the staff of the Royal College of Science, also at South Kensington, became an Assistant Professor in 1909, and was elected to the newly created chair of physical chemistry at Imperial College in 1913 at which time the department which Armstrong had directed was closed down. Philip had broad interests in the physical chemistry of solutions, including the solubilities of gases, in alloys, in adsorption and in infrared spectroscopy. He was a first-rate teacher and a very respected figure in chemical administration.

Because the Royal Institution and the British Association for the Advancement of Science planned to have, in September 1931, a grand celebration of Faraday's discovery of electromagnetic induction, Council decided to have only one General Discussion that year, in April (*Minutes*, 22nd May, 1930). Quite probably, too, it was alarmed by the size of the recent reports of Discussions which had been hovering around 200 pages while one, that for Molecular Spectroscopy in 1929, had been 339 pages, at a time when the number of pages required for ordinary papers was rising rapidly, viz. 1927, xxiii, 379 pages; 1928, xxiv, 416 pp.: xxv, 373 pp.: 1930, xxvi, 468.: 1931, xxvii, 603 pp. There is no mention in the Minutes of the Publications Committee of a discussion on this point at that particular time but, as we shall see later, the size of the Discussion Reports was a constant worry to that Committee. Council felt that this 1931 meeting should if possible be in Oxford and 'preferably be of greater industrial interest than some of the Discussions which have recently been held. "Physicochemical Aspects of Fuel Utilization" was briefly considered'. The Oxford members of Council (Hartley and Egerton) were invited to report as to the nature and possibilities of such a meeting. Somehow on 1st October 1930 Council decided that the meeting should be held in Liverpool, the topic being 'The Chemical Action of Light' later amended to 'Photochemical Processes'.

At all events the Liverpool meeting was very timely. The nature of primary photochemical processes was one of the four sub-topics and the introductory paper by Professor R. Mecke, then at Heidelberg, shows how startling was the improvement in understanding since the

1925 meeting, because of the development of spectroscopy. There was now appreciation of the effect of the quenching of fluorescence in gases by collisions, of the importance of long-lived triplet states in sensitization by e.g. mercury atoms and in the activation of some molecules, of the power of the Franck–Condon principle to explain dissociation of molecules into atoms, and of the nature of 'pre-dissociation'. Professor M. Bodenstein (Berlin) described the stationary state method for analyzing secondary reactions and considered its power and its limitations. Professor A. Berthoud (Neuchatel) introduced the section on photochemical changes in liquids and solids and Professor E. C. C. Baly (Liverpool) that on photosynthesis.

By now these General Discussions normally attracted the eminent foreign authorities in the subject: there were ten such at this meeting. The President (Robert Mond) entertained all of them and the speakers very handsomely at the Adelphi Hotel: Marlow's genius for alcoholic entertainment was given full scope. D. W. G. Style comments on this (*Trans. Faraday Soc.*, 1953 **49**, 554) and E. J. Bowen (personal communication) says that he was on that occasion introduced to 'Black Velvet' which the writer believes in this context to mean a mixture of Guinness's stout and champagne. A photograph taken on the last day shows 70 participants (see plates section) so this was not one of the very large meetings.

The one new face on Council for 1931–32 was that of F. I. G. Rawlins, who was only 36. He had gone to Trinity College, Cambridge, after being educated privately, and then worked for a while at the University of Marburg. He returned to Cambridge and turned to crystallography, becoming Supervisor in that subject at Fitzwilliam House in 1929; but he also was part-author, with A. M. Taylor, of a book about *The Infra-red Analysis of Molecular Structure* (1929). Later (from 1934) he acted as a scientific advisor to the Trustees of the National Gallery, eventually becoming Deputy Keeper.

2.7. 1932–1934

There were once again two General Discussions in 1932. The first was to be at Oxford, so an Oxford sub-committee was invited to propose a topic. They suggested 'Weak Electrolytes' but on Council H. S. Taylor, 'of Princetown' (sic), proposed 'Adsorption of Gases at Surfaces', and this was supported by an Oxford member, A. C. G. Egerton. Sidgwick and Hartley accepted this, so another very timely Discussion, re-named 'The Adsorption of Gases by Solids' took place. Taylor, who gave the General Introduction, was fizzing with the concept of activated adsorption that he had proposed about a year previous and was surprised, even pained, that it was not accepted universally at once. A survey of experimental work by Rideal showed, however, how difficult it then was to decide quite what happened to an adsorbate – how much remained on the surface and how much penetrated into the adsorbant through pores or cracks or even through the lattice – and how several such penetration processes could require activation energy. Moreover, Volmer described observations which showed that molecules adsorbed on a solid surface have a mobility that is associated with a heat of activation. Polanyi discussed theories of adsorption and again the experimental difficulties of deciding how these applied to particular cases were evident. About 170 members and visitors attended the opening session, bur five visitors from Germany – Professors Bonhoeffer, Freundlich, Mark, Polanyi and Volmer – who had been expected were prevented at the last moment from coming to England. Although Hitler did not become Reichskanzler until January 1933, already in 1932 the shadow of Nazi power was affecting the German Universities. Social life for the meeting centred on Balliol College, where an informal dinner was held, 100 being present.

The second General Discussion of 1932 was held in Manchester, the first there since 1921. It was also the second one organized by the Colloids Committee and was on 'The Colloid Aspects of Textile Materials and related Topics', so it satisfied Council's wish for a Discussion of greater industrial interest. It was the first to be so oriented since 1926. Nevertheless it had academic interest too, for the debate about the size of the constituent molecules in cellulose, silk, rubber and polystyrene continued and is well illustrated in this Report. Professor H. Staudinger put his view, based on viscosity measurements, that cellulose chains contained an average of about 750 dextrose residues, corresponding to a stretched length of about 3900 Å and a molecular weight of some 120000. He quoted high values also for polystyrene (about 6000 monomer units) and caoutchouc (about 18000 units). W. N. Haworth and E. L. Hirst concluded, from end-group analyses of methylated celluloses, that the chain contained only 100–200 residues. Hermann Mark, who had been warned to leave Germany and had gone to Vienna, was able to come this time. He claimed that recent X-ray determinations showed that the crystallites in cellulose were about 600 Å long, i.e. about 120 residues. He criticed Staudinger's very bold extrapolation of his empirical viscosity law from relatively small molecules and also the assumption on which it rested, viz. that the molecules were present in solution essentially as straight chains. The size of crystallites was, however, determined from line broadening in the X-ray diffraction pattern, which was not accurate and was not very relevant if the chain could extend from crystallites into amorphous regions and so on, a point which was not then appreciated; so crystallographic evidence could not necessarily give an answer. Examinations in solution, whether by viscosity, osmometry or centrifuging, required that the material be dispersed in a solvent: the technique used for examining cellulose was mostly to make a derivative, e.g. to acetylate, nitrate or methylate, and there was a risk that such chemical treatments would cause degradation into smaller molecules. Staudinger emphasized this risk. The other parties tended to discount it. There was a lively controversy, but the disagreement remained unresolved. As for the origin of the Staudinger method, what had happened was that in 1929 he had asked the Nothilfgemeinschaft der deutschen Wissenschaft for money to buy a Svedberg ultracentrifuge but had been refused (Olby, *op. cit.*, p. 14). Therefore he was driven to find an alternative and devised his viscosity method. It was, admittedly, essentially empirical and hence he found difficulty in convincing people of the value of his results. In a book written much later, K. H. Meyer (*Natural and Synthetic High Polymers*, Interscience Publishers Inc., New York, 1950 Chapter A.I.2, p. 31), who had headed the group at Ludwigshafen, remained very critical of the crudities of Staudinger's arguments which had led him to some incorrect conclusions; but at least Staudinger was less wrong than were his opponents of 1932, for subsequent osmotic studies showed that his values for molecular weights of these very large molecules were too low! [Even later Debye, Huggins, and others produced theoretical treatments of solutions with very long solute molecules which fortified Staudinger's method. M.M.D.] There was, of course, much more to the meeting, e.g. in the same section on raw materials Astbury described his early work on the structure of keratin. The 'Production, Deformation and Degradation of Fibre Particles' were extensively discussed: swelling was one phenomenon which got attention and in this context the importance of hydrogen bonding was becoming appreciated.

The meeting was a major one in every way. The total number of members of the societies involved together with visitors was 260. The Report filled 368 pages – a record. Once again the President entertained the overseas members and visitors to dinner, a party of 30, at the Dorchester Hotel. It included R. O. Herzog, Mark and Staudinger. This was the first meeting at which arrangements for the entertainment of the ladies who were guests of the Society are specifically mentioned.

Dr. N. V. Sidgwick, F.R.S., succeeded Mond as President in 1932. He, like Lowry, was interested in structure. Since the early years of the century he had been responsible for experimental research, of which perhaps the most important was the demonstration of the reality of hydrogen-bonding in some ortho-substituted phenols. It was, however, as a writer that he had a major influence on chemical thought. His book *The Electronic Theory of Valency*, published in 1927 soon after he joined Council, had very quickly brought him fame relatively late in life for he then was 54. In 1933, during his term of office, he was appointed to a personal Chair of Chemistry at Oxford. He was one of the most notable Oxford characters of his time. He was very clear-minded. He had great integrity and dedication. He had no inhibitions about biting the woolly or the slipshod speaker. He had an acid wit (see Section 2.4 for literature references).

There was a good deal of re-shuffling of Vice-Presidents which, with retirements, led to three new members of Council being elected namely Drs. L. L. Bircumshaw and R. G. W. Norrish, and Professor R. Whytlaw-Gray, F.R.S. Bircumshaw was a metallurgist at the N.P.L. who was particularly concerned with studying the oxidation of non-ferrous metals. Norrish, 35 years old and a former pupil of Rideal, had just become Humphrey Owen Jones Lecture in Physical Chemistry at Cambridge in succession to Rideal and had made a reputation for original research on reaction kinetics and mechanisms. Whytlaw-Gray, who was 55, was a delightful, gentle and very interesting person. He had a passion for very accurate physico-chemical measurement and a flair for manipulation which found expression in his classic work on the limiting density of gases. During the 1914–18 war he became engaged in work on particulate clouds and he made some brilliant, fundamental investigations e.g. of particle size, numbers of particles in unit volume and the lifetime of such clouds. His early career was varied. It included failing the examination for entrance to the Army because he was not interested, studying engineering at Glasgow, hearing Ramsay lecture there and then going to work with him at U.C.L., becoming Assistant Professor at U.C.L., being a temporary science master at Eton during the war, and then in 1923 becoming Professor of Inorganic Chemistry at Leeds. He also became administrative head of the chemistry department, not because he was interested in administration but apparently because he was manoeuvred into it (see E. G. Cox and J. Hume, *Biog. Mem. Fellows R. Soc. London*, 1958, **4**, 327).

The first General Discussion in 1933 was on 'Liquid Crystals and Anisotropic Melts': it was held at the Royal Institution in London, Sir William Bragg welcoming the Society. The Council Minutes of 24th May 1932 do not state who suggested this topic, but the odds are that it was Rawlins. He would have known of the work by Bernal and Dorothy Hodgkin (née Crowfoot). He and Bernal were appointed to the sub-committee set up to organize the meeting.

In 1933 this subject was one of purely academic interest and was rather esoteric at that. Very little work on it had been done in this country. Several of the papers presented at the meeting became standard references. The nature of the smectic phase was understood in essentials: from the optical properties the conclusion had been drawn that there is a mobile, multilayer structure, each layer consisting of long molecules stacked side by side. This model had been supported by pre-1933 X-ray diffraction studies. The nematic phase was less well understood but several workers concluded that there were swarms of long molecules arranged in roughly parallel threads rather than in layers, i.e. without the longitudinal order found in the smectic phase but retaining, more or less, the lateral order, G. W. Stewart found that X-ray diffraction photographs showed the same pattern as that for the isotropic liquid phase of the same substance, although the intensity was greater than for the latter. This type of pattern, which he had found also to be common to isotropic liquids and very dense gases, he attributed to the presence of small, partly ordered groups of molecules – the 'cybotactic' groups. Zocher proposed a model in which there is a greater degree of order than would correspond to numerous independent swarms, and with some degree of elastic response to external perturbations. The cholesteric phase was not understood, but the structure of cholesterol had been established only in 1932, partly in Bernal's determination of unit cell size and space group. At the meeting there was much discussion of the nematic phase. Ornstein and Kast reviewed the evidence for the swarm theory. This included orientation by magnetic and electric fields, as indicated by changes of transparency, dielectric properties, and X-ray diffraction patterns. Such observations certainly supported the swarm concept, since only by cooperation in swarms could sufficient torque be produced in the relatively low fields: estimates of swarm size were attempted. There were, however, a number of puzzling phenomena for which no satisfactory explanations were offered. Professor F. Rinne of Freiburg-im-Bresgau, who died shortly before the meeting, adduced evidence for liquid crystal-like properties of biological materials. Bernal and Dorothy Crowfoot's contribution was to examine by X-ray diffraction one smectogenic, solid crystalline material and several nematogenic ones, and also crystals of cholesteryl chloride and bromide, hoping thereby to gain some ideas about the structures of the liquid crystalline phases. They found that in the smectogenic material, the molecules of which have 'heads' and 'tails', there is a multilayer arrangement in which any head is surrounded by tails in the same layer and is opposite a tail in the next layer. They found also that the molecules in the nematogenic materials, though parallel, were not in layers but were 'imbricated' i.e. overlapping. Finally they found roughly plate-like structures for the cholesteryl compounds. In the crystals the molecules are arranged with their planes parallel, tilted but hardly imbricated. They commented that the iridescent colouring observed in cholesteric liquids, which formally is a consequence of layers 500–5000 molecules thick, although of fundamental importance is not paralleled by crystal structure and may be of hydrodynamic origin. It seems, then, that the properties of liquid crystals which have since made them industrially important were largely known and partly understood at that time but there were no uses for them. Later changes in technological needs suddenly made them important.

The Society was honoured by the presence of Professor R. Schenck of Berlin, then President of the Deutsche Bunsen-Gesellschaft, and by other eminent foreign authorities in the field. Events in Germany had not yet had any very obvious effects on such visits. From Council Minutes it appears that members of Council made voluntary contributions to meet the cost of entertaining the visitors.

The second General Discussion of that year was on 'Free Radicals' and was at Cambridge. It was peculiarly timely because free radicals had quite suddenly become fashionable. It had been deliberately proposed by Lowry who, in his introductory paper, pointed to the swings in fashion between polar and non-polar views of bonding and of reaction processes and emphasized that the concept of electron pair bonding had made it easy to envisage either a polar fission of a bond or a non-polar one. Evidence for the latter had gradually accumulated. It came in various ways, from the discovery in the early years of the century of the stable, aromatic free radicals, from the stable inorganic oxide radicals, from the paramagnetism of oxygen, from studies of gas reactions – first photochemical and then thermal ones – especially the phenomena of retardation and inhibition of reactions by added substances sometimes called 'negative catalysis', from the recent demonstrations by Paneth and by Polanyi that alkyl radicals really have a free if brief existence, and from the rapid developments in molecular spectroscopy which had made it possible to detect radicals by absorption or emission. Moreover, Aston's 'canal-ray tube' made it possible to infer the existence of radicals from the ionic fragments produced in gas discharges. The realization that radicals could reasonably be postulated in thermal reactions had made chain mechanisms much more attractive, for radicals were the very type of high energy intermediate that such mechanisms called for, having high potential bonding energy instead of kinetic energy and therefore having a much higher degree of specificity in reaction. The time was ripe for a review of all these developments and this Discussion provided an outstandingly good one. C. E. H. Bawn (*Trans. Faraday Soc.*, 1953, **49**, 558) rightly described it as a landmark. As he pointed out it was on this occasion that F. O. Rice submitted his scheme for the chain mechanism of the thermal cracking of hydrocarbons, and E. Rabinowitch presented a paper jointly with J. Franck in which their now well known effect of primary recombination of the radicals produced in solution by photodissociation was postulated.

The social centre for the meeting was Pembroke College where, at the usual Guest Night Dinner, Lord Rutherford informally proposed the toast of the Society. There were 27 foreign guests in all, an unusually large number. Paneth and Polanyi were both present but they were now refugees, the former described as of 'Königsberg and London', the latter as of 'Berlin and Manchester'. There were a number of other refugees at that meeting, namely Professor F. Arndt, Drs. L. and A. Farkas, Professor H. Freundlich, Drs. P. Harteck, Gertrud Kornfeld and A. Weissberger. The President (N. V. Sidgwick) welcomed them. These and many other losses from an intellectual élite which had grown up over almost a century were to have dire effects on German science that were to last for at least another 20 or 30 years; but they also led to a great enrichment of intellectual life in, especially, the United Kingdom and the United States of America.

The total effect of having two very large Reports of General Discussions and an increasing number of ordinary papers was to push up the number of pages for the 1933 *Transactions* (29) to an estimated 1358 (the actual final total was 1336) and this in turn led to an expected deficit of about £400. Consequently, at a Finance Committee meeting on 7th November 1933, recommendations for economies were made to Council, which were accepted in December. Nevertheless it was agreed that, despite the difficulties, the Society should continue to publish all the material accepted by the Publications Committee as being of the proper standard and that there should not be a limitation of the size of each issue. There were recommendations to avoid duplication of treatment of the same material and to reduce prolixity

The Inter-war Years, and the Second World War

which we will consider in more detail when we come to review publication policy as a whole.

There were only two newcomers to Council in 1933, namely Dr. H. J. T. Ellingham and Mr. C. N. Hinshelwood, F.R.S. Ellingham, born in 1897, was an Imperial College man. He graduated during the 1914–18 war, being a pupil of H. B. Baker and J. C. Philip, and after it was over he became a Demonstrator there at a salary of £230 p.a., a humble level from which he soon rose. In 1933 he was a Lecturer. He is remembered with affection by one pupil for his 'beautifully organized and witty lectures on thermodynamics'. He wrote an excellent text book in collaboration with Allmand (1923) on applied electrochemistry and later devised the 'Ellingham diagrams' which show very simply the temperature dependence of the free-energy changes occurring when metals are oxidized or oxides (or sulphides) are reduced. Later still (1945) he became an important figure in chemical administration as Secretary, and then as Secretary and Registrar, of the Royal Institute of Chemistry (*Chemistry in Britain*, 1976, **12**, 322).

Hinshelwood, also born in 1897, was a College Tutor in 1933. Had there been no war he would have come into residence as a Scholar at Balliol College, Oxford, in 1916; but instead he was sent to the Queensferry Royal Ordnance Factory where he showed amazing ability at solving chemical problems and where he became interested in chemical kinetics from studies of the slow decomposition of solid explosives. After the war he came at last to Oxford and took a shortened honours course during which he showed such outstanding knowledge and judgement that his Tutor, Brigadier (Sir) Harold Hartley, was able to persuade Balliol to elect him to a Fellowship in 1920. One year later a Tutorial Fellowship fell vacant next door, at Trinity College when D. H. Nagel died, and he was elected.

He soon turned from the decomposition of solids to gas reactions. One of the major problems to concern him was the mechanism of first-order reactions. At first he was more attracted by the simple theory put forward by Lindemann at the 1921 Discussion (Section 2.1) than by the chain reaction theory proposed by Christiansen and Kramers with its seemingly arbitrary and improbable requirement that reaction energy should be conveyed specifically from product molecules to reactant. While he was one of the authors of radical chain mechanisms for some thermal reactions that were explosive under certain limited conditions of temperature and pressure, e.g. the hydrogen–oxygen reaction, he was sceptical about their wider significance until convinced by the work of his own group. In 1933 he was gay, irreverent and witty. As yet not much burdened by administrative cares he was able to concentrate on research and teaching. His research, which was full of excitement, was going very well: at the early age of 32 he had been elected to the Fellowship of the Royal Society. He was a superb tutor, and the personal relations inherent in the local tradition of tutorial teaching brought him great satisfaction and happiness. According to his friends these years were the happiest of his life (E. J. Bowen, *Chemistry in Britain*, 1967, **3**, 534; Sir Harold Thompson, *Biog. Mem. Fellows R. Soc. London*, 1973, **19**, 375).

2.8. 1934–1936

The first of the two General Discussions in 1934 was on 'Dipole Moments', which at that time meant the electric dipole moments of molecules. In 1912 P. J. W. Debye had put forward a classical theory which related this quantity to the bulk properties dielectric constant, density and temperature. He used P. Langevin's treatment of magnetic dipole orientation in a field and Clausius and Mosotti's simple treatment of the internal field problem; so it was a good approximation for a dilute gas but a more dubious one for a liquid or a solution. At that time (1912) the techniques for measuring dielectric constants had not developed sufficiently for this theory to be of much practical use, but the invention and application of the electronic triode valve made possible a very great improvement in ease and accuracy of measurement by the 1920s. A few pioneers (see P. Debye, *Polar Molecules*, Chemical Catalog Co., Inc., New York, 1929) then started making measurements of electric dipole moments of molecules which soon showed that they are of great importance to chemists, being directly related to molecular symmetry. They were, indeed, the first molecular quantity derivable from bulk properties which could be rigorously related to structure. They give also information about electron distribution within the molecule and even, with some assumption and limitations, within bonds and groups. This was of particular value in organic chemistry, where ideas of polar effects in reactions had long been current. They made possible informed discussions of molecular interactions through electrostatic fields. The only one of the real pioneers who attended seems to have been Debye himself, but there was a good attendance from the second generation, including Dr. A. E. van Arkel, Dr. E. Bretscher, Professor J. Errera, Dr. O. Hassel, Dr. H. Sack, Dr. C. P. Smyth and Dr. J. L. Snoek. The papers and discussion were excellent, covering as they did the determination, the interpretation and the application of electric dipole moments. Among the topics covered may be mentioned rotation about a single bond, including a theoretical treatment by W. G. Penney and G. B. B. M. Sutherland showing that the stable conformations of hydrogen peroxide and of hydrazine are skew ones and not the symmetrical, non-polar ones, and the use of electric dipole moments as an additional test of Pauling's hypothesis of 'resonance' in molecules. Debye talked about the very important new concept of dielectric loss by dipole relaxation, and used the lighted cigar which he habitually smoked to illustrate the behaviour of a dipole in a field. At the end of the meeting Marlow gave him a huge new one 'to replace wear and tear on the dipole'. During the meeting several members and visitors became aware that the President had, very characteristically, compiled an exhaustive card index of all the known dipole moments, and they expressed the wish that it might be more generally available. This was prepared for the press by two members of Sidgwick's group, Dr. G. C. Hampson and Dr. R. J. B. Marsden (the writer was at the time in Pasadena) and was published as an 86-page Appendix to the Report of the meeting. For further comments on the meeting see the writer's account (*Trans. Faraday Soc.*, 1953, **49**, 547).

The second General Discussion in 1934 was on 'Colloidal Electrolytes'. It was the third one organized by the Colloids Committee and the last of the series originally planned; but the Society was invited to hold further meetings under the auspices of that Committee and agreed to do so. It was also the first one to be held after the death of Sir William Hardy, the Committee's founder, so tributes were paid to him. University College, London, where Donnan still held sway, was the venue (see plates section). The topic was a central one in colloid chemistry for three reasons. It related to the origin of surface charge on lyophobic colloid particles and to the way in which this stabilized colloidal solution. It had the more special meaning which McBain had given it, of the behaviour of the salts of the longer chain carboxylic or sulphonic acids and of some dyestuffs in forming micelles, as shown by their anomalous thermodynamic properties and electrical conductivities in aqueous solution. It related also to the behaviour of lyophilic colloids, such as the carbohydrates or proteins, which have numerous hydrophilic groups some of which could undergo dissociation into ions. By 1934 there was a reasonably good understanding of these matters due in some measure to structural studies by X-ray diffraction and to the growing appreciation of the nature and importance of hydrogen bonding, together with other polar molecular phenomena. Moreover, the topic was of burgeoning industrial importance because the range of surface-active agents and detergents was increasing concurrently with the understanding of their modes of action. There were papers relating to all these aspects.

The meeting was a sizeable one: the group photograph shows 126 faces. There were 23 foreign guests present representing the majority of distinguished foreign workers in the field, in Europe and in Russia. Professor H. Freundlich, the grand old man of colloid chemistry, was living in London as a refugee, as already noted. Wolfgang Ostwald was there but W. Pauli was not. Professor J. W. McBain could not come from Stanford but Mrs. Laing McBain did. The report set a record for length – 422 pages. There is no mention of any financial support from the other societies represented on the Colloid Committee, so the whole cost of printing this huge report presumably fell on the Society. It helped swell the cost of printing Volume 31 and thereby to give rise to a deficit of about £300 for 1935, as was reported to the Finance Committee on 4th March 1936. As we have seen, the Committee had already, in 1933, made recommendations to Council for reducing the cost of publication, but the pressure was felt to be lessened in 1934 when news came of a very handsome bequest from the late Colonel J. J. Bourke which, in the words of the Committee's Minute, 'should enable the Society to print all papers of its high standard of merit'. In

The Inter-war Years, and the Second World War

1936, therefore, the Committee contented itself with calling for more stringent limits on the number of contributions invited for General Discussions and asking again that authors be urged to write more concisely.

The Council for 1934–35 contained no really new faces at all. Rintoul became President. There were some switches between Vice-Presidents and ordinary members of Council, and Dr. R. Lessing and Sir Robert Robertson returned as ordinary members. Robertson had just been extruded by his seniority from the group of five ex-Presidents who were *ipso facto* Vice-Presidents; so apparently it was considered highly desirable that he should still have an official voice in the affairs of the Society, and he was content to return to Council in that humbler capacity. As it turned out, because of Lowry's untimely death in November 1936, Robertson soon rejoined the select band of elders. This incident stimulated the writer's desire to know more about this kind of switch and how long some people had an influential voice in the Society, with interesting results. Thus, Robertson proves to have served continuously, in one capacity or another, from 1920 to the end of the second world war and, after a break, yet again as an ordinary member of Council. Donnan was elected to Council in 1913, became a Vice-President 1914–20, was off for a year but became a Vice-President again 1921–24 prior to becoming President 1924–26, and then was an ex-Presidential Vice-President: because of the death of Rintoul (who was Robertson's brother-in-law) in office and because of the war Donnan continued in that capacity until 1946. Another position of influence was membership of the Publications Committee: Donnan was a member from 1924 onwards. There were others who served for long periods. A. W. Porter served altogether 20 years, from 1913 to 1933, including his time as President. C. H. Desch served in different capacities from 1915 to 1948 with a break of only one year; and T. M. Lowry served similarly from 1906 to his death in 1936 with only two breaks of one year each. M. W. Travers was another President who served for a long time at various levels, from 1928 to 1953 with one break of two years, as was E. K. Rideal who served from 1921 to 1957 with one break of two years. A. J. Allmand was on Council 1919–22 and 1931–36: he was on the Publications Committee from

1932 until 1947, being Chairman from 1938, and moved direct from that position to being President (for one year only) and then to being a Vice-President until his death in 1951. Some who did not become President also served for long periods. Thus, W. C. McC. Lewis did so from 1920 to 1941 with three short breaks. Emil Hatschek joined Council in 1913 and was switched forth or back from ordinary membership to Vice-President six times, so that he served on Council until 1937; and even then he remained until 1941 a member of the Publications Committee to which he was first appointed in 1914. J. R. Partington was on Council from 1919 to 1938 with one break of three years, and he was on the Publications Committee from 1921 to 1939, being Chairman 1933–38. George Senter was on Council from 1910 to 1924 and continued on the Publications Committee to 1934. Thus there were several periods of service, and of possible influence, of 20 years and a few of 30 or more years.

It is interesting to see what use the ex-Presidential Vice-Presidents made of their opportunities. Neither Hadfield nor Sidgwick attended a single meeting of Council after they ceased to be President, and Sir Robert Mond attended only two. All three men had many other things to do: Sidgwick, for example, became President of the Chemical Society in 1935. However, Porter, Robertson, Donnan, Desch and Lowry attended with varying degrees of regularity. The consequence was that on several occasions in the early 1930s three or even four ex-Presidents appeared at the same time and formed from about a fifth to a third of those members of Council present. If they had chosen to act as a *bloc* then, clearly, they could have been very influential on decisions at those meetings. After 1936 there were never more than two ex-Presidents present. As we saw in Chapter 1 the Society survived its infancy through the long and devoted service of a faithful few. The continuing presence and influence of such a group gave continuity and stability in its middle years, and obviously the Society engendered a close, personal loyalty in many members of Council; but, of course, in a time of rapid change this carried a risk of ossification, of the Society having a limited outlook and acquiring the reputation of being run by a clique, all of which are well known institutional ills. We shall have to watch for such possible effects.

In February 1934 suggestions were made on Council for General Discussions in 1935. Lowry wanted to have one on 'Unsaturation and Conjugation'. Donnan suggested four topics, viz. 'Chemical Effects of Electrical Discharge in Gases', 'The Shape and Quantum Dynamics of Small Molecules', 'Reaction Kinetics of Liquid Media', and 'Equilibria and Thermal Dynamics (sic.) of Technical Processes'. However, Council decided to wait until the new President (Rintoul) took office. In those days it was still possible to organize such meetings quickly. In the event none of these topics was chosen. The first of the two General Discussions in 1935 was in fact on 'The Structure of Metallic Coatings, Films and Surfaces': it was the first time for some years that the Society had turned again to metallography and the electrometallurgical aspects of electrochemistry which had been two of its original major interests (see Appendix C).

There were, indeed, new and important things to say. The Discussion covered work on the structure of coatings or films made by condensation or sputtering, electrodeposition, hot-dipping or spraying, and also of metal surfaces generally. The methods considered were the conventional metallographic ones and the new techniques of X-ray or electron diffraction. Professor G. P. Thomson, J. J.'s son, in 1927 while at Aberdeen, had shown the diffraction of an electron beam by thin films, and this technique had been quickly taken up and applied, especially by Professor G. I. Finch at Imperial College, London, in about 1930 (see M. Blackman, *Biog. Mem. Fellows R. Soc. London*, 1972, **18**, 223). Ellingham, who also was at Imperial College, suggested this Discussion topic to Council in June 1934 and was asked to develop preliminary plans: he also persuaded the Rector of Imperial College to invite the Society to meet there. The information about surfaces gained from electron diffraction, which was the basis of the first half of the meeting, proved to be the main novelty. Some decades earlier, Sir George Beilby, who was a founder member of the Society, had made the brilliantly original and very important postulate that on polished metal surfaces there is a non-crystalline layer (see F. I. G. Rawlins, *Trans. Faraday Soc.*, 1953, **49**, 582). Finch was able to show that there is indeed a layer which is not polycrystalline or microcrystalline but is effectively a supercooled liquid layer into which other metals will dissolve in a way that they would not into the crystalline basis metal. Unfortunately Beilby had died in 1924, much too soon to see his hypothesis verified. By this new technique it was also possible to detect films of oxide, selenide or chloride after various treatments. Cases of induction by the basis metal of an abnormal crystal structure in a film deposited on it were demonstrated. The nature and structure of lubricating layers of graphite were shown and the development by wear of such a layer on cast iron explained the virtue of that material for bearing surfaces. Beside such novelties there were a good number of sound reports concerned particularly with technological advances and the science behind them.

In summing up, Desch emphasized the important effect on the mechanical characteristics of surface layers or inclusions of oxide, hydroxide or other compounds and the difficulty of getting clean metal surfaces, although even he probably did not realize at that time how very difficult it would prove to be, and indeed that it was almost impossible with the vacuum techniques then available. The meeting included much social activity of a kind then possible in London at reasonable cost. There was the usual Guest Night Dinner to honour the overseas guests, who numbered eleven, but they also attended the Anniversary Dinner of the Chemical Society and a Friday Evening Discourse at the Royal Institution, given by Lord Rutherford, on 'The Neutron', an entity discovered only three years previously.

In October 1934 Council had agreed with the proposal by Lowry and Donnan that the Society should continue with the colloid meetings and so with the Colloids Committee. It was this Committee that organized the second meeting in 1935, on 'Phenomena of Polymerization and Condensation'. The original suggestion was made by Rintoul who, because of his work on nitrocelluloses, had long been interested in colloids and large molecules, and its acceptance shows how colloid chemists were now tending to think more in terms of large molecules and not only of aggregates (see *Trans. Faraday Soc.*, 1936, **32**, 1485). Rintoul presided at the meeting, which was held at Cambridge and was a brilliant occasion with the Vice-Chancellor welcoming the Society and a very large number, about 35, of overseas guests and visitors. Some of these are identified on the relevant figure in the plates section. They included not only the early giants – Staudinger, Mark, Kurt H. Meyer – but also younger men like W. H. Carothers, who at the Dupont laboratories had produced nylon. The meeting combined scientific interest and industrial relevance to a remarkable degree, for this was the time of very rapid development in the study of reactions that produced very large molecules, and the beginning of major growth in the plastics industry.

The Report was 412 pages long, and this time the cost caused repercussions, as we shall see later. H. W. Melville, in looking back at this meeting (see *Trans. Faraday Soc.*, 1953, **49**, 565), stressed how clearly it mapped out the regions of study for the next decade or two. The old controversy about the nature of large molecules was still smouldering. Kurt H. Meyer put what may be called the Herzog–Meyer view that substances like cellulose or rubber consist in the solid state of bundles or micelles of quite, but not very, large molecules. The latter are 'primary valence chains', held together by ordinary chemical bonds: the micelles cohere by secondary bonds or van der Waals forces. When treated with a suitable solvent the micelles first swell and then, at least to some extent, disintegrate as solution occurs. Staudinger believed that the primary molecules in solution of such substances as the polystyrenes are very large and are not micelles. Meyer said that 'Staudinger's alternative is expressed as if one were to discuss whether a house is built up of bricks or walls', a remark which shows the weakness of analogies. Staudinger might have retorted that there is no absolute need to associate bricks with building a house, which can be done by pouring concrete. So the argument rumbled on. One difficulty was that there still were few determinations of molecular weight other than those by Staudinger. By 1935 there was appreciation of the possibility that some methods of molecular weight determination would give number-average values while others would give weight-average values. The means of producing one-, two- or three-dimensional polymers were clearly recognized: this was the main theme of Carothers's paper. He and some others were also quite exercised about terminology and the semantic confusions that arose e.g. from the polycondensations for which he had become famous.

There were several discussions of the mechanism of polymerization of ethylene derivatives, all based on the idea of the double bond opening up to give a biradical. It was clearly realized that chain reaction mechanisms were involved but, although the catalytic action of metal and of peroxides was known, the idea of the reactive, growing monoradical was not yet enunciated. Mark and also Bawn tried independently to set up a system of equations for the kinetics of linear polymerization and to compare predictions with observations, but found the task difficult. Mark was puzzled by the speed of the reaction. Staudinger mentioned that ethylene could, with difficulty, be polymerized to a liquid of quite low molecular weight: in the ensuing discussion Dr. E. W. Fawcett, of the I.C.I. laboratories at Winnington, described how a white solid, insoluble in acetone, moderately soluble in benzene, of composition $(CH_2)_n$, and molecular weight about 4000 could be obtained by heating ethylene at about 1000 atmospheres pressure to 170 °C (*Trans. Faraday Soc.*, 1936, **32**, 119), but nobody seemed immediately interested. [It was, of course, the first report of the I.C.I. high-pressure method of producing polyethylene, i.e. polythene, which acquired major importance in the 1939–45 war. M.M.D.] There was considerable discussion of the relation of mechanical properties, e.g. elasticity, to structure. As a whole, therefore, the meeting focussed attention on all the topics that were immediately important. Rarely has the Society succeeded better in its ambition to be a catalyst.

Council in 1935 certainly did not suffer from staleness, for there were no less than five newcomers viz. Mr. U. R. Evans, Dr. Samuel Glasstone, Dr. C. F. Goodeve, Mr. A. McKeown, and Professor M. Polanyi. There was also a translation, of Hinshelwood, to be a Vice-President and two returns of old hands, W. C. McC. Lewis and J. R. Partington to Vice-Presidencies. Evans, who was then 46, went as an undergraduate to King's College, Cambridge, of which College he eventually became an Honorary Fellow. He studied electrochemistry in Wiesbaden and then for a while he worked with E. K. Rideal's father, Samuel Rideal. After serving in Signals throughout the 1914–18 war he returned to Cambridge and became an authority on the corrosion of metals. His interests therefore fitted closely to the Society's early tradition. By 1935 he had produced two major books. He was gifted as a linguist and wholly dedicated to his subject. Glasstone, born

1897, after graduating worked with Allmand at King's College, London, for his M.Sc. and Ph.D. degrees. For seven years he was a Lecturer at the University College of the South-West (now Exeter University): then he went to Sheffield and was there in 1935. He was researching actively in several problems of electrochemistry, such as oxidation reactions at electrodes, polarization and overvoltage. He was to emigrate to the U.S.A. quite soon (1939) to take up a research associateship in Taylor's department in Princeton, and eventually to become famous as a writer in physical chemistry and nuclear technology. Goodeve was a very lively young Canadian, only 31 years old at the time. After graduating at Manitoba he came to U.C.L. to work in Donnan's department, returned to Manitoba for a while but soon came back to U.C.L. as a Lecturer. He was to spend the rest of his working life in this country. In the 1930s his research interest was in molecular spectroscopy: he had contributed substantially to General Discussions. We shall meet him again. McKeown was an Ulsterman who, after graduating from the Queen's University, Belfast, came to W. C. McC. Lewis's department in Liverpool in 1920 to work for a Ph.D. degree, but soon gave up this plan when he was appointed an assistant Lecturer. By 1935 he was 35 years old and was a Lecturer. His interests were in solution properties and in chemical kinetics, including those of photochemical reactions, and his most important contribution was probably the book which he wrote, on *Photoprocesses in Gaseous and Liquid Systems* (1929), with his colleague R. O. Griffith (member of Council 1929–32). Polanyi we have met already. He quickly settled into the Professorship at Manchester, where he went in 1933, and soon had a lively research group established. With his outstanding originality and breadth of interest, his drive and intellectual power he quickly became a major figure in physical chemistry in this country.

2.9. 1936–1938

In October 1934 Council had referred two topics to the Colloid Committee for future consideration: these were 'Smoke and Dust' and rather surprisingly, 'Flame and Explosion'. That Committee did not take up these suggestions, but a year later Council approached

Whytlaw-Gray, who had just gone off Council, to ask if a meeting on the former topic could be held in Leeds. Almost certainly he had been the originator of the idea, and although Lessing was also interested, welcome came from Leeds. What emerged was a General Discussion on 'Disperse Systems in Gases; Dust, Smoke and Fog', held in the Spring of 1936, and the first to be held at Leeds. It included a paper on 'The Inflammation of Dust Clouds', so there was some flame and explosion.

To one unfamiliar with this field it is most interesting to see how problems were identified and how they were treated both theoretically and experimentally. Such matters as the process of formation of smokes and fogs, counting the number of particles per unit volume, measuring their size, shape and other characteristics such as electric charge, the rate of disappearance of clouds by coagulation and precipitation or by evaporation, the importance and nature of nuclei, the techniques for clearing fog or removing dust, including the effects of sound and ultra-sound waves, were all discussed systematically. The industrial aspects and associated public issues were also thoroughly considered. The economics and practicability of clearing an airfield runway of fog by thermal means was one such aspect, especially interesting in view of the Fog Investigation Dispersal Operation which was developed for war-time use soon after. The discussion of smoke abatement was enlivened by questions from Charlie Goodeve, asked with a degree of bluntness and youthful impatience which caused him to be slapped down by his elders: in this he showed the interests in technological problems and matters of public policy which were later to become his major concern. The detailed description of the Howden–I.C.I. flue-gas washing plant was nostalgically reminiscent of papers in the Society's very early days.

The Vice-Chancellor welcomed the Society. Unfortunately the President (Rintoul) was ill at the time, so Robertson took the Chair and introduced the dozen distinguished foreign visitors, all but one from the continent of Europe. It was another very successful meeting, realizing perfectly the early aim of the Society to be a bridge between science and technology.

The second General Discussion in 1936 was also out of London. At the invitation of Professor James Kendall, F.R.S., of whom more later, it was held in Edinburgh. His proposal of a Discussion on 'Molecular Interaction in Liquid Systems' was considered and accepted by Council in 1935, an appropriate sub-committee being set up. Because of the distance between the capitals and the shortcomings of correspondence this sub-committee found difficulties in working, so Council decided in March 1936 to urge the Edinburgh members to come to London for a meeting and even to offer to refund their expenses. In response to Council's wishes the title was made more specific, becoming 'Structure and Molecular Forces in (a) Pure Liquids and (b) Solutions'. It was most timely because there had been great advances due both to the theoretical developments and to new experimental techniques based especially in X-ray diffraction and absorption spectroscopy. Fritz London, who by then had fled from Berlin to Paris, gave a classic paper in which he explained how dispersion forces arise, how they have the right qualitative characteristics, e.g. of additivity, to contribute to the so-called van der Waals forces, and how quantitatively they give reasonably good fits with observation in appropriate cases but leave much, or even most, to be explained by other forces in many cases such as water. Bernal attempted to describe the properties of a liquid assembly of spherical molecules in terms of a scaling parameter, a coordination number that could vary continuously, and an irregularity coefficient. This was prompted by the information derivable from X-ray diffraction studies on liquids such as had been made by Prins and Stewart. It was remarkably successful; for example, he claimed that it predicted a sharp melting point. There were other outstanding contributions, including one from that delightful person Edmond Bauer and his very lively, polyglot pupil Michel Magat, and also from Frenkel and from Eyring.

Joel Hildebrand, then a mere lad of 55, gave the introductory paper on solutions and in this section much attention was given to the conditions which must be met for 'ideal' solutions to be formed and to the forces which cause negative anomalies to Raoult's law. There was much argument about their possible nature, whether it was merely a matter of shape or whether 'chemical' forces were involved and how this could be established. The hydrogen bond was recognized as arising in some conditions. It was a splendid occasion, notable for the liveliness and high quality of the discussion as well as for the set pieces. [After one of Bernal's contributions to the Discussion, Hildebrand asked the chairman whether Bernal could be obliged to leave the room immediately, to write out in full what he had just conveyed to the audience. M.M.D.]

The meeting was marred by the very sad, but not unexpected news of Rintoul's death in late August at the age of only 66. This was the second time that a President of the Society had died in office, but this time, at least, he was near the end of his term of office. Donnan and Kendall took the Chair in turn at the Edinburgh meeting; and the new President, M. W. Travers, took office on 1st October. Travers we have already met. When he became President he was 64 and was still an Honorary Professorial Fellow and Nash Lecturer at Bristol University. A further loss to the Society in November of that year was the death of Lowry. He was aged only 62, and was still an ex-Presidential Vice-President. He had been one of the most devoted and influential founder-members of the Society.

There were four newcomers to Council in 1936, viz. J. D. Bernal, H. J. Emeleus, Professor A. Ferguson and Professor James Kendall, F.R.S. Bernal, as we have already seen, was showing remarkable versatility and enterprise in exploring molecular structure by diffraction methods and in developing the theory of liquids, both these with only dispersion forces and those with chemical forces like water about which he had already written a classic paper with R. H. Fowler. He was highly articulate, and excellent in discussion. Much later, Sir Lawrence Bragg said of him 'Of course we always got ideas from Bernal. Bernal is the most fertile and the greatest brain of all the people working in this field. He is far more known for starting a new idea than for completing the work on it. He lets other people do that.' (*Chem. Br.*, 1970, **6**, 149).

Emeleus was an Imperial College man, born 1903, who after spending some time at the Technische Hochschule in Karlsruhe and then two years at Princeton, where he worked with H. S. Taylor, returned to London in 1931 and became a member of the staff of his old College. He was interested in photochemical reactions, and in the mechanism of reactions as revealed e.g. by the phosphorescence which occurs in some oxidations, both of organic and inorganic compounds, and by the kinetics, as in the oxidation of silanes. Ferguson was a man of striking size, being 6ft 6in tall and weighing 21 stone (for metrical posterity, 1.98 m and 133 kg). In appearance he was not unlike Dr. Samuel Johnson (J. H. Awberry, *Proc. Phys. Soc.*, 1952, **65**, 998) but the manner was different. Born in 1880 in Entwhistle he had started work as a pharmacist's assistant and started to study for a London science degree at a local institute. In 1902 he entered the University College of North Wales at Bangor as an Exhibitioner. After graduating he stayed there as an assistant lecturer in physics until 1919. He then went to the Manchester College of Technology as a Lecturer, but he soon (1921) moved to Queen Mary College where he spent the rest of his working life, becoming an Assistant Professor of Physics in 1934. In 1936 he was still Papers Secretary of the Physical Society despite having had serious eye trouble. He was also a member of the Board of the Institute of Physics; and before the end of his term of service with the Faraday Society (he was a Vice-President 1937–43) he was an editor of the *Philosophical Magazine* and of *Nature*, and President of the Physical Society (1938–41). From about 1920 he had given the major part of his time and effort to organization and committee work at which he was very good; so all in all he was a powerful recruit to Council.

Kendall had succeeded Sir James Walker, F.R.S., as Professor of Chemistry in 1928 when he was 39. Although he was born in Surrey he was an Edinburgh graduate because Kendall's father thought that son James might attempt the Indian Civil Service Examination and Edinburgh men had just taken the first three places (N. Campbell and C. Kemball, *Biog. Mem. Fellows R. Soc. London*, 1980, **26**, 255). After graduating in science and arts he decided to seek a career in chemistry, so he spent a year abroad working part of the time with Arrhenius. Then, having been invited by Professor Alexander Smith, a former Edinburgh man, he emigrated to the U.S.A. in 1913 to take a post at Columbia University, where he had a most fruitful time as a researcher. During the war he became one of the very few Englishmen to hold a commission in the U.S. Naval Reserve, eventually attaining the rank of Lieutenant Commander. Soon after the return to Columbia Smith died, and Kendall was urged to undertake the major task of revising his old chief's series of textbooks. Also, in anticipation of a possible call to Edinburgh to succeed Walker, in the long tradition there of pupil succeeding master, he went in 1926 to the Washington Square College of New York University in order to gain administrative experience as head of department. Early on he discovered his great gifts for exposition, both by the spoken and the written word, and later he realized that he was an excellent administrator. So the distractions increased steadily: and when the call to Edinburgh came in 1928 he had almost given up research. In his fruitful period he worked very actively as an experimenter on several problems of solution chemistry, e.g. the formation of loose compounds in binary systems, heats of vaporization of associated liquids, solubility products, conductivities of electrolyte solutions, and the separation of elements and isotopes by ionic migration. One of his last projects was an attempt to separate the hydrogen isotopes by electrolytic fractionation which was, however, overtaken by work in the U.S.A. He was a very effective provider for the Edinburgh department; and he had the sympathy and enthusiasm which were most encouraging to young researchers. He continued to write very readable books about chemical ideas and the history of chemistry, and to give very stimulating introductory lectures. He was a great character, i.e. he was unorthodox. Campbell and Kemball tell how it fell to Kendall publicly to present the Nichols Medal, in 1928, to H. S. Taylor and E. K. Rideal, who had written the well known book 'Catalysis in Theory and Practice'. He said 'There has always been discussion as to who wrote which chapters of Rideal and Taylor's book. I can give a sure way of finding out. When you find a split infinitive, that had been written by Taylor; when you find an equation that does not balance, that had been written by Rideal.' The audience was much amused: the authors were not. He was shrewd. He was obviously right in urging that a Discussion be held on liquids and solutions in 1936. It is likely that one reason for his election on Council was a desire to strengthen the Society's links with the Scottish Universities. It may be remembered that these had existed from the earliest days when Professor Crum Brown was a founder member and Lord Kelvin was President; but in the 1926 list of members there were only 19 personal or institutional members with addresses in Scotland.

The Inter-war Years, and the Second World War 117

As the newly elected General Secretary of the Royal Society of Edinburgh, Kendall had influence; and he obviously wanted to lessen the remoteness felt by chemists in Scotland.

In 1934, when Council agreed to the continuation of the Colloid Committee, it made two specific recommendations for future General Discussions to that Committee, one of which was used in 1936, and one vague one, viz. 'A subject of interest to biologists' which it felt should be held 'in the reasonably near future'. In April 1937 there was such a one, on 'Properties and Functions of Membranes, Natural and Artificial'. It was held, very appropriately, in U.C.L. In the first part there were indeed numerous contributions from biologists such as J. Z. Young who described the structure of nerve fibres, from biochemists such as Ancel Keys who talked about the permeability of the capillary membrane in man, and from biophysicists such as Kenneth S. Cole who reported on attempts to analyse the dielectric characteristics of aqueous suspensions of sea-urchin eggs. The Discussion illustrated the difficulties which physical scientists must feel in getting to grips with biological problems, namely that of finding phenomena to which their relatively simple ideas and principles can be usefully applied and of deciding, if they find them, whether they are of fundamental importance, only secondary, or even trivial. Professor A. Krogh, for example, urged that the Discussion should be restricted to membranes which are passive, i.e. not those across which material is transported with the consumption of energy, because in the latter it was not clear what is membrane and what is pump or, as Krogh himself said, 'because it is too difficult to distinguish the membrane from the cell of which it is an integral and perhaps variable part'. Thus, he included membrane phenomena which could be treated by the chemical ideas of thermodynamics and kinetics but excluded those which are the very stuff of biology. Not surprisingly, objections to this view were voiced. Already, in those days, there were suggestions that such transport takes place by the material being complexed on the input side of the membrane, so as to make it compatible with the membrane and cause it to flow across under a concentration gradient, and being de-complexed on the output side from which it cannot flow back through the membrane because of its incompatibility. In this way the membrane functions as a double non-return valve, while the 'pump' is the complexing and decomplexing mechanism operating with the consumption of energy. Ideas as to what these mechanisms might be were, however, extremely vague. The second part, on artificial membranes, was led by Kurt H. Meyer and it dealt exclusively with what we have called the passive type of membrane. The topics discussed were essentially the specificity of their permeability in relation to their chemical nature, and the thermodynamics and kinetics of membrane processes. There were frequent references to Donnan's treatment of membrane equilibria, which showed its continuing value. With only early electron microscopes to show fine structure, membranes – even the much simpler artificial ones – were still mysterious. The recent developments from new valency theory of the old concepts of polar, hydrophilic and lipophilic groupings and interactions, including the specific one of the hydrogen bond, were much in use. No dramatic advances could be expected from such a meeting. It was one step in the long dialogue between biological and physical scientists.

The second General Discussion in 1937 was another that has come to be regarded as a classic. It was on 'Reaction Kinetics', and it was the first occasion on which the theories of Eyring and Polanyi received full, public discussion in this country. The topic had been proposed by Polanyi, who played a leading part in organizing the meeting and arranging for it to be held at Manchester. His sub-committee imposed a more definite pattern than had hitherto been customary, by inviting a number of authors to prepare reports on particular aspects of the subject. It is possible that the sub-committee had had some prompting, for over a long period and particularly in 1936 the Finance Committee had urged in Council that the organization of General Discussions should be tighter in order to increase comprehensibility and, by avoiding unnecessary overlap, to save space. The cost of printing the Reports was proving to be a heavy burden.

This time the pattern consisted of a general review of the theoretical methods of treating the reaction process, then one of experimental evidence bearing on these theories, and finally a consideration of reactions involving protons. The treatment of the reaction process by calculating a potential energy surface, and the assuming that there is a quasi-thermodynamic equilibrium between normal reactant molecules and an activated complex or transition state defined by the saddle point, with a translational mode along a path of minimum energy, instead of one vibrational mode, was developed during the early 1930s by Eyring and Polanyi and their pupils. The potential energy calculations were based on the treatment of atomic interactions by Fritz London and expounded by him at the 1936 Discussion. Claims were made that this theory resolved certain difficulties into which the earlier 'collision theory' had run, particularly those of explaining why some reactions go more slowly and others more rapidly than is calculated from the collision number. It was agreed that if both treatments could be applied rigorously they would give the same answer, but it was argued also that when only an approximate treatment could be made then the activated complex treatment gives a better answer. On the other hand, in order to go beyond using thermodynamic functions and to gain a physical picture of what was happening, it was necessary to interpret partition functions in terms of atomic motions and degrees of freedom. Then, too, the Eyring and Polanyi treatment required that methods for calculating potential energies, which were known to be approximate, be pushed to or beyond their limits, and that they be laced with several bold assumptions. There was, therefore, great scope for fundamental discussions between the eminences who were there in force. A few examples may be given.

Edward Guggenheim, of whom more later, is on record as saying in discussion that the transition state theory was 'entirely empirical' (*Trans. Faraday Soc.*, 1938, **34**, 27) but, according to R. P. Bell who was there, what he actually said was that it should be described as '*Sesqui*-empirical' because three assumptions had to be made in order to get two results. R. H. Fowler rebuked sin as manifested by sloppiness in the definition of models. Hinshelwood stressed the need for building up a physical picture of the reaction process and of the factors which could affect it. He emphasized the importance of time. He and Polanyi disagreed about the part which multiple degrees of freedom could play. He took the view that they, together with a long interval between activation and reaction, could explain the high rates of first-order reactions; whereas Polanyi argued that degrees of freedom were irrelevant and that the cause of the high rate must be a loosening of bonds in the activated complex. E. Rabinowitch made lively contributions in discussion about the effect of time lags on reaction velocity and about reactions in condensed systems. In introducing the third section R. P. Bell raised the possibility of effects of proton tunnelling, years before they were to be demonstrated. As D. W. G. Style remarked later (*Trans. Faraday Soc.*, 1953, **49**, 555) 'an unusually high proportion of the twenty-one contributions . . . were papers of fundamental importance. In every way it was an historic meeting.'

The dispersal of German Jewish and anti-Nazi scientists was no longer a matter for comment but was an accepted fact. It may be noted that, apart from Polanyi himself, of those participating E. Wigner was then in Madison, Wisconsin, while E. Rabinowitch and A. Wassermann were in London. K. F. Bonhöffer, brother of the theologian Dietrich, remained in Leipzig.

In 1937, Garner returned as a member of Council and there were two newcomers, viz. Dr. D. W. G. Style and Professor S. Sugden, F.R.S., Style, who was only 33, was a spectroscopist and photochemist. He had come to King's College, London, as a student in 1923: he stayed to work for his Ph.D. with Allmand. After spending one year abroad in Polanyi's laboratory in Berlin he returned to King's as a Lecturer and spent the rest of his working life there, later succeeding Allmand. He was quiet, and rather aloof in manner; but he was a very clear-sighted and independent person who had a love and a flair for doing his experiments with simple, home-made apparatus. He was a dedicated scientist, and he was much loved by those who got to know him (*Chem. Br.*, 1980, **16**, 379). Sugden, who was 45, had just graduated at the Royal College of Science when was broke out in 1914. After service with the R.A.M.C. he was released to join the Research Department of the Royal Arsenal, Woolwich, to work under Sir Robert Robertson, where he did a useful job; but he was happy to return to academic life, becoming a Lecturer at Birkbeck College in 1919. He quickly made his mark, conceiving the parachor, a quantity which aimed to give molecular volumes in liquids corrected to the same internal pressure (L. E. Sutton, *Obit. Not. Fellows R. Soc. London*, 1951, **7**, 493). This was one of the last attempts to derive structural information from empirical, additive constants: [Some colleagues in physical chemistry renamed it the metachele. M.M.D.] Methods based soundly on physical theory were, as we have seen, developed in the late 1920s and early 1930s, and these proved vastly more powerful. Sugden took up some of these methods himself. In 1937, following Donnan's retirement, he became a Professor of Chemistry of U.C.L. He was small of stature, very lively and very cheerful – a delightful person.

2.10. 1938–1940

The first General Discussion in 1938 was held at Bristol and was on 'Chemical Reactions involving Solids'. this was proposed to Council in April 1937 and it is highly likely that the idea came from Garner (see Section 2.4) who had by 1937 published a number of papers on adsorption, especially on heats of adsorption, heterogeneous catalysis and the rôle of nuclei in the decomposition of solids. Basic advances had become possible because of the new understanding of crystal structure which had led to the concepts of lattice imperfections, and the development of wave mechanics in generalizing ideas about energy levels. By far the greater part of the experimental work and of the ideas in this field had by then been done in, or had come from continental Europe, The pioneers included Bloch, Frenkel, Pohl, Schottky, Carl Wagner, Jost, de Boer and Brillouin. They had explored the effects of various kinds of lattice defect on the absorption spectra of inorganic crystals and on diffusion processes including electrical conduction by ions. They had also showed how irregularities can be produced thermally or by irradiation with light. These studies had led to the beginnings of an understanding of the processes of reactions in solids. The band theory of energy levels in solids had been established but not much developed, and N. F. Mott, then at Bristol, was one of the people currently developing and using it. Clearly, therefore, the topic was one which needed publicising in this country and at that time. Moreover, with both Garner and Mott at Bristol, that was obviously the right place for it. A section was devoted especially to solid carbon, as suggested by Desch; and there was a brief one on photochemical processes which included reviews of the ideas about the photographic process for which a sound theoretical basis was just beginning to emerge. Once again, most of the leading authorities contributed: many of them were present, and there were excellent discussions.

The second Discussion that year was on 'Luminescence' and was held at Oxford. How it originated is not clear, but possibly Hinshelwood, who was a Vice-President in 1937 when the decision was made, relayed a suggestion by E. J. Bowen. It was of more limited interest than the two preceding Discussions, but was already of considerable industrial importance. Cathode-ray tubes were coming into common use, e.g. the B.B.C. had started television broadcasting in the London area in 1936, and fluorescent lighting tubes were being manufactured although they were still a novelty. There were three sections to the Discussions, the first two on luminescence induced by radiation, in liquids or vapours and in solids, respectively, and the third on chemiluminescence. The second section was therefore linked with the previous (April) Discussion. Not surprisingly there proved to be less understanding of condensed systems than of mono- and diatomic gases. Although general theoretical schemes could be adduced there were many phenomena for which rather *ad hoc* explanations were offered. Altogether it was a good meeting.

The new President in 1938 was E. K. Rideal, who by then was John Humphrey Plummer Professor of Colloid Science at Cambridge: he was 48 and was in his prime. Sir Frederick (now Lord) Dainton, F.R.S., feels that Rideal, whom he knew well in those days, had two remarkable characteristics that should be stressed. One was the fertility of his mind. 'There were no subjects on which he did not have an idea, indeed, very many ideas. Of course many of them were of little use, and he was often not discriminating. That the range was enormous is shown by the papers which he published touching, e.g. on all aspects of reaction kinetics in homogeneous and heterogeneous systems, on spectroscopy, on the formulation (with Fowler) of the effect in gas reactions of the number of degrees of freedom on the fraction of molecules activated by collisions, on colloid stability, the shape and size of molecules as determined by the film balance and by sedimentation.' This is supported by the writer's memory of a visit that he made to Rideal's laboratory in 1929 in company with the late Professor Klaus

Clusius, who then was still with Eucken at Göttingen (probably as his Assistant). Clusius was amazed by the variety of work going on and wondered how Rideal kept control of it. Dainton adds that Rideal's lectures were 'chaotic but very stimulating for those who had already graduated in chemistry'; and Rideal's second outstanding characteristic was 'his great generosity which, allied to his natural friendliness, made him a marvellous person for encouraging younger people'.

Emile Mond continued as the Honorary Treasurer, but at the Council meeting on 14th December he was reported as being seriously ill: he died on the 30th day of that month (*Trans. Faraday Soc.*, 1939, **35**, 301). By a sad coincidence his predecessor and relative, Sir Robert Mond, F.R.S., had died only about two months earlier (*Trans. Faraday Soc.*, 1938, **34**, 1369). Dr. R. E. Slade, who first joined Council in 1926 and had become a Vice-President, was nominated and elected Honorary Treasurer by Council at their next meting in January 1939. He was then Controller of Research for I.C.I. Ltd. The new faces were those of Professor R. H. Fowler, O.B.E., F.R.S., who became a Vice-President, Mr. E. J. Bowen, F.R.S., and Dr. F. D. Miles. A. O. Rankine returned as a Vice-President and so did Whytlaw-Gray. R. Lessing returned to Council after an absence of only one year. Fowler was born in 1889, went to Trinity College, Cambridge, became a Fellow in 1914, and from 1932 was John Humphrey Plummer Professor of Mathematical Physics. He was a man of rare intellectual distinction. His field was statistical mechanics. He and Edward Guggenheim must at the time have been writing their classic book on 'Statistical Thermodynamics' which appeared in 1939.

Bowen we have met before in connection with the General Discussions in photochemistry in 1925 and 1931. He and Hinshelwood were near contemporaries at Balliol College, Oxford, and he was one of the few people with whom Hinshelwood felt at ease. His enthusiasm and verve always made him stimulating company. Sometimes known as 'the Empirical Master' he was a fertile and ingenious experimenter. Like another photochemist, D. W. G. Style, Bowen delighted in building his own apparatus out of old cocoa-tins, odd lenses and prisms. There is an apocryphal story that his old tutor, (Sir) Harold Hartley, F.R.S., when inspecting the former Balliol Laboratory, opened a drawer, found a set of discarded false teeth, and immediately said 'I wonder what apparatus Bowen will build these into?' A real dental episode occurred when the new Physical Chemistry Laboratory at Oxford was officially opened in 1941. Bowen showed publicly, by ultraviolet light, that the then Vice-Chancellor had false teeth!

Miles, then aged 53, was an industrial chemist with the Nobel Division of I.C.I. Ltd., where, probably, he was by 1938 head of the physical chemistry section of the research department which Rintoul had set up. He was a rather withdrawn, very scholarly man who might have flourished in an academic setting but who did in fact prove very effective in an industrial one. He was the author of two books, one a major, very learned work on the physical chemistry, formation and use of nitrocellulose, and the other a history of the Division's research department at Ardeer, written with great insight and sensitivity. His knowledge was great, both in range and in depth; and it was always freely available to his colleagues.

The first General Discussion in 1939 was on 'Hydrocarbon Chemistry', held in London at Imperial College. Although the origin is not clear from Council Minutes, the suggestion is likely to have come from Rideal, for his group had done very good work on some of the catalytic reactions. There was a welcome return to an older custom of having a summarizing introductory paper, this one being given by Rideal. He emphasized the industrial importance of the topic. The fuel oil industry had had to find out how to convert one hydrocarbon into another; and, furthermore, 'the various hydrocarbons are likely to become of increasing importance since they can serve as the raw materials of a number of the newer chemical industries, an economic and scientific field which is as yet almost virgin soil.' Coal was still the major industrial raw material for the organic chemical industry in this country. It was generally agreed that the homogeneous gas reactions used by the fuel industry involved free-radical chains; but a quick glance through papers and discussions confirms Rideal's remark that there were wide divergences of opinion about the chain mechanisms, e.g. between F. O. Rice and E. W. R. Steacie, while Hinshelwood and his pupils thought that there were also non-radical reactions. The gas reactions based on heterogeneous catalyses, such as the Fischer–Tropsch syntheses, the direct hydrogenation of coal, and ring closure were known. There were several papers, some by British authors, which threw a good deal of light on the mechanisms thereof.

The different schools of thought were well represented, and there were vigorous debates. Olefin polymerizations were again discussed, as they had been in 1935. A new view about the kinetics thereof was advanced by R. G. W. Norrish who argued that it was not the initiation reaction that controlled the rate of polymerization, as Staudinger, Mark and others believed, but the propagation reaction. Lord Dainton comments that neither view showed sufficient appreciation of the importance of the rate of termination of reaction chains. It may be noted that although M. W. Perrin, who had been largely responsible for the 're-discovery' of polyethylene at the Alkali Division of I.C.I. Ltd., contributed a paper he did not say much about this new polymer. It was patented in 1936 but still was of minor commercial importance – only ten tons were sold in 1939. It was thought of as an expensive material to be used only when outstanding dielectric properties were required, as for the highly secret RDF system of aircraft detection, later re-named RADAR. Large-scale uses had not yet been foreseen. The mechanical properties of high polymers, including the elasticity of rubber, were beginning to attract attention. They, too, received some discussion; but ideas were still rudimentary. H. W. Melville has stressed some of the highlights of this Discussion (*Trans. Faraday Soc.*, 1953, **49**, 568).

Once again there was good international representation, 27 guests and visitors coming from overseas. The Society had had yet another meeting which was exceptionally timely and brimful of interest.

The April Discussion was to be the last full one before war broke out on 3rd September. There was to have been another under the auspices of the Colloid Committee, on 'The Electric Double Layer', and in recognition of Rideal's Presidency it would have been held at Cambridge; but this was abandoned. Rideal persuaded Council that, since many of the contributors had completed their manuscripts in spite of the disturbed conditions, all the papers that were available should be published. These make a thick volume of 319 pages without any discussion. They cover electrophoresis, streaming effects and surface conduction, electrocapillarity, the theory of the double layer and colloid stability, and the biological and technical applications such as the separation of biological materials by electrophoresis which Tiselius had developed. A 22-page collection of written contributions to the discussion was published later in Volume 36.

In the 1939 Council three old hands returned as Vice-Presidents, namely W. C. McC. Lewis after only one year off, and J. E. Coates and C. C. Paterson after intervals of eleven and thirteen years, respectively. There were four newcomers. One was C. R. Bury, born 1890, who had graduated with first-class honours in chemistry at Oxford in 1911. As a member of Trinity College he was one of D. H. Nagel's pupils and, moreover, one of the early products of the research school in the Balliol and Trinity laboratories, where he worked for a year with (Sir) Harold Hartley. He then went to Göttingen and in 1913 was appointed as Lecturer in physical chemistry in the University College of Wales at Aberystwyth, where he remained until 1943 when he joined the research department of the Billingham Division of I.C.I. Ltd. Most of his papers are about the phase rule, and solution properties; but by far the most important one was on the electronic structure of atoms and ions (*J. Am. Chem. Soc.*, 1921, **43**, 1602). He showed very clearly that by modifying Langmuir's fourth postulate he could give electron allocations to main groups which fitted, and therefore were justifiable by, the chemical properties of the elements. He was able in this way to rationalize the occurrence of the inert gases, of the transition elements, and of the rare earths. Also, he could make some predictions about the properties of the transuranic elements. His electron allocations are little different from those now accepted. On the basis of his 'correspondence principle' Niels Bohr anticipated some of Bury's conclusions; but he was not nearly so clear and the physical basis that he offered was vague. Professor Mansel Davies wrote: 'Not of all Nobel chemists can it be said that every college student of chemistry is taught the content of their work'. Yet Bury got little credit. In an obituary notice (*Nature*, 1969, **222**, 805) Mansel ends 'His directness of manner and independence of mind doubtless contributed to the absence of public recognition of his merits.'

The second newcomer was Dr. J. J. Fox, who succeeded Sir Robert

The Inter-war Years, and the Second World War

Robertson as Government Chemist in 1936. It had been intended that he should go into business, but because of his mother's influence he was allowed to follow his own interest in science. He first joined the Government Laboratory in 1896 when he was 22, attended courses in chemistry at the Royal College of Science, and then went to evening classes at the East London Technical College (now Queen Mary College) eventually gaining a B.Sc. by research in 1908 and a D.Sc. in 1910. He became a member of the permanent staff of the Laboratory in 1904 and won steady promotion. He was an expert on analytical methods generally, but he was particularly interested in the application of physical methods to analysis and was one of the pioneers. He published papers on the relation between chemical structure and ultraviolet spectra with Sir James Dobbie, who was Government Chemist before the first world war, and then on infrared spectra with Sir Robert Robertson when he succeeded Dobbie in 1921. He was round, cheerful, kindly and very helpful – 'a very lovable as well as a respected man.' (see *Obit. Not. Fellows R. Soc. London*, 1945–48, **5**, 141; *J. Chem. Soc.*, 1945, 719). Dr. W. J. Shutt, Lecturer in electrochemistry at Liverpool University, was the third new man. He was about 43 years old, a Manchester graduate who had specialized in electrochemistry from early on, and who joined the staff at Liverpool in 1919. There is a reference to 'his triumph over a life-long infirmity' in a tribute paid to him by the Chairman of the Faculty of Science after his untimely death in 1943 (aged 47). He published sparingly and mostly about the anodic passivization of metals. The fourth man was Dr. O. J. Walker, who was born in 1899, took his B.Sc. and Ph.D. at Edinburgh by 1925, and then worked for a while with Kasimir Fajans in München. In 1939 he was a Lecturer in Donnan's department at U.C.L., where he worked on electrode reactions, on spectroscopy and photochemistry.

2.11. 1940–1945

Council had planned to have a General Discussion in the spring of 1940 on 'Processes of Isotopic and Molecular Separation' and, provisionally, the autumn one on 'Interatomic Distances', the former to be held in London and the latter somewhere in Wales. A 'colloid' meeting was also contemplated for 1941. In November 1939 Council realized that all these plans had to be abandoned, and suggested a two-day meeting on 'Current Problems of Fundamental Importance' which was the vaguest title yet suggested. What actually happened was that Rideal arranged a short, one-day meeting in Imperial College, London, on 'The Hydrogen Bond', on 17th May 1940. On this occasion, asked to review the role, if any, of the hydrogen bond in protein structure, Astbury said that since the detailed structure of the proteins was largely unknown all he could do was to describe the various hypotheses about their structure in which hydrogen bonds played a part. E. L. Hirst and two co-workers concluded that the sub-units of the starch molecule, each of about 24–30 glucose sub-sub-units, are held together by normal covalent bonds and not by hydrogen bonds. The starch macromolecules can, however, associate by hydrogen bonding to form large particles. G. B. B. M. Sutherland, J. J. Fox and A. E. Martin, J. M. Robertson and Rogie Angus described applications of the new physical methods – infrared spectroscopy, X-ray diffraction and diamagnetic susceptibility – to investigating hydrogen bonds.

In January 1941 a Special General Meeting was called, at which members agreed to a temporary alteration in the rules, whereby the Officers and ordinary members of Council could continue to serve during the war, beyond the periods provided in those rules. Thus the Society did in the second war as it did in the first one, and the effect was again to give a very lively, hard-working President and an admirable Secretary great influence on its life.

There were two one-day meetings in 1941 and one in May 1942, all held at the Hotel Rembrandt instead of in the customary academic accommodation. These were respectively on 'The Oil/Water Interface' (held in conjunction with the Physical Society), on 'The Mechanism and Chemical Kinetics of Organic Reactions in Liquid Systems', and on 'The Structure and Reactions of Rubber'. The first of these might appear to be the brainchild of the Colloid Committee, but in fact there are no Minutes for any meeting of the Committee between January 1938 and December 1945; so Rideal probably chose it. The meeting was brief (11 a.m. to 4 p.m.) and of limited interest, but there were valuable contributions, e.g. on the permeability of cell membranes, on the role of protein and of other 'amphipathic' substances in such surfaces and in the stabilization of oil or lipoid emulsions in water, on the electrophoresis of oil droplets, and on the adsorption of surface-active agents at such interfaces. The discussion was obviously lively.

The second was only slightly less brief (10 a.m. to 4.30 p.m.) but it was of wide interest, was well attended, and became a valuable source (see D. W. G. Style, *Trans. Faraday Soc.*, 1953, **49**, 556). In his introductory paper C. K. Ingold said that he understood it to have been planned originally at the 1937 Discussion on 'Reaction Kinetics' and he commented that the profound effect of that meeting was shown by the extent to which the concept of the transition state was now used. This concept, and the effect generally of the improved understanding of valency, including the polarity of covalent bonds, were indeed influences which had stimulated development, but there were also experimental factors especially the use of isotopes to elucidate mechanisms. By way of background it may be remarked that at that time the major school of study of the mechanisms of organic reactions in solution involving polar processes, as distinct from free radical processes, was that of Ingold and E. D. Hughes at University College, London (temporarily evacuated to Aberystwyth).

There was a substantial British tradition of research in this field. K. J. P. Orton and Arthur Lapworth were among the pioneers; and Ingold pointed out that half the material of the 1941 Discussion came from former pupils of Orton, the most senior of whom was H. B. Watson of Cardiff. Robert Robinson had developed the ideas of Lapworth, his revered master, and had produced theories of polar reaction processes based on simple electronic interpretations of valency; but although he described them in lectures he published little about them, and when he did he tended to publish in rather inaccessible journals, so he failed to popularize them (Lord Todd and Sir John Cornforth, *Biog. Mem. Fellows R. Soc. London*, 1976, **22**, 465). Ingold, who was seven years younger, developed theories which in some respects were very similar. Robinson believed that Ingold had borrowed from him without acknowledgement. He felt this very strongly to his dying day and made no attempt to hide his feelings, so relations between the two men were very strained. Ingold published freely in accessible journals and introduced his own nomenclature, e.g. 'electrophilic' instead of 'cationoid' and 'nucleophilic' instead of 'anionoid', which became generally accepted. Moreover, with E. D. Hughes he examined in detail a wider range of thermal reactions in solution than Robinson did (C. W. Shoppee, *Biog. Mem. Fellows R. Soc. London*, 1972, **18**, 349). Leaving aside the question of whether he consciously borrowed from Robinson, the answer to which will never be known, his whole approach was different. Robinson was a very great organic chemist who used his theories brilliantly in syntheses and elucidations of structures that were of major importance: he developed cruder theories dealing with a wider range of phenomena than did Ingold. For example, he speculated very fruitfully about biosynthesis, and about the origin of oil. He was very sure of his own ideas. When, in about 1931, the writer showed Robinson that electric dipole moment observations supported his theories of electron drift between the benzene ring and various attached groups Robinson said 'Well, of course, we always knew that that was so', and then, after a pause, 'but it is quite pleasing to see it confirmed by physical methods'.

Ingold did many more quantitative kinetic studies than did Robinson. His range, which included incursions into molecular spectroscopy, was of a different kind. He was much more a physical chemist in methodology, although he once said, teasingly, to the writer that physical chemistry was not a subject in its own right: it existed only to serve organic and inorganic chemists. Several schools of chemical kinetics had been, or were currently of influence in this field. Such had been H. M. Dawson's at Leeds: such were the Polanyi–Evans school, those of Hinshelwood and of R. P. Bell at Oxford, and that of E. A. Moelwyn-Hughes at Cambridge. These provided basic ideas and techniques; but the kineticists regarded organic chemistry as providing convenient tests of their theories, they were not interested primarily in its problems as were Robinson and Ingold. It could be that Ingold had Hinshelwood, or Polanyi and Evans in mind when he said in his introductory remarks that 'we may note with gratification that all the papers have avoided ... the tendency to be observed among physical chemists to drift obscurely into the misty regions of the general theory of reaction rate.' The papers at the Discussion certainly dealt with a considerable variety of standard organic reactions, viz. substitution in aliphatic and aromatic compounds, elimination reactions, ester hydrolysis, condensation of carbonyl compounds, anionotropic reactions, addition reactions with olefins, Friedel–Crafts reactions, the Cannizarro reaction, and ring closure. They mostly dealt with processes involving ions or polar processes, but a significantly different contribution was that of W. A. Waters on radical reactions. Appreciation of the importance of these latter in organic chemistry had been growing since the work of Kharasch in the U.S.A. and of D. H. Hey and Waters in this country, during the 1930s. The Report was 204 pages long. It was a remarkable effort for wartime and was to prove a landmark for a long time to come.

The one-day Discussion held on 29th May 1942, again at the Hotel Rembrandt, was on another subject dear to Rideal's heart, viz. 'The Structure and Reactions of Rubber'. The papers presented showed how rapidly thought had developed after the acceptance of Staudinger's concept of macromolecules. Apart from this breakthrough, the new physical methods – electrical dipole moment measurements, thermal measurements, and especially the diffraction studies and molecular spectroscopy – were making available much more detailed and precise information about the structure of molecules – even large ones – and about intra- and inter-molecular energetics. This new information made possible the use of thermodynamic principles which had been enunciated quite a long time earlier but had not been widely applicable at that time. So it became possible to connect changes of entropy with changes of molecular motion and freedom in a much more definite, purposeful and fruitful manner. Details were still unclear and debatable, but the general outline of an explanation of elasticity, or the lack of it, in high polymers was becoming established. There was a strong demand from industry for a better understanding of all the mechanical properties of polymers. Progress was at last being made also on the problem, which was of great practical importance, of why and how rubber perished by oxidation. The isolation of the primary hydroperoxide and the acceptance of radical chain reactions were the keys thereto. This subject was very suitable for a domestic discussion because a number of very bright young men were working in this country; and several of them were pupils of Rideal, including G. D. Coumoulos, R. M. Barrer, D. D. Eley, G. Gee and R. F. Tuckett.

In September 1943, during daylight hours, another brief, wartime Discussion was held at the Hotel Rembrandt on yet another topic in which Rideal was particularly interested, namely 'Modes of Drug Action'. Sir Henry Dale, who had recently retired as Director of the National Institute for Medical Research and had succeeded Sir William Bragg as Director of the Davy–Faraday Laboratories at the Royal Institution, and who was President of the Royal Society, in a general introductory address warned physical chemists to beware of thinking that they could find easy, general solutions to problems in this field and emphasized the highly specific nature of drug action. For example, the ease of adsorption on a surface might be a necessary quality but it was much too general to explain how the action ultimately occurred or even to explain how effectiveness varied in a homologous series. Rideal and some of his pupils, including J. H. Schulman, had been studying these general factors, and Dale's remarks may have been directed at them; but in his introductory address to the physico-chemical section Rideal gave much thought to the ways in which specific activity could arise, i.e. how intermolecular attachments could occur. J. A. V. Butler says of this Discussion that 'Many of the types of interaction between drugs and substrates, which have since been the subject of much research, were fully discussed.' (*Trans. Faraday Soc.*, 1953, **49**, 576). For example, Rideal and others suggested that biological activity might require some combination of factors, viz. a more general one such as differential solubility in the cell lipoids or adsorption at cell interfaces, with some specific ones due to a combination of polar effects, such as hydrogen bonding, with van der Waals non-polar interactions. This could mean that the drug would have the general power to pass through a cell membrane and the specific power to react with one particular enzyme when it gets through.

The possible importance of specific stereochemical relationships had long been appreciated. There was considerable argument about whether the concept of receptor sites or centres on or in cells was a valuable one: this arose particularly from a paper by H. R. Ing. Inorganic chemical reactivities were certainly recognized, but those which were then known were not very specific, e.g. the blanket effect of the mercuric ion in combining with –SH groups. On a different note Hinshelwood and some of his pupils described experiments in the modification of bacterial growth by change of medium, and the addition of disinfectants or sulphonamides. His general interpretation of these and similar effects, that the enzyme systems in all the bacterial cells adapted gradually to the external influences, was to become highly controversial. Despite the wartime limitations, this was a good Discussion of substantial long-term value.

Notwithstanding the decision of 1941 there were some changes on Council in 1943. At a Council meeting on 12th August 1943, attended only by the President, Dr. R. Lessing, Professor S. Sugden and the Secretary, Lessing and Sugden said that they felt that a change of President was undesirable but that there should be changes of Vice-Presidents and ordinary members. This view was eventually accepted at a Special General Meeting held during the Discussion in September, by which time W. J. Shutt had died, making one replacement essential. As a result, at the Annual General Meeting held on 11th December, Council proposed a re-shuffle of Vice-Presidents and four new faces on Council, namely those of T. Beacall, O.B.E., Professor M. G. Evans, Dr. R. W. Lunt, and Professor H. W. Melville, F.R.S. Thomas Beacall was an industrial organic chemist who, with three co-authors, had written a book about 'Dyestuffs and Coal-Tar Products: their Chemistry, Manufacture and Applications' which was first published in 1915 with the aim of providing some of the systematic, basic knowledge of which the British chemical industry was then desperately short. In 1938 his address was 'The Patent Office'.

Meredith Gwynne Evans, born in 1904, after a brilliant undergraduate and graduate career at Manchester, spent a happy year working with H. S. Taylor at Princeton on the photochemical effects of replacing protium by deuterium in some simple molecules. He would certainly have had opportunities of seeing and talking with Henry Eyring, and this may have been how his interest in the 'transition-state' theory of reaction kinetics was initiated. When he returned to

Manchester in 1935 as a Lecturer Polanyi was established there and the two of them began a very fruitful and exceptionally harmonious collaboration. According to Professor John Polanyi, F.R.S., Eyring and his father had already published a paper in 1931 describing the calculation of a potential surface (see *Biog. Mem. Fellows R. Soc. London*, 1977, **23**, 423). Next, Pelzer and Wigner in 1932 calculated the properties of an 'activated complex' or 'transition state' using this surface; and then in 1935 Evans and Polanyi, and Eyring independently, generalized this treatment. So the theory was born. Dainton has commented that it, the Michaelis–Menten equation, the Lindemann mechanism for first-order reactions, and the Langmuir theory of heterogeneous catalysis all have in common the idea of reaction taking place via an intermediate state, the rate of leakage from which is much less than the rate of formation so that it can be treated as being either a special kind of molecule or a complex with an enzyme or a surface, which is in equilibrium with ordinary reagent molecules; thus, they are all elaborations of the theme first proposed by Arrhenius. In passing we may note that Polanyi's contributions to this theory marked the climax of his career as a physical chemist. In 1939 Evans became Professor of inorganic and physical chemistry at Leeds. During the war he was very busy with practical researches and administrative problems (H. W. Melville, *Obit. Not. Fellows R. Soc. London*, 1953, **8**, 395).

Dr. R. W. Lunt, 45 years old at the time, had graduated with high honours as a chemist from Liverpool University and then joined Donnan at U.C.L. where he worked largely on chemical effects in gas discharges. He was a member of a group at U.C.L. which was supported by I.C.I. Ltd., and when this broke up, about 1934 or 1935, he, having developed an interest in electrical engineering, took employment with Callendars Cables Ltd. Professor H. W. Melville, only 35 years old in 1943, was an Edinburgh graduate who, as an 1851 Exhibitioner, had travelled the high road south to Cambridge where he worked with Rideal. He became a Fellow of Trinity College in 1936. In 1940 he was appointed Professor of chemistry at Aberdeen in succession to Alexander Findlay, but because he had been asked to undertake research for the Ministry of Supply it was agreed that Findlay should carry on for a while. By 1943 Melville was recognized as an outstanding researcher and administrator. He had made many contributions to the Society's Discussions, especially to those on polymers.

The Society managed to have a brief Discussion in January 1944. It, too, was on a subject of immediate interest to Rideal's group, viz. 'Molecular Weight and Molecular Weight Distribution in High Polymers', and it was held jointly with the Plastics Group of the Society of Chemical Industry in the Hall of the Institute of Mechanical Engineers. H. W. Melville reviewed the available methods for measuring large molecular weights, explained the distinction between number average and weight average values, explained the need to know the distribution and stressed the effects it could have on different properties. G. Gee showed, from a statistical analysis of the entropy term for a solution, that the osmotic method is valid in the limit of infinitely dilute solutions; but he also emphasized the limitations of Staudinger's method of determining molecular weights from viscosities. Clearly, Staudinger had had a fair degree of luck in arriving at an essentially right answer from a dubious method. The discussion was wider than the title indicated for, in trying to explain how different distributions could arise several authors, namely G. Gee, E. F. G. Herington and H. W. Melville, tackled the fundamental and difficult problem of the kinetics of polymer formation. The thermal degradation of polymers was considered by H. H. G. Jellinek who showed that it was necessary to assume that all bonds are not equally strong but that there is some kind of distribution of weak links. The peculiar phenomena arising in the melting of high polymers which are only partly crystalline were discussed by two of Rideal's pupils, E. M. Frith and R. F. Tuckett. Altogether it was an admirable meeting.

At an Annual General Meeting, very belatedly held on 2nd January 1945, more changes were proposed for the Council which was nominally to hold office for 1944–45. The reason for the long delay was probably the day and night bombardment of London first by flying bombs from June to September 1944 and then by rockets. Even by January 1945 the rocket bombardment had not been ended, but presumably it had been sufficiently contained to make a meeting in London an acceptable risk. There was one newcomer as a Vice-President, namely Professor Alexander Findlay of Aberdeen, who then was President of the Royal Institute of Chemistry. Born in 1874 he lived most of his early years in Aberdeen. He graduated from the University and at first worked on an organic chemical topic with Professor F. R. Japp there; but in 1898 he went to Leipzig with an 1851 Exhibition to work for a Ph.D. with Wilhelm Ostwald. This experience gave him a settled interest in physical chemistry. He came back to an appointment at St. Andrew's but soon moved to work with Ramsay at U.C.L. He then took a post at Birmingham, and in 1911 became Professor of Chemistry at Aberystwyth. In 1919 F. S. Soddy moved from Aberdeen to Oxford and Findlay realized his great ambition by succeeding him. However, his hopes of building a lively research school were frustrated by continuing poor facilities and by his own conscientiousness as a teacher and administrator, so that he did not live up to his early promise as a researcher. Nevertheless he became a very considerable figure in physical chemistry in the United Kingdom. He wrote a number of textbooks and general books on science which were well regarded. He was very clear-minded, sympathetic and courteous. Characteristically, when after years of struggle he had finally persuaded the University to commit itself to building a new chemistry laboratory, he resigned so that his successor could plan it to his liking; but, as we have seen, he was persuaded to carry on until Melville could be partly freed from his wartime tasks (R. B. Strathdee, *Chem. Br.*, 1967, **3**, 24).

New faces as ordinary members of Council were those of Drs. Alfred Caress, H. W. Thompson, and O. H. Wansborough-Jones. Caress, 41 years old, was another pupil of Rideal's: by 1945 he was a fast-rising star in I.C.I. Ltd. He had been associated with plastics ever since the company, rather reluctantly, decided to acquire a foothold in that market by buying first a controlling interest in Croydon Mouldrite Ltd. in 1933, and then full control in 1936. By 1938 he was research manager of Mouldrite, and when in that same year the Plastics Group was formally constituted Caress was appointed a Director. By 1945, largely because of the war, the Plastics Group was important and successful even though it had not been able to gain control of polyethylene to which Alkali Group was clinging tenaciously against the efforts of Caress and others (see W. J. Reader, *I.C.I., A History*, Vol. II). Thompson, then 37, was an Oxford product, a Fellow of

St. John's College. He had been a pupil of Hinshelwood as an undergraduate and started research with him. He made some important advances in the kinetics of gas reactions, but soon left this field and, after a period in Berlin where he worked for his Ph.D. with Haber he took up infrared spectroscopy, still in its infancy, and applied it with great vigour. He had a second passion – the encouragement of Association football. He had been a Blue, and quickly became very influential in the game: already by 1941 he was a member of the Football Association Council. Wansborough-Jones, then 40, was a Cambridge man. He had worked with Rideal and then also with Fritz Haber in Berlin. In 1930 he became a Fellow of Trinity Hall. During the war he was given an Emergency Commission in the Army and emerged as a Brigadier in 1945. For several years thereafter he was Scientific Adviser to the Army Council. We shall meet him again.

The very last Discussion meeting during the war, on 'The Application of Infra-Red Spectra to Chemical Problems', was held in January 1945 at King's College, London, after the belated Annual General Meeting. Sir John Fox, F.R.S., who was to have opened the Discussion, had recently died, so Rideal and Ingold gave introductory talks. The two leading schools at that time were at Cambridge, where work had been initiated by A. M. Taylor and F. I. G. Rawlins and was now led by G. B. B. M. Sutherland, who had returned from a sojourn with the University of Michigan infrared school, and at Oxford where work had been initiated and was still led by H. W. Thompson. Sutherland and Thompson gave a most interesting review of recent developments, from which it was clear how advances in theory and technology had combined to give very rapid progress. This was of two kinds: the accurate, detailed analysis of spectra to obtain molecular parameters for small molecules, and the relatively simple, more empirical use of the spectra of larger molecules to obtain limited structural information or to make chemical analyses. The basic theory of rotation–vibration spectra, including the selection rules, was well established although development was needed. The classical theory of simple vibrations had been applied to molecules by Niels Bjerrum as far back as 1914, and more generally by D. M. Dennison in 1931. The concepts of a central force field, and of a valence force field with its relation to bond force constants, were also put forward in 1914 although applied only to carbon dioxide. However, the practical task of correctly assigning observed frequencies and deriving force-field parameters was proving a formidable one even for triatomic molecules. The need for isotopic replacement was appreciated and it had been used, but as yet only a beginning had been made. During the War effort had been largely channelled into analytical applications. There had been some work done, with difficulty, in the 'long-wave' region, i.e. up to about $250\,\mu$ or $40\,\mathrm{cm}^{-1}$; and there was clear realization that development in this region was very desirable so that rotational parameters for heavy molecules could be obtained. However, at a cursory glance there seems to have been no mention in this Discussion of the possibility of microwave spectroscopy, although the absorption by water vapour at some microwave frequencies must have been known. Probably this was because it was still wartime and such matters were official secrets.

2.12. RETROSPECT

So ends our review of the Society's General Discussions and our record of the people involved in them, from the end of one war to the end of the next. It was deliberately done in some detail because the history of the Society cannot be properly understood without knowing how physical chemistry and chemical physics developed in that era. These aspects of the science of chemistry did advance to an amazing degree which was certainly comparable with that in the previous phase of great development, when thermodynamics and kinetic theory were first applied to chemistry in the late nineteenth century. At the same time there was a transformation of the Society from a small, modest and rather amateurish one, struggling for recognition, to one with world-wide fame and influence. It was mainly through the General Discussions that the Society achieved its high standing. They became international scientific occasions because of their unique organization, their informal atmosphere, and the apt, timely choices of topic. In this country there was virtually no competition with them. At the Annual General Meetings of the Chemical Society of that era the President would give an address, and occasionally there was in addition a lecture.

There were ordinary meetings, much like those of the Faraday Society, at which papers were read and discussed; and there were extra meetings when an eminent chemist would give a lecture that would be followed by a discussion. However, there was nothing resembling the two-day Faraday General Discussion Meetings. Therefore the Faraday Society had a great opportunity and rose to it, aided by luck in the historical development of the form of its General Discussions, but mainly because of the opportunism and energy of successive Councils. The amount of physico-chemical work being done in this country had likewise increased and its standard improved, so that it could, by the end of the inter-war period, bear comparison with that in continental Europe or the U.S.A.

We can see several causes for all this. From the mid-1920s onward the revolution in quantum mechanics gave enormous impetus to those parts of chemistry which depended directly on quantum phenomena, especially to molecular spectroscopy where many rigorous, quantitative treatments had become possible. In addition it had much wider, more qualitative effects on chemical thought, e.g. by giving much clearer insight into the nature of chemical bonding, by starting people thinking in terms of quantum energy levels where they had not previously done so – as in treating metals, semi-conductors or insulators, junction phenomena and electrode reactions – and by introducing entirely new concepts such as tunnelling or those derived from electron spin and nuclear spin. We cannot aim to explain in any fundamental way why quantum theory developed as it did when it did; but we can at least say that its development provided a fine example of scientific method, of the interplay between theorists who put up cock-shies and experimenters who knocked them down and so stimulated the theorists to put up better ones.

The power of the experimenters grew with the improvements of theory; but it grew also because of the appearance of new or improved techniques. Most of these were due to applications of physics but were not necessarily, or not wholly dependent on quantum theory, e.g. the diffraction methods, determinations of electric or magnetic dipole moments of molecules, and the isolation of stable isotopes. These vitally important new tools were often linked to wide technological advances, e.g. the development of better X-ray tubes, of thermionic valves and valve circuits, of better vacuum systems including pumps and seals, of better radiation detectors, of electromechanical calculating machines – indeed a myriad of such detailed advances, many of them the fruits of invention or of imaginative engineering rather than of new science. It is, however, worth stressing that much, probably most physical apparatus used by chemists was still built either by the researchers themselves or in their laboratory workshops, provided that their laboratory had one. There were exceptions to this generalization, e.g. optical instruments such as spectrometers for the visible and ultraviolet regions, galvanometers, precise electrical measuring instruments, cathode-ray tubes and a few electron microscopes. For such things there was a reasonable market, so this factor, together with the need for having a group of specialized skills to produce them economically, had encouraged commercial production. In general, however, the instrument engineer and the commercial builder of apparatus did not yet dominate the scene. Chemical research had not yet come to be really expensive in terms of money.

These factors were general and had worldwide effects: there were others which had more particular effects in this country. One such was the flight of refugees from central Europe; but this had an even greater effect in the U.S.A. Another was the increase of public awareness of science, which became a much more usual subject in secondary schools where it was aided by the recruitment of an excellent body of teachers in the years following the 1914–18 war. Parallel with this was the increase in financial support, especially by central Government but also by the chemical industry, for academic research. These changes in public attitude came from the general acceptance of a belief that fundamental research led directly to survival in war and prosperity in peace. In the previous chapter we noted the setting up during the 1914–18 war of the Department of Scientific and Industrial Research. In 1920 its gross expenditure was £330 176, while for 1938 it was £872 127 (see Russell Moseley, *R. Soc., Notes and Records*, 1980, **35**, 167); but it is clear that by far the greater part of this expenditure was on the running of the National Physical Laboratory and the Research Association establishments, where the emphasis was on applied research, rather than on research in the Universities. The total expenditure, for the physical sciences, on grants to research students and on grants to established research workers either for personal support or for the support of their research, was £12 661 for 1918–19, rising to a maximum of £47 905 in 1922–23, then falling slowly until 1932–33 to around £26 000 where it stayed, more or less, until the outbreak of war. Judging from the number of grants of the three kinds made to chemists, chemistry received about 40–50% of these sums. Even when multiplied by 20 these seem small by today's standards, but at the time they were sufficient to transform research. At the same time the University Grants Committee and private benefactors were providing money for building and equipping new laboratories. It may be noted that as Government money became more freely available so there developed quite bitter struggles for the control of it: this is well illustrated by Russell Moseley's paper on 'Government Science and the Royal Society: the Control of the National Physical Laboratory in the Inter-War Years' (*loc. cit.*). From that article it also appears that Governments seemed to find it impossible to follow a consistent long-term financial policy for research. Periods of relative generosity were interrupted by periods of bitter economy, as when Mr. Geddes swung his 'axe' in 1921.

The very rapid development of chemistry was matched by that in physics and in the science of metals, subjects which the Society was originally founded to promote, at least in the borderland areas where they overlapped with chemistry. By 1930 the Society found itself effectively limited to coping with two General Discussions *per annum*, certainly because of the cost of publishing but probably also because it lacked both sufficient scientific man-power to provide the *ad hoc* committees and the office support essential for organizing any greater number. Activity in one of the Society's original spheres of interest, namely in electrochemistry, decreased in the inter-war years. Nevertheless the need to make choices in the land of overlap became more pressing, and one of the earlier interests had to be sacrificed. Between 1929 and 1939 there was only one General Discussion that was mainly concerned with the properties of metals, namely that of 1935 on metallic coatings, films and surfaces, whereas from 1919 to 1928 inclusive there were about nine which were mainly or at least substantially concerned with metals or with metallurgical processes. This happened the more readily because the societies that earlier had been concerned with the practical aspects of metal science had begun to take much more interest in the theoretical aspects.

An obvious question which then arises is whether the Society covered all of the newly developing fields as well as it should have done. One of these became known broadly as 'chemical physics'. Now one of the Society's original aims was to promote 'chemical physics', but in 1903 nobody seems to have been quite sure what this meant; and in 1919 this aim was changed to promoting 'physical chemistry'. By the 1930s 'chemical physics' had acquired a reasonably definite meaning: one could say that it had become jargon, probably crystallized by the founding in 1933, by the American Institute of Physics, of the *Journal of Chemical Physics*, which quickly exerted much influence. The term had come to mean the application of quantum theory and statistical mechanics to chemical problems, and also the use of the new physical methods for the determination of molecular structure. It implied an emphasis on molecular properties rather than on the bulk properties studied in classical physical chemistry.

There were several Discussions on such new matters in the 1930s, but there were two gaps. One was valence theory, last discussed in 1923. The other was the diffraction methods of determining molecular structure: the results of such investigations but not the methods themselves were last discussed in 1929. Certainly there were many applications of the ideas and information which came from these sources. The use of electron diffraction for the study of surface layers was treated in 1935. Rather similarly some of the methods used in valence theory were applied in the Discussion on 'Reaction Kinetics' in 1937. There were, however, no wide, General Discussions of basic theory and results, although these topics would seem to have been well within the purview of the Society and Discussions could have been very useful.

Were these omissions accidental or of deliberate choice? The question is one that should be asked because we have seen that two or three 'Old Boy' networks had considerable influence in Council and

The Inter-war Years, and the Second World War 143

we need to know if these caused any limitations of aim. Such networks were probably inevitable when the scientific community was still small, and they have great advantages for some purposes but not for all. The Council Minutes do not help much to provide an answer; but after careful thought the writer has come to the conclusions that the omissions were probably accidental. Because there were so many topics that needed discussing, and the choices were determined by rather randomly fluctuating pressures from various groups, a perfect sequence could hardly be expected.

There were not many people on Council during the 1930s who were primarily interested in structure or in valence theory. Sidgwick and Lennard-Jones certainly were. Lowry, who suggested the 1929 Discussion of 'Crystal Structure and Chemical Constitution'; suggested one on 'Unsaturation and Conjugation' for 1935. Rawlins, Bernal, Sugden and Goodeve were all working on structure; and Donnan suggested a Discussion on 'The Shape and Quantum Dynamics of Simple Molecules' before 1935, probably on Goodeve's behalf. By its decision in 1929 to hold a Discussion on a topic related to colloids every two years the Society had tied one quarter of its available Discussions to this field, and although the commitment was sometimes interpreted liberally, or even ignored, and produced some excellent Discussions, it can be deemed to have produced a bias toward the study of bulk properties. On the whole the indications are that the majority of Council members were more interested in the use of the ideas and information from chemical physics for explaining the phenomena of physical chemistry than they were in how these ideas or that information were derived; but this did not affect choice every time, and the actual series of General Discussions, even if not perfect, was certainly very fine.

To conclude this Section two remarks may be made about the war and the Society. The first is that in the second war, as in the first one, the Society was lucky to have a forceful and very hard-working President and a most loyal and effective Secretary: between them they kept the flag flying. It was only to be expected that for the time being Rideal would mould the Society round his interests as Hadfield had round his: Council met relatively infrequently so the President and

Secretary had to make many of the decisions. The second is that the Society did not benefit from the second war in quite the same way as it did from the first one. This time the Government was alive from the start to the importance of organizing scientists for national service, and with the aid of the scientific societies this was quickly done. The Chemical Society had set up an Advisory Research Council which included Donnan, Hinshelwood, Slade and Rideal among its members; so the Faraday Society asked them and W. L. Bragg to be its Advisory Research Committee. Although many members of the Society made notable contributions as individuals to national survival, the Society as such was not directly involved in organizing research as it had been in the first war. Moreover it is roughly true to say that whereas in the first war chemists were the glamour boys, in the second one this role was taken by the physicists with radar, the nuclear bomb, and a vital share in Intelligence and in operational research as their major contributions.

It remains now, in our survey of this period, to look at the main events in the administrative, publishing and financial activities of the Society, and this will be done in the next chapter.

Sir Joseph Swan
President, 1903–1904

Lord Kelvin
President 1905–1907

Sir William Perkin
President 1907

Sir Oliver Lodge
President 1908–1909

Sir James Swinburne
President 1909–1911

Dr R. T. Glazebrook
President 1911–1913

Sir Robert A. Hadfield
President 1913–1920

Prof. A. W. Porter
President 1920–1922

Sir Robert Robertson
President 1922–1924

Prof. F. G. Donnan
President 1924–1926

Prof. C. H. Desch
President 1926–1928

Prof. T. M. Lowry
President 1928–1930

Sir Robert Mond
President 1930–1932

Prof. N. V. Sidgwick
President 1932–1934

W. Rintoul
President 1934–1936

Prof. M. W. Travers
President 1936–1938

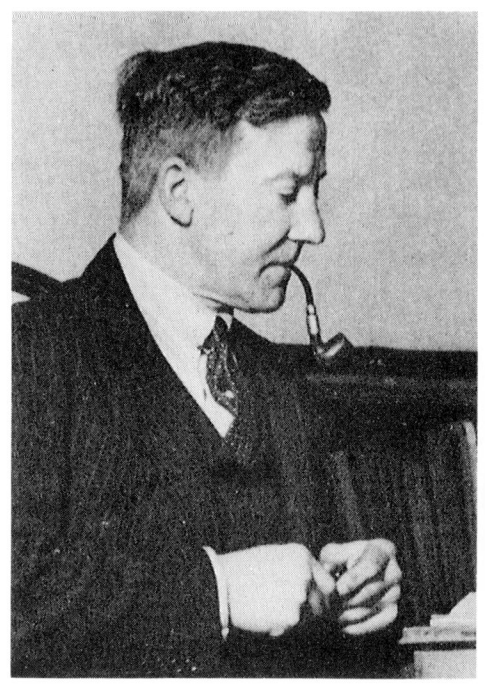

Prof. E. K. Rideal
President 1938–1945

Prof. W. E. Garner
President 1945–1947

Prof. A. J. Allmand
President 1947–1948

Prof. Sir John Lennard-Jones
President 1948–1950

Sir Charles Goodeve
President 1950–1952

Prof. F. S. Dainton
President 1965–1966

Prof. C. E. H. Bawn
President 1967–1968
© James Bacon & Sons

Prof. G. Gee
President 1969, 1970
© Godfrey Argent Studio

Prof. J. W. Linnett
President 1971

F. S. Spiers
Secretary 1903–1926

G. S. W. Marlow
Secretary 1926–1948

F. C. Tompkins
Secretary 1948–1971

Group photo taken at Liverpool, 1931 at the General Discussion on *Photochemical Processes*. Taken from *Trans. Faraday Soc.*, 1931. Vol. 31, facing p. 358. Those seated in the front row (reading from left to right) are Dr Kondo, Professor Baly, Professor Berthoud, Mr Emile Mond, Professor Bodenstein, the President, Professor Mecke, Sir Robert Robertson, Miss Wallace, Dr Frankenburger, Professor Eggert and Professor Allmand. There will also be seen seated on the ground Messrs. Batley, McMillan, Eltenten, Leathwood, Williams, Belton and Childs. First row standing: Messrs. Kirkbride, Griffith, Willey, MacDonald and Norrish, Mrs Kerridge, Major Freeth, Professors Donnan, Spencer and Porter and Dr Lochte-Holtgreven. Second row standing: Messrs. Emeleus, Griffiths, Marlow, Childs, Style, Topley and Bailey, Professor Partington and Dr Ludlam. Third row standing: Messrs. Angus, Cassie, Smith, Stein and Goodeve.

Group photograph taken at University College, London, 1934 at General Discussion 61 on *Colloidal Electrolytes*. Taken from *Trans. Faraday Soc.*, 1935, Vol. 31, facing p. 1. Seated in the front row, reading from left to right are: Dr G. F. Davidson, Dr C. Robinson, Prof W. C. M. Lewis, Dr O. Quensel, Mr Macfarlane, Dr G. S. Adair, Mme. A. Roche, Prof. E. Gorter, Prof. E. K. Rideal, Prof. A. Frumkin, Prof. E. Elöd, Miss D. Jordan–Lloyd, Dr F. A. Freeth, Mr W. Rintoul, Prof. H. Freundlich, Prof. H. R. Kruyt, Prof. F. G. Donnan, Dr F. Eirich, Prof. I. Traube, Mrs Laing–McBain, Prof. Wo. Ostwald, Prof. E. J. Bigwood, Fraulein Krüger, Prof. A. J. Rabinovitch, Dr Ph. Gross, Mrs J. W. Goodeve, Fraulein G. Kornfeld, Dr E. Valkó, Prof. W. D. Treadwell, Mr E. Hatschek, Fraulein L. Werner, Miss G. Wakeley and Prof. C. K. Ingold.

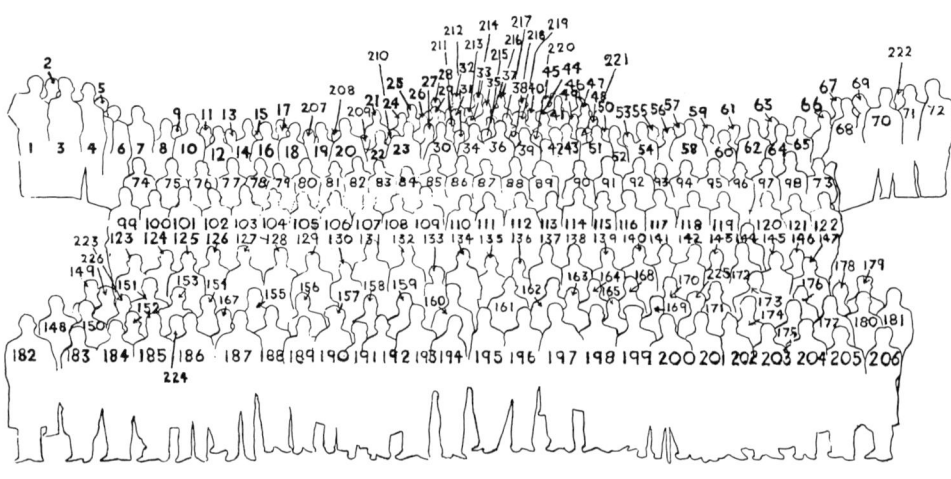

Group photograph (and accompanying identification key) taken at University College, London, 1945 at the General Discussion on *Oxidation*. Taken from *Trans. Faraday Soc.*, 1946, Vol. 42, facing p. 99. See key on opposite page.

Alphabetical list of those appearing opposite

Abel, E. 203
Abnett, Miss E. A. 89
Adam, N. K. 168
van Arkel, A. E. 174
Atkinson, B. 74

Baldwin, R. R. 86
Bangham, W. H. 156
Barker, J. T. 151
Barrett, J. W. 11
Baxendale, J. H. 123
Benesch, R. 3
Bevington, J. C. 139
Bircumshaw, L. L. 13
Blease, R. A. 113
de Boer, J. H. 199
Buchner, E. H. 195
Burnett, G. M. 59
Bury, C. R. 172

Carter, S. R. 141
Chamberlain, G. H. 87
Chambers, W. 43
Cockbain, E. G. 180
Cornish, G. R. 96
Coulson, C. A. 41
Coumoulos, G. D. 181

Davies, C. W. 34
Dear, W. N. 107
Donnan, F. G. 196
Dymock, J. 54

Eirich, F. 149
Elce, H. 51
Eley, D. D. 179
Evans, A. G. 1
Evans, M. G. 165
Farkas, L. 204
Farmer, E. H. 182
Fordham, S. 95
Frank, G. C. 178
Freling, E. 155

Gaydon, A. G. 162
Gee, G. 17
Gorter, E. W. 225
Gray, V. R. 104
Greenhalgh, R. K. 83

Hall, R. H. 52
Hamer, W. E. 9
Harris, A. 58
Hartley, G. S. 69

Herington, E. F. G. 62
Hewitt, J. T. 183
Hilton, F. 222
Horton, L. 136

Ingold, C. K. 30

Jellinek, H. H. G. 144
Jones, T. T. 29
Jungers, J. C. 187

Kenner, J. 36
Kitchener, J. A. 88
Kögl, F. 198
Kornitzer, Miss B. 133
Kramers, W. J. 97
Kruyt, H. R. 193

Laffitte, P. 189
Le Fevre, Mrs C. 5
Le Fevre, R. J. W. 2
Letort, M. 188
Lewin, S. 112
Lock, S. 70

Magat, M. 167
Mardles, E. W. J. 223
Marlow, G. S. W. 160
Marriott, J. V. R. 146
Marshall, J. 92
Martin. 10
Masson, C. R. 63
Matalon, R. 191
Mathieu, M. 190
Matthews, J. B. 171
Mazurkiewicz, A. G. 173
Meares, P. 119
Mering, J. 224
Mikolajewski, E. 226
Monceaux, P. 152
Mooney, R. B. 77
Mullins, B. P. M. 164
Mund, W. 175

Macdonald, J. Y. 106
McGrew, F. C. 185
Mackay, Miss M. 60
McLeavy, G. 111

Naylor, R. F. 55
Newall, H. E. 37

van Ormondt, J. 202

Park, G. S. 134
Payman, W. 145
Peters, L. 110
Pirani, M. 66
Polanyi, M. 169
Prettre, M. 192

Rhodes, Miss H. 42
Richards, R. B. 8
Richer, G. 159
Rideal, E. K. 194
Robertson, A. J. B. 109
Robertson, A. R. 127
Robinson, C. 68
Rushbrooke, G. S. 47

Schweitzer, H. 161
Shepherdson, A. 153
Siedroyc, W. 176
Skinner, H. A. 40
Slade, R. E. 184
Small, P. A. 94
Stern, J. 118
Street, J. C. 105
Style, D. W. G. 82
Sully, B. D. 14
Szwarc, M. 205

Theorell, H. 200
Travers, M. W. 197
Turck, K. H. W. 90
Twigg, G. H. 91

Ubbelohde, A. R. 142

Vargas Eyre, J. 130
Vernon, E. J. 163
Verwey, E. J. W. 166
Vichnievsky, R. 154

Walsh, A. D. 45
Waterman, I. 201
Watson, W. F. 61
Weeks, G. A. 132
Weiss, J. 23
Whyte, R. B. 46
Williams, G. 76
Wiltshire, J. 108
Wodak, Mrs E. F. 206
Wood, L. B. 71
Wynne Jones, W. F. K. 143

Group photograph taken at Cambridge, 1935 at General Discussion 63 on *Phenomena of Polymerisation and Condensation*. Seated on ground, from left: Garner – – – – – – – D. G. Davies – Allsop. First seated row, from left: T. C. James, W. H. Mills – – – – – Lowry, W. H. Carothers, E. C. C. Baly – – – E. Mond, Standinger – – – – Sidgwick, Rideal, Mark, Lennard-Jones, de Boer(?), Houwink(?) – – – – – . First row standing, from left: – – S. R. Carter, G. Cox, Astbury – – – – – – Dunstan, Cadman, Norrish, Travers, C. R. Bury – – Goodeve, C. Le Fevre, Mrs Goodeve – Marlow – – – – – – – E. C. Bailey – Eirich. Second row standing, from left: – – – – – – – – Salomon – R. Le Fevre – – – – – – Hartley – – – – – – J. K. Roberts – Allmand – – Wansborough-Jones. Third row standing, from left: – – D. Crowfoot, Megaw, Bawn – – Boys(?) – – – – F. E. T. Kingman – – – – Bryce, Spence – – N. K. Adam – – – – – F. I. G. Rawlins(?) – . Back row, from left: – – – – – – – – – – F. C. Frank, M. M. Davies, J. B. Marsden – – Moelwyn-Hughes, Gee, A. E. Alexander, Wynne-Jones – – . Individuals deduced as being present at the meting (from the report in the *Transactions*) but not identified: H. W Melville (Cambs), J. E. Carruthers (Cambs), S. C. Gran (Cambs), F. C. Wood (Manchester), E. E. Walker (Manchester), A. Wasserman (London), E. Heyman (London), P. Lewis Dale (London), A. M. Taylor (St Andrews), N. J. L. Megson (Teddington), L. C. Vernan (Teddington), W. P. Pepper (Liverpool), M. W. Perrin (Northwich), D. Findlayson (Long Eaton), J. Boulton (Braintree), R. S. Morrel, W. R. Davies, J. Farquharson, H. I. Waterman, J. J. Leendertse, A. Abkin, S. Medvedev, H. Dorstal (Vienna), R. Freundenberg (Heidelberg), Dr & Mrs J. R. Katz (Amsterdam), K. H. Meyer (Geneva), G. Stafford Whitby (Ottawa), W. Herer (Freiburg), E. Husemann (Freiburg), H. B. Weiser (Houston, TX), W. O. Milligan (Houston, TX), J. C. Patrick (Yardville, NJ), G. Walter (Vienna), Dr & Mrs D. Signer (Bern), P. N. Kogerman (Tartu), E. Bergman (Rehevoth), D. L. Atwegg (Paris), T. F. Bradley (Linden, NJ), Dr & Mrs R. H. Kienle (Boundbrook, NJ), Fraulein D. D. Kruger (Berlin), M. Mathieu (Paris), R. Meyer (Paris), J. H. van der Monne (Amsterdam), Orthner (Frankfurt), H. Pringsheim (Prague), E. Proskauer (Leipzig), Sauter (Freiburg), T. Takei (Tokyo), P. von Tafel (Delft), A. Weidenger (Amsterdam), F. Weigert (Leipzig).

CHAPTER 3

Publication, Finance, and General Administration 1918–1945

3.1. THE GENERAL STATE OF AFFAIRS IN 1918

WHEN WE LEFT the more mundane affairs of the Society in Chapter 1 (Sections 1.3, 4, 5) they were in a difficult but hopeful phase. To recapitulate briefly: in its early years the Society had undertaken a very ambitious programme of publications, of *Transactions* and *Proceedings*. The custom of independent publication of original, primary papers was one that by then had long been established for learned societies; but the *Proceedings*, though only an auxiliary journal, was initially elaborate and expensive (Section 1.1). In 1905 it cost £148 6s 2d to print as against £139 9s 11d for *Transactions*, so Council soon chose to whittle it down. With this done, the Society eventually managed to show a succession of credit balances on each year's working, and in the Council Report for 1910 the Treasurer was able to say 'This is the first time in the history of the Society that a credit balance has appeared in the Balance Sheet': it was no less than £40 7s 0d. There must have been some guarantee fund to carry the initial series of debit balances, but the writer has found no mention of who provided it. We can only guess that one or more of the original members of Council did this, and that he or they preferred to remain anonymous.

This state of modest prosperity lasted until 1918 but then the Society found itself losing money year after year. In the first part of this chapter we shall concern ourselves with these and later financial problems and with the factors behind them viz. publication policy and membership policy. We shall also consider some of the more scientific aspects of these policies; e.g. how well, by its publications, the Society served the field that it professed to cover, and especially how it reacted to the changes and developments therein. We shall look for changes in the character as well as in the size of the membership and how far this may have related to, or have affected publication policy. Inter-relations will prove complex: distinctions between hen and egg may be difficult to make. We shall also review some of the other administrative activities of the Society, especially those involving relations with other scientific bodies. Finally we shall look further at the human and the personal side of the organization.

First it seems appropriate to make some general, if rather trite, remarks about the way in which the Society regarded money. Although a learned society, in contrast to an industrial or commercial firm, assesses its success by the usefulness of its services and products to its consumers, rather than by the money profit that it makes from providing these services and products, it must obviously have sufficient regard to the money profitability of its operations to remain solvent. Therefore it is often subject to internal tensions between those who want to press solvency to the limit in their concern to achieve their intellectual and social aims for the society and those, usually including the Treasurer, who would like to see a larger safety margin and who therefore try to apply a brake to the spending faction. The Faraday Society will provide good examples. The Society was fortunate in being small in staff and so, although it was not a particularly generous employer in money terms, its labour relations were outstandingly good, as we shall see. Moreover, it did not have to be deliberately a job creator or a job preserver so it was exempt from the further tensions and the confusion of aims to which some large corporations are prone. It had simple aims. Its problems were to find the means to realize them and to keep these aims in step with evolving needs.

3.2. THE FINANCE COMMITTEE AND THE PUBLICATIONS COMMITTEE

Our task requires us to look at the roles played by the Finance and the Publications Committees in the life of the Society, so we need to refresh our memory about the early history of these Committees. The Society did not have a Finance Committee until the Council appointed one on 20th October 1903 with the specific task of considering the financial aspects of the exchange, then proposed, of original publications with the American Electrochemical Society. The members were Messrs. Alexander Siemens, Sherard Cowper-Coles and the Treasurer (Dr. F. Mollwo Perkin). They must have met because they reported back to Council on 5th November 1903. They certainly met on 29th February 1904, because Minutes of that meeting have been preserved, and they reported back a second time on 13th April 1904. No Minutes of later meetings are available until a new volume starts with those for 16th January 1922; so for a long time the writer presumed that those for meetings between these dates were lost in the conflagration of 1941, although it seemed odd that one or two volumes of such a series should have had a separate fate. In the Council Minutes from April 1904 to September 1921 there are no references to a Finance Committee, but only to the Treasurer presenting financial statements, yet there are several references to the Publications Committee in its various metamorphoses which we shall presently note, and one of these was to the 'Editing and Financial Committee' which suggests a conflation (Council Minutes, 15th May 1906). On re-reading the Council Minutes of 21st September 1921 the writer realized that a Finance Committee was then appointed *de novo*, as one of four actions agreed to cope with a fresh financial crisis, and this new Committee first met on 16th January 1922; so it now appears that there was no separate Finance Committee between 1904 and 1921, and that no volume of its Minutes has been destroyed or lost.

At its very first meeting on 11th February 1903, Council appointed an Editing Committee 'to consider the question of Publication of Proceedings and to report to the Council.' Some account of its activities has been given in Chapter 1 (Section 1.1), but because this is rather scattered a brief recapitulation is useful. The original members were Messrs. B. Blount, W. R. Cooper, O. J. Steinhart and the Secretary (F. S. Spiers). There is no record in Council Minutes of any change in membership until a 'new Editing and Publishing Committee' was constituted on 4th April 1905; but in the meantime Mollwo Perkin and Cowper-Coles attended meetings, no doubt because it was in this period that the decision was made to cease publishing in *The*

Electrochemist and Metallurgist and for the Society to publish its own *Transactions* and *Proceedings*. Indeed, on 16th October 1904, those present were only Cowper-Coles, Perkin and the Secretary; and on 8th December 1904, only Perkin and the Secretary. Such informality of procedure was what might be expected in a society newly founded by a group of friends; but from 22nd July 1904, after Council had started to consider seriously the major change of publication policy, the Committee did at least keep Minutes of its proceedings and these fortunately are all preserved. Huntington acted as Chairman for most of the time from 1905 until his death in April 1920. He was succeeded by W. R. Cooper, another founder member.

With one exception, which we shall soon note, no further instruction about this Committee's duties seems to have been recorded in Council Minutes: the standard formula was that the Committee was reappointed 'with the usual powers'. The Committee underwent several changes of title and some of function. It became, as already noted, 'the new Editing and Publishing Committee' in April 1905, the 'Editing and Financial Committee' in May 1906, the mere 'Editing Committee' again in November 1906, the 'Editing and Executive Committee' in June 1907, and the 'Editing Committee' yet again in June 1909: it eventually settled down as the 'Publications Committee' in June 1915. The 'Editing and Financial Committee' of May 1906 was empowered to 'pass new members on behalf of Council', an executive function which, as its Minutes show, it certainly performed. Indeed, as was mentioned in Section 1.1, in 1913 it 'passed' two new members who had been elected by Council two months earlier. This is an extreme example, but in general the working relations between the Committee and Council are not very clear: as we have seen the duties and powers of the Committee were not clearly defined. Much of what the Committee did was what we would expect to be delegated to it, i.e. the running of the refereeing system and the other sequential steps of publication, arranging programmes for Ordinary Meetings and special lectures, recommending printers and advertising agents; but sometimes there is no obvious link between business that occurs in the Minutes of the Committee and of Council when this might be expected. Sometimes, as in 1904, it is the financial people or the Finance

Publication, Finance, and General Administration 1918–1945

Committee rather than the Publications Committee who take the initiative, especially about economies in publication. The early Minutes do not give a clear idea of how the publication system was run, for so few papers are actually mentioned that it seems certain that many of them – probably those with favourable referees' reports – were dealt with by the Chairman and the Secretary. Perhaps groups of them were formally authorized for publication as 'issues of *Transactions*', although the writer could not convince himself that even this was always done systematically. The refereeing system is not described, so we do not know when the Society's custom of having two entirely independent first referee's reports began.

3.3. THE IMMEDIATE CAUSES OF THE POST-WAR CRISIS

Even before the crisis at the end of the war the Society's finances had been delicately balanced (Section 1.3), so it is not surprising that from time to time suggestions were made for keeping down the cost of publication e.g. by somehow persuading authors to be briefer. Thus, in March 1914 the Editing Committee discussed the 'bulk of *Transactions*' having been prompted to do so by Council. The only suggestions that emerged were that papers should be limited to 5000 words and that overlapping should be avoided in General Discussions by pointing out to authors that there was always a General Introduction. Papers for General Discussions obviously presented special difficulties. Contributions were invited by *ad hoc* Special Committees, so the ordinary system of refereeing papers could hardly be applied; and this difficulty was probably aggravated by some authors failing to meet deadlines so that the organizers were often pressed for time. Furthermore it would be only normal human behaviour for each Special Committee to want its General Discussion to be bigger as well as better. Yet some kind of control was proving necessary especially because the number of Discussions per year was increasing as they became better established. The attempts to achieve control form a long and rather surprising story which we shall review later in this chapter. Our present concern is with 1918 and the years immediately following.

The 1914–18 war had inevitably caused inflation; so the printing bill, which was the major charge on income, had increased sharply. The nominal cost per page set had at least trebled (Section 1.4). During the war the increase had been mitigated by the rough constancy of the number of pages, for although the number arising from the increasingly successful General Discussions tended to rise, those from the ordinary papers tended to fall. Nevertheless the total cost of printing had practically doubled by 1918 (Council Report, 1918) and was clearly going to increase much further because of prospective price rises, a fast-rising membership, and an increase in the number of papers to be expected after the end of hostilities. All this was aggravated by the sudden need to rent an office, at 10 Essex Street, Strand, at a capital cost of £87 17s 7d for equipment and alterations (Council Report, 1919) and an annual cost of about £106 for rent, cleaning, heat, light and telephone. The cost of an assistant to the Secretary had risen to £182 p.a.: in 1917 only £39 p.a. had been voted, but possibly this was for part-time help only. The whole of these costs did not fall on the Society for, initially, Spiers was to be the Secretary both of the Faraday Society and of the newly founded Institute of Physics, so the latter body agreed to pay £160 p.a. for one year. There was, also, a sudden increase in the cost to the Society of the publications exchange with the American Electrochemical Society (Section 1.5), about which long and continuing story more will be said later. The total result was that after a very small credit balance for the year's working in 1917 there came a large deficit of £202 15s 5d for 1918, and similar losses in 1919 and 1920. By the end of 1919 the cumulative effect was sufficient to convert 'the reserve in the Treasurer's hands in 1918 to a final deficit of £73 2s 10d, this being the extent to which the Society was liable to its Life Membership Fund of £400 on December 31st last.'; so the Society was in a state of mortal financial sin. The Council tackled this emergency vigorously and variously. It will be best for us to consider the several strands of policy one by one.

3.4. ATTEMPTS TO INCREASE INCOME

The first step was to seek an immediate increase of income. On 9th December 1919 Council decided to ask all members to pay a

'Voluntary Levy' of 10s (50p) each. The response was very good and eventually paid off almost the whole of the 1918 deficit. Council repeated the call in September 1921, with similar success. The relevant Minute (29th September 1921) continues 'It was agreed that Members be informed that this step was taken in preference to raising the subscription and that consideration was being given to means by which such a step might be obviated in the future.' As we shall see, the unwillingness of Council to increase the annual subscription long remained part of its ethos; and so when A. J. Allmand suggested an increase, in January 1922 'The Council decided to adhere to its former decision.' It is noteworthy that the members' subscription stayed at £2 p.a. until 1950.

Another step was to try to persuade the Government to help. In the Council Report for 1921 we read:

'*Government Support for Scientific Publications*
The Board [i.e. Council] gave its support to the action of the Conjoint Board of Scientific Societies [recently formed, in 1917] in appealing to the Chancellor of the Exchequer for financial assistance to such Societies as would be obliged to curtail scientific publication in view of the increased costs of printing and postage. It was proposed that the aid should take the form of an increase in the grant made to the Royal Society in aid of publication, the Royal Society administering that increase in accordance with the needs of the Societies requiring grants.

The Chancellor has expressed his regret at being unable to contemplate any increase to the existing grants at a time when the reduction of public expenditure is of paramount importance.'

Both statement and phrasing sound remarkably familiar. Nevertheless, in 1924 the President (Sir Robert Robertson) informed Council that 'through the good offices of Professor Thorpe' he had applied to the Royal Society for a grant. That Society was able to make one of £100 toward the cost of publication in 1923, though whether from Government or private benefaction was not stated. In 1924 Council applied for a further grant of £200 and actually got £150 in 1925 from 'the Publications Fund'. A third grant of £150 made for 1925 was certainly from the Brunner, Mond Publication Fund.

A device to increase effective income, which was remarked upon previously (Sections 2.1–2.3), was to arrange General Discussions in cooperation with other bodies that would bear some of the cost of publication. This was done for about seven, from 'The Examination of Materials by X-Rays' (April 1919) to 'Physical and Physico-Chemical Problems relating to Textile Fibres' (April 1924) and 'The Physical Chemistry of Igneous Rocks' (October 1924), with substantial savings.

The longer-term way of increasing income, which the Council favoured above all others, was to increase the membership. Council realized very clearly, and said from time to time in its Reports (e.g. 1918, 1919, 1924 and 1936), that if the Society was to serve and, indeed, to change the scientific community as it would like to do, the membership had to increase. Achievement of this clear, missionary aim might have been hindered by an increase of subscription rate, and we can guess that this is why Council remained so determined to avoid such an increase. Apart from not taking this negative step, it took positive ones. One such was to propagandize. A booklet describing briefly the objects and work of the Society was compiled and circulated among members and others (Council Report, 1918). Certainly the membership increased rapidly about that time, but how much the booklet had to do with this we do not know. Later, in 1923, a 'notice card' explaining the advantages of Student Membership was prepared and circulated to 33 universities and colleges. On 29th January 1924 it was reported to Council that the result of sending out 66 of these had been negligible. In 1925 'A proposal to appoint a special Propaganda Committee was not proceeded with.' After 1921 numbers changed little for the next ten years, but then they rose rapidly again. An analysis of the way in which this happened proves interesting and rather surprising, as we shall see later (Section 3.11).

Some of the easings of rules about membership which were made from 1913 onwards were described in Chapter 1 (Section 1.4). An entirely new kind of easing started in October 1920 when specific proposals from the Institute of Physics for reduced joint subscriptions were considered and agreed to. They were (1) that entrance fees should be waived for a member of the Institute who was already a member of one 'participating society' and who wished to become a member of another such society: these were, presumably, the societies which cooperated to found the Institute viz. the Faraday Society, the Optical Society, the Physical Society, the Royal Microscopical Society and the Röntgen Society (see Section 1.5); (2) that reductions of annual subscriptions should be made for members of the Institute who were also members of two or more participating societies, these to range from 15% for membership of two, to $33\frac{1}{3}$% for membership of four participating societies. The attraction for the Faraday Society was that this scheme might bring in more members, but the disadvantage was the degree of reduction of subscriptions. Another possible attraction was that more physicists might be induced to join, which Council said it would like to see happen (Council Report, 1926). This scheme did not have a marked, or even a perceptible effect on membership. In 1923 the Secretary (Spiers) suggested that this pattern should be extended to membership of the chemical societies, and the President was asked to discuss the possibilities with the President of the Federal Council for Pure and Applied Chemistry; but nothing came of this.

There was a different extension, for in 1926 the Council of the Faraday Society agreed that the Royal Societies in Australia should be admitted as 'participating societies', with the rather cryptic assurance that 'the abatements in subscription to the Society, if any, arising from this proposal would be negligible'; so its purpose was not clear. Only a year later Council considered a recommendation of the Finance Committee that the whole scheme be reviewed, because only 33 members were affected, with a total reduction in their subscriptions of less than £14, 'whilst the Society is put to trouble and delay in receiving subscriptions.' In the spirit of the scheme, Council decided in 1930 that members of any of the societies represented on the Colloid Committee would not have to pay an entrance fee to join the Faraday Society.

A more ambitious, international scheme was that agreed with the Deutsche Bunsen-Gesellschaft in 1929, which was that a member of either Society could join the other without formality or entrance fee, and could pay both subscriptions to his first society, in the currency of his own country, less a 10% discount on the total; but Council was anxious to make clear that this 10% discount was not to be superimposed on the discounts already available through the Institute of Physics scheme. It may be remembered that a rather different form of collaboration was under discussion in 1914 until the outbreak of war killed it (Section 1.5). In Council Minutes only 29 members are specified as joining through the D.B.G. up to the end of 1930, though it may be that they were not separately identified thereafter.

At the same time Council intended that the scheme should be extended to the Société de Chimie Physique, the American Electrochemical Society and the Nederlandsche Chemische Vereniging if those societies so wished. There were some complications, e.g. the French society wanted a much greater reduction than 10%: already in 1929 members of the Faraday Society were enabled to purchase the *Journal de Chimie Physique* at reduced rates that were not specified (Council Report, 1929). The Dutch society offered *Recueil* with more than a 10% reduction. Accord with the American society was quickly reached and took effect from the beginning of 1930. Agreements with the French and the Dutch societies were reached in 1930 (Council Report, 1930). Fifty-four members of the American society are recorded as joining the Faraday Society in 1930 and they, together with the influx from the D.B.G., produced a minor peak for that year. These international schemes ran into trouble when Britain came off the gold standard; but of that more later.

The actual changes, not only in the number of members but also in the kind, are very interesting and deserve further discussion; but to do this now would be to interrupt our present preoccupation too much; so they will be treated later as a separate topic (Section 3.11).

We now have to consider the last of the ways in which Council tried to increase the Society's income, namely by increasing the sales of *Transactions* to non-members. As may be seen from its Reports, Council rightly attached importance to such sales while, at the same time, it sought to widen the scope for membership. Thus, university librarians had, from 1909, been allowed to become members on behalf of their libraries as is shown by the membership list for 1911 which includes three; and from 1918 onwards commercial firms were eligible for corporate membership (Section 1.4). In 1924 there was an important development. Messrs. Gurney and Jackson, Ltd., wrote

Publication, Finance, and General Administration 1918–1945

proposing that they should become the Society's sole agents for the sale of its publications, in return for a special all-round discount of $33\frac{1}{3}\%$. The Finance Committee deliberated, and in May advised against acceptance, which advice Council followed. However, in November the firm came back with a fresh offer. This prompted further bargaining by the Committee out of which came final acceptance by Council in December 1924. The main points were that the firm would have a 20% discount on all retail sales and a 10% discount on sales through the trade, a trade discount of 10% being allowed to booksellers and wholesale agents. This would not apply to the distribution of *Transactions* to members of the Faraday Society or the American Electrochemical Society under the existing arrangements. The Council Report for 1926 said that 'the arrangement has continued to work satisfactorily. The net amount received for sales of *Transactions* and Reports of General Discussions reached the large figure of £1023 10s 3d as against £791 18s 7d in 1925.' This rate of increase could not be expected to continue but there was a good, steady income from sales, and there was explicit praise for the arrangement annually until 1935 by which time, no doubt, the benefits had come to be taken for granted. The arrangement lasted until the end of 1949. This same firm also helped by taking over the agency for advertisements in *Transactions*, after another firm had failed to give satisfaction, and for 1927 increased the income from £14 4s 8d to £66 9s 10d.

3.5. ECONOMIES, MINOR SAVINGS

In addition to these attempts to increase income, Council tried also to reduce costs. Although, as we have seen (Section 3.4), it was very unwilling to increase the members' annual subscription it was not so averse to giving less service for the same subscription. It was very anxious to reduce printing costs, so it decided to stop publishing the slimmed down *Proceedings* from the end of 1920. This was only a minor economy, because the cost of printing *Proceedings* in 1919 was but £26 7s 0d. By then the contents were mainly rather chatty abstracts of the papers and of the contributions to the discussions, both for General Discussions, which it may be noted were described and numbered as

'Ordinary Meetings', and what may be called ordinary 'Ordinary Meetings' at which the earliest tradition was followed of having papers on a miscellany of subjects. There were also official announcements, the Council Report for 1919, and an account of the Annual General Meeting of 13th December 1920 at which Hadfield made his retirement speech. Thus the *Proceedings* were pleasant though not essential; but it is a matter for regret to the historian that the Council Reports were subsequently published as rather insignificant pamphlets which most recipients did not bother to keep systematically. There is now no complete set: if the Society had one it must have been burnt in 1941. The writer has found few reports of the Annual General Meetings: they appear occasionally and tantalisingly in bound volumes of *Transactions*.

Other minor economies were not to make the *Transactions* page the same size as that of the *Proceedings of the Royal Society*, because of the extra cost of 'employing leaded type for the longer line' (1922), not to publish a list of members in 1924 but to keep it in the office on a 'Bizada' filing system, not to provide covers for authors' reprints (1924), not to have any gaps between papers so as to reduce the cost of paper used by about 4% but, obviously, not the cost of type-setting (1927). Council also decided not to have a professional reporter at Ordinary Meetings but to invite authors to send in their comments: in 1903 Spiers had agreed to report discussions so as to save the cost of a professional, so it seems that by 1925, probably because of pressure of work and possibly because of the deterioration of his health, he had ceased to do this. Further changes were for the Society to be less generous to authors by charging them more for reprints over the number supplied *gratis* (1928), and later (1933) to allow *gratis* reprints only to members and those invited to contribute to General Discussions; and to levy a realistic charge on industrial and quasi-industrial organizations that wanted large numbers of reprints. From 1933 the experimental matter of papers was printed in smaller type than the main text.

In a further attempt to contain printing costs the Society changed its printer. Messrs. Unwin Brothers, who had been printers from the beginning, submitted a revised estimate to the Publications Committee in 1919; but that Committee unanimously accepted one from the

Electrical Press Ltd. This arrangement held only briefly, for in November 1920 the Committee, again unanimously, accepted the estimate of the Aberdeen University Press: that arrangement held for the rest of the life of the Society as an independent body. The result of these changes is that Part 1 of Volume 15 was printed by Unwins and the other two parts by the Electrical Press; while in Volume 16 the first two parts were printed by the Electrical Press, the third by the Aberdeen University Press, and a large Appendix, reporting the 1920 'Colloids' Discussion, by Messrs. Eyre and Spottiswood Ltd. for His Majesty's Stationery Office, because the Department of Scientific and Industrial Research had agreed to publish it. There is an interesting variety of fonts in these two volumes. None of the printer's estimates is quoted in the Publication Committee's Minutes, so we do not know how substantial the saving was. What we do know is that in 1926 the Aberdeen University Press agreed to allow a 5% discount if its bills were settled within a month. The Finance Committee reckoned that for 1926, if money to do this were borrowed from the Bank at 6% p.a., which was 1% over Bank Rate, this borrowing would cost about £10 and the saving would be about £50. Accordingly the Honorary Treasurer arranged for an appropriate increase in the permitted overdraft. There was a further, unspecified amendment of the contract in 1932 which meant that 'the cost of printing the *Transactions* in 1932 exceeded that in 1931 by only £12 12s 8d, although the number of pages printed was 928 instead of 828' (Council Minutes, 19th May 1933).

Although many of these economies could be described as good housekeeping, some of them verged on penny-pinching. It is, of course, in accordance with one of Parkinson's laws that there were two other, major economies to be made which Council faced with some apparent reluctance. We will now examine these.

3.6. THE EXCHANGE WITH THE AMERICAN ELECTROCHEMICAL SOCIETY

By 1918 the cost of the scheme for providing the *Transactions of the Faraday Society* to members of the American Electrochemical Society became unacceptably high. It had always tended to be expensive to the Faraday Society, which had to supply many more copies of its *Transactions* than did the American society; but from 1909 there were payments from the latter society recorded in the Faraday Society accounts which roughly balanced the entries for the costs of the exchange. Suddenly, in 1915, the American society ran into financial difficulties (Section 1.5) so it wanted to terminate the free exchange and, instead, to make members of either society pay a small sum extra to their subscription if they wanted the *Transactions* of the other society. This seemed to distress the President (Hadfield), who at first offered to make a personal contribution to the American society. Since about the latter part of the eighteenth century – certainly from the early nineteenth century – learned societies had realized that their publications were a prime and efficient means both of spreading enlightenment and of gaining prestige and influence (see e.g. R. S. Cahn, *Survey of Chemical Publications*, The Chemical Society, London, 1965, Appendix 1), so it may well have been for such reasons that the Faraday Society Council did not want to see the distribution of the Society's *Transactions* diminished, as might happen if the members of the American society had to make a conscious decision to pay for them.

Nevertheless, the scheme agreed in late 1916 was a variant of the American society's proposal. Unfortunately it proved far more costly than had been hoped, even though the American society was generous in making some refunds to the Faraday Society. The provision to members of one society of the *Transactions* of the other was arranged at a price of $5.00 for the two annual volumes of the American Electrochemical Society, The Faraday Society charging the same for its *Transactions*.

There was one surprising hitch. In May 1930 Council heard that members in Australia and in the U.S.A. were having difficulty in getting *Transactions* through their Customs. The Secretary was asked to take up this matter with the appropriate contacts; but there is no subsequent Minute, so we do not know why this difficulty arose or how it was resolved.

3.7. CONTROLLING THE COST OF *TRANSACTIONS*, 1915–1924

The most important task facing Council in its fight to control expenditure, from about 1918 onwards, was to end the state of anarchy which prevailed in the publication of *Transactions*. This has already been mentioned in Chapter 2 (Section 2.4) because most of the trouble came from the General Discussions and it was referred to earlier in this chapter (Section 3.3). We need now to review events thoroughly. The necessary recapitulation and the detail may at first seem tedious; but the story proves to be a surprising one, so the reader is urged to have patience.

During the 1914–18 war the General Discussions became the source of from half to three-quarters of the material for *Transactions*, and they continued to be so for the years immediately after the war because although the absolute number of pages for ordinary papers began to increase again so too did those for General Discussion material. The General Discussions of that era were one-day, or sometimes only evening events. The number *per annum* was variable and sometimes large: thus, in 1919 there were two major ones, in 1920 and 1921 there were three major and one minor, in 1923 four major ones – including the first two-day meeting – plus a minor one, while in 1924 there were again four major ones and one minor one. More important are the numbers of pages generated by these meetings which, for the immediate post-war years, were:

1920, 593; *1921*, 529; *1922*, 193; *1923*, 603; *1924*, 394

Consequently, although the number of pages from the ordinary papers was relatively steady over that period, the size of volumes fluctuated wildly. What Council needed to do, if prudence was to be its guide, was to control the average amount of material so that it matched resources, and to make the flow more even so that forward financial planning might become possible. It may be recalled (Section 2.4) that attempts at control went as far back as 1915, for in June of that year Council renamed the Editing Committee the Publications Committee and widened its function. Council agreed (Minutes, 29th June 1915) to a motion moved by Dr. George Senter, a member of that Committee viz. 'That the members of Special Committees to organize General Discussions be appointed by Council as hitherto, and that the Chairman or a nominee of the Publications Committee should be an ex-officio member of such Special Committees which should make their Reports to the Council in concert with the Publications Committee.' An amendment, which would have empowered the Publications Committee to appoint the organizing committees, was proposed but, although it was seconded by the Chairman of the Publications Committee (Professor Huntington), it was lost.

What happened subsequently? The answer is surprising! From the Minutes of the Publications Committee and of Council there is no evidence that the prescribed procedure was fully and properly followed for any General Discussion before the one on 'The Examination of Materials by X-Rays' held, nearly four years later, on 20th April 1919. This was, indeed, proposed by Hadfield to a Publications Committee meeting on 13th January 1919 when it was approved and referred to Council; and on 7th April the Committee approved the programme. Before this, for three meetings, the Committee had done no more than approve papers a week or two beforehand, when it could not effectively control the programme: for three other meetings it even accepted them after the event; while for five General Discussions there is no evidence that it took any action at all. The effort of following the prescribed drill in 1919 seems to have proved exhausting. Another three years passed before the Committee approved, on 18th January 1922, a programme as much as two months ahead of a General Discussion, and this was for a rather minor one about 'Some Properties of Powders' held on the evenings of 9th and 23rd March 1922. In between, as before, the Committee accepted papers when it was too late to exercise control, or did nothing at all. Two of the special committees appointed by Council were even told to report back to it and not to the Publications Committee. Council continued to supervise the organization of the Discussions quite carefully, to judge by the number of reports made to it. From all this evidence one is driven to conclude that the Publications Committee was proving not to be a suitable body to control the Discussions; but speculation about the possible reasons for this will be left until later in the story.

On the same day, 18th January 1922, that the Publications Committee showed rather unusual activity by approving a programme in advance (for the Discussion on 'Powders') the Finance Committee met and decided to intervene. Porter, Cooper and the Secretary attended both meetings: we do not know which one came first. It recommended to Council that, because of the amount of material already going through the press (450 pages) 'not more than 200 pp. of new matter should be accepted for reading or publication before June 30th 1922.' Council agreed. Perhaps because of this, the output from Discussions was kept low for 1922 by having only one major one in addition to the minor one mentioned above; but there is no evidence to show that this was in fact a consequence of action by the Publications Committee. That Committee asserted effective control for only one of the four General Discussions which took place in 1923, i.e. it reverted to its old habits and consequently there was a record-sized Volume 19 (1923–24) of 949 pages. For the Discussions of 1924 the Publications Committee exercised control over the first two, a minor and a major one, but did nothing whatever about the other three. In July 1924 the Finance Committee became alarmed by the estimated cost of Volume 20, which was £1100, i.e. more than 50% greater than that of the previous Volume (£700) which itself had been a record. Therefore it urged that Volume 20 should be limited to 600 pages and that not more than two General Discussions should be held in 1925. Council agreed. Later the Finance Committee asked that specific page limits be imposed on the Reports for two General Discussions: that on 'Base Exchange in Soils' was allotted 70 pages and that on 'The Physical Chemistry of Steel-making', 100 pages. These limits were accepted with some reluctance but were roughly observed.

In June 1925 Council itself took the initiative in control for, being evidently concerned about the number of papers expected for the Autumn General Discussion on 'Photochemical Reactions in Liquids and Gases', it requested the organizing Committee to report concisely to the Publications Committee about its plans. Nevertheless, the Minutes for that Committee record only that it accepted the papers after the event; and that Committee had taken no action at all about the Spring meeting in 1925.

3.8. CONTROLLING THE COSTS OF *TRANSACTIONS*, 1925–1933

W. R. Cooper had had to resign as Chairman of the Publications Committee at the end of 1924, because of what seems to have been the quite sudden onset of ill health (he died in 1926 aged only 58). He was succeeded by Professor A. W. Porter, who already was Chairman of the Finance Committee and who therefore should have been able to coordinate the two committees. This appears to have been in his mind, for in October 1925 he 'invited the Council to define precisely what should be the functions of that Committee in regard to the material contributed to General Discussions organized by special "ad hoc" Committees appointed by the Council.' 'It was decided that such special committees should be appointed by the Council as heretofore but that the programme of General Discussions should be referred to the Publications Committee for ratification before taking action and that subsequently all matter should be passed for press by the Publications Committee.' Considerable formal power was thereby restored to the Committee; and after this sally by its Chairman it did indeed have another bout of activity. It approved the programme for the Spring meeting in good time, and then it made a major attempt at control by advising drastic changes in the organization of the projected Autumn Discussion on 'Physical Phenomena at Interfaces'. The organizers were proposing to be very generous to potential authors. After an exchange with the Publications Committee a compromise was reached and resulted in a modest Report of 68 pages.

This worry about overlapping in General Discussion papers was an old one, expressed for example in 1914 (Section 3.3), and it was a continuing one. It led the Publications Committee, under Porter's chairmanship, to generalize. A Minute for 9th November 1927 reads 'The Committee after some discussion expressed the opinion that the General Discussions of the Society would serve a more useful purpose if not more than three or four comprehensive papers were circulated in advance so that adequate opportunity can be given for those papers to be read and digested. The Committee also felt that the many distinguished visitors who have hitherto been invited to contribute papers should be invited to contribute to the discussion, so that the discussion rather than the primary papers would constitute the greater part of the Reports.' These views were conveyed without recorded comment. They were never adopted in a Discussion, so they must have been regarded as too extreme. Therefore they may have diminished the repute and influence of the Committee. Nevertheless, the underlying worry continued. For example, in December 1928 Council agreed that for the 'Molecular Spectroscopy' meeting of September 1929 there should be a special introduction to each of the four Sections, in addition to a general one for the whole Discussion, and that other authors should be advised of this.

The recommendation of the Finance Committee in 1924 that not more than two General Discussions should be held in 1925 was heeded, and indeed there were never more than two a year thereafter; but from the famous one about 'Strong Electrolytes', in the Spring of 1927, they tended to become two-day instead of one-day events and so for each to generate Reports of some 200–400 pages instead of the 100–200 pages for earlier Discussions. This change was a major one, yet there is no record to show that it was discussed as a matter of general principle and agreed after full consideration of the consequences, either by Council or by any of its Committees: it just happened in a series of *ad hoc* moves. That it was a sly trick to outwit the Finance Committee is unlikely. Rather, it was a natural response to the expanding needs of chemistry which, by that time, had started the rapid development described and emphasized in Chapter 2. Be that as it may, the cost of publishing the Reports was obviously increased again. By 1930 the change was fully established, and it set the pattern of General Discussions for the rest of the Society's independent existence.

Another change that affected costs may be mentioned here. In 1922 there were only three issues of *Transactions*, but Council resolved that thereafter there should be four. Then in 1926 Council agreed to a proposal by Porter, as Chairman of the Finance Committee, that *Transactions* should appear every two months 'in order to expedite the publication of original papers', and that 'papers might be published before being read at a Meeting.' Only a year and a half later, after careful discussion, Council agreed that 'for the more speedy publication of contributions to the Society' there should be monthly parts each of about 48 pages, except that for those General Discussion Reports which extended over more than one part, all the parts might be published simultaneously in one cover, thereby reducing the cost. It must be remembered that at that time these Reports formed part of *Transactions* although from early on they also were sold as separates. This change was going to increase costs for the next Volume by up to £150, and already an overdraft of £257 was expected by the end of the year; but the Finance Committee was of the opinion 'that the increased and regular rate of publication will enhance the prestige of the Society and will lead to increased receipts from membership subscriptions and from subscriptions from non-members' for whom the charge a year would still be £2 10s 0d (£2.50) plus postage. Monthly publication began in 1928, and the results eventually justified the Council's hopes.

Returning to the saga of the General Discussions, we find that for a period of two years, from January 1926 to February 1928, the Publications Committee did supervise them actively; but then again it began to approve rather late, though still before the event, and after 1930 it did nothing. Indeed, for the last three years of Porter's Chairmanship, viz. 1931–33, it met only twice a year, which precluded active supervision and which must mean that Porter had abandoned hope that this Committee could be effective in that role. He may have felt rebuffed by the lack of response to the Committee's draconian recommendations of 1927. Under his successor, Professor J. R. Partington, the Publications Committee, at its only meeting in 1934, 'decided that it was not necessary or desirable to meet at regular or stated intervals, but that they would meet two or three times a year as the exigencies of the work required.' These exigencies could have included the supervision of General Discussions, but in fact did not. Even more than before, the routine supervision was left to Council: it received a series of progress reports from most special committees, but not from all of them. Some appear to have acted independently once they were appointed: the Colloid Committee did this twice and both of the resulting Reports were very large. We may note that the Editor reported at the time that the number of papers of high quality then coming in was increasing, and it can be seen from Fig. 3.1 that this was when membership was beginning to increase rapidly.

A year later (17th November 1933) there was alarm on the Finance Committee when it was found that because Volume 29 had been lumbered with two Reports totalling 574 pages it was going to be 1350 pages as against 928 for Volume 28, and that this would reduce the Society's reserve to about £100. With the aim of avoiding such deficits in future the Committee expressed certain views and made some recommendations:–

(a) They approved of the Editor's publishing all the material passed by the Publications Committee and they did not recommend that a limited number of pages be prescribed for each number of the *Transactions*; but they did emphasize the need of urging authors to be as concise as was consistent with clarity.

(b) They considered that there could be less material for General Discussions without serious disadvantage; so they recommended that 'the greatest care should be exercised both in the selection of the authors and in 'the condensation of their contributions'. Therefore, they recommended that organizing committees 'be required to report to the Council all their recommendations as to persons invited to contribute to discussions.' The aim was to reduce not only the number of contributions but also of visitors expecting hospitality.

(c) They recommended that authors, particularly at General Discussions, should be charged if they required excessive and costly corrections to their papers.

These recommendations, hardly notable for their novelty, were accepted by Council on 15th December 1933. Council also deemed worthy of attention a suggestion by Lowry that one of the two annual General Discussions might be on a less ambitious scale than recently: it will be recalled that by 1933 they had become two-day instead of one-day events. It considered, further, that organizing committees should take greater care to eliminate duplicate matter from the papers and to avoid the 'republication in full of papers which have recently appeared elsewhere.' So things seemed set for yet another attempt to improve control of the General Discussions; but before the next meeting of Council there was a dramatic, completely unforeseen change of circumstances. A founder member and life member of the Society, the late Lt. Col. J. J. Bourke, who had just died, had made the Society his residuary legatee. The exact value of the benefaction was not yet certain but it was obviously going to be substantial. Ultimately it proved to be ca. £15 500. Therefore Council was informed at its next meeting that the President, the Honorary Treasurer and the Secretary had taken no action to implement the restrictive decisions reached at the December meeting.

3.9. RENEWED EFFORTS TO CONTROL COSTS, AND EVENTUAL SUCCESS, 1934–1939

In its Report for 1933 Council commented on the increased popularity of *Transactions* with authors, and said that despite the increased burden that this imposed they felt it in the best interests of the Society not to limit the number of pages to be printed, and expressed the hope that as a result of the Bourke legacy and various economies that were to be made – mainly increasing charges to non-members – they would be able to print all the papers that came up to their high standard. The Report also said that Council had urged the Publications Committee to insist that 'Authors write their papers with utmost conciseness consistent with clarity, and to take steps to reduce the bulk of material contributed to General Discussions.' What did the Publications Committee do in response? As may be recalled it met in June 1934 and decided to meet again only when the 'exigencies of the work required.' It did agree to publish N. V. Sidgwick's tables of electric dipole moment data, an expensive project which arose from the Discussion held in the previous April. Otherwise, at that meeting and at two in 1935, there was not even a mention of specific General Discussions. So it ran true to the form that had become established. It did nothing. By then, too, it had effectively delegated control of the publication of ordinary papers to the Chairman and the Editor, acting on reports by the referees. In fairness, the writer feels he must emphasize that on the rare occasions when there were difficulties with authors the Committee still gave as much care and thought to arriving at a fair outcome as it ever had.

In March 1936 Council was again worried by an estimated deficit of £300 for 1935. The Finance Committee recommended that savings be made insisting authors write more concisely and that the size of the General Discussion reports be more stringently limited. Council accepted these recommendations and then added to them restrictions that were to apply to the Edinburgh meeting in September 1936, and thereafter. They were:–

'(a) Immediately after the decision to organize a General Discussion, the Publications Committee should decide what space in the Transactions should be allotted to the Report (including all the papers and the discussion thereon); for the Edinburgh meeting, the Council suggest to the Publications Committee that 250 pages should be the maximum allotted.

(b) That in deciding the provisional list of contributors, the sub-committee should make up its mind as to the subject on which every contributor should be invited to write.

(c) That the sub-committee should thereafter allocate the available space between the contributors, and that the contributors should be informed that the maximum number of words of their contribution should be 3000, or whatever number in each case might be decided.

(d) That every contributor be invited as speedily as possible to send a synopsis for the assistance of the sub-committee in preventing duplication, but not for publication.

(e) That so far as possible introductory matter should be included in a General Introduction to a group of papers rather than in individual papers.'

The Publications Committee subsequently decided that referees cannot always make recommendations about shortening papers and the Editor reported he was controlling the length of papers for the forthcoming Leeds and Edinburgh Discussions.

There is nothing in later Minutes to show that the Publications Committee acted to implement paragraph (a) above. The new policy of controlling the pages allocated for Discussions was, however, presented at the Annual General Meeting in Edinburgh in the context of the Society's financial problems.

The Publications Committee of 4th October 1935 first decided that polemical papers should not be encouraged and then agreed to incorporate into their Minutes a statement which had appeared in the *Transactions* for some months.

'Written contributions by way of discussion on the above may be submitted to the Editor. The question of their publication in *Transactions*, together with the authors reply (in any) will be decided by the Publications Committee.'

No earlier authorization of this announcement can be found in the Minutes either of the Committee or of Council, and when it came to the attention of Council at their next meeting they decided that this notice should not continue to appear.

Perhaps it was in an effort to tidy up things that on 27th July 1938 Council decided to promote some amendments of the Rules whereby, for the first time, the Publications Committee should be specifically mentioned therein, the Chairman of that Committee would become *ex officio* a Member of Council, and Council would be enabled to delegate to that Committee any of their powers relating to the approval and publication of papers. These provisions hardly had time to come into effect before the outbreak of war in 1939 brought very abnormal conditions for publication.

In looking back over this strange story, the first question to ask is how effective were these repeated attempts at regulating the output from General Discussions? Some of the relevant data are: *Transactions* volume pages peaked in 1920–22 (592 pp.: 529 pp.), in 1923–24 (603 pp.) and 1929 (582 pp.) and 1933 (573 pp.). Each peak was followed by a valley, and there were later peaks in 1935 and 1936 (670 pp.: 672 pp.).

All the peaks except the one in 1929 were followed by attempts at regulation. There were four attempts (in 1915, 1925, 1933 and 1936) to arrange that the Publications Committee should be the regulatory body; but none of them worked for long. The size of the Discussion Reports seem to vary both with the nature of the topic and with the inclinations of the organizing committee. 'Molecular Spectroscopy' (1929) may have merited 339 pages; but it is less clear that 'Colloidal Electrolytes' (1934) needed 422 pages.

The repeated failure of the Publications Committee in this context may have been structural. It sought, in the earlier years, the necessary power, but never could wield it for long. Certainly it would have had to meet more often: this it seemed disinclined to do. That its members were not even paid third-class travelling expenses did not help. There were also differences in the views of the Publications Committee and of Council: there are intrinsic difficulties in the refereeing of invited papers, and these seem to have been compounded by imperfect organization.

The only method of control that seemed to work was for Council to allot directly to organizing committees block grants of space, as it did in 1936 for three Discussions. By that time the Society seems to have arrived somehow, by some kind of collective experience, at a settled view of what size the average Report should be, and a willingness to abide by it. The oscillations diminished and the annual output from Discussions settled to the figure which became the long-term mean, i.e. about 500–600 pages, i.e. for two Discussions, except for some violent fluctuations during and immediately after the 1939–45 war.

The Society's administration over this period can but be regarded as technically unimpressive: it shows the difficulty that an amateur organization can have in trying to apply a consistent, long-term policy. Nevertheless, what was much more important was that the Society survived and learned from experience. As was emphasized in Chapter 2, the General Discussions were as a whole a remarkable scientific success; and it is relatively of minor importance that some Reports were overblown and some cut unduly.

By the time the output from the General Discussions had been stabilized, i.e. about 1936, it had become the minor part of *Transactions*. From about 1930 the pages required for ordinary papers increased rapidly; and, with a lag of about two or three years, so did the membership. Dr. R. S. Cahn wrote 'publications double, in geometrical proportion, approximately every ten years' (*Survey of Chemical Publications*, The Chemical Society, London, 1965, p. 7). He probably used papers rather than pages as a measure; but we may conclude that the Society's *Transactions* followed a growth pattern that was general for that period. After the hiccup caused by the 1939–45 war, the rate of increase settled down to a lower value, of a doubling in about 17 years.

3.10. THE GENERAL FINANCIAL PICTURE

We must now look briefly at the result of all the moves that we have considered to increase income and to contain expenditure. This is most simply done by looking at the gain or loss on the successive years' workings. Balance sheets were struck and accumulated balances were shown annually; but in the period under review the basis for reckoning them was changed about four times in all, so far as the writer can see, first by amateur auditors and then by professionals. Consequently the annual balances seem not to be exactly comparable, but near enough to be illuminating. The annual figures are shown in Table 3.1.

These data show how precariously the Society lived throughout the inter-war years. The figures quoted show an accumulated deficit of £951 by the end of 1935. They also show vividly the recurrent, alarming lurches which prompted attempts to control publication costs. Only after 1936 did the Society stabilize its finances, when it was able at last to control the output from the General Discussions and when, also, the membership was rising rapidly. The income from subscriptions in 1932 was £930 4s 3d, in 1936 was £1395 18s 8d, and in 1939 was £1672 8s 8d. During the later years of the 1939–45 war there were some large credit balances because *Transactions* was abnormally small. These were largely invested to provide against the anticipated post-war needs but, as was usual in those days, the investments were of a type that offered no hedge against inflation.

The Bourke legacy worked no miracle: it did not prevent losses in 1934 and 1935 although interest payments had begun to come in. Indeed, as we saw earlier, it had the immediate effect of causing restrictions on publication to be eased and so, perhaps, of contributing to the creation of a deficit. The gross return from the capital sum of

Table 3.1

Year	Loss	Profit
1918	£202 15s 5d	
1919	£173 15s 5d	
1920	£231 12s 7d	
1921	£46 13s 0d	
1922		£242 8s 9d
1923	[no accounts available]	
1924	£293 7s 4d	
1925	£187 4s 9d	
1926		£20 1s 0d
1927		£167 13s 3d
1928		£541 14s 10d
1929	£87 5s 7d	
1930		£213 18s 2d
1931		£172 4s 3d
1932	£130 5s 9d	
1933	£480 11s 10d	
1934	£70 16s 3d	
1935	£406 12s 2d	
1936		£238 2s 5d
1937		£592 19s 6d
1938		£535 15s 11d
1939		£567 14s 11d
1940		£631 9s 7d
1941		£531 2s 7d
1942		£739 14s 9d
1943		£1152 11s 8d
1944		£1145 1s 0d
1945		£1739 13s 6d

about £15 500 was ca. £490 p.a., including recoverable income tax, which is about 3.2% and reasonable at that time.

In retrospect it is clear that this period was an heroic one for the Society: it had to cope with great and rapid expansion if it was to fulfil its destiny. Despite the administrative shortcomings that can be seen with hindsight, it succeeded. From 1920 to 1939 *Transactions* increased in size roughly three- to four-fold. The scientific quality improved almost beyond recognition. There were limitations in scope that were commented on at the end of Chapter 2 and which we shall consider again later in this chapter; but within them the Society had gained a first-class international reputation.

3.11. MEMBERSHIP AND CHANGES IN THE CHARACTER OF THE SOCIETY

We have already reviewed the administrative actions taken to increase the membership of the Society: it is appropriate now to consider the changing character of the membership which itself directly relates to that of the Society.

The membership increased sharply between 1915 and 1921, remaining almost constant until 1933, then rising rapidly until World War II brought a check. Insights are provided by six lists of membership – those of April 1911, 1913, for January 1917, September 1919, October 1926 and January 1938. From these it is possible to divide the membership into three classes:

1. The academic group, being staff members of universities or similar institutions.
2. Those working at research institutes, often of a national or governmental character.
3. The industrial group, which includes those in industry, commerce, and the ancillary professions, e.g. consultants, patent agents, etc.

To each of these categories of individuals, corporate members or libraries are added.

From the counts based on these groups some significant results emerge. In the early years of the Society the industrial members outnumbered the remainder by more than two to one. Up to 1919 the growth of the academic group was slow, but the annual loss of those discontinuing membership was greatest for the industrial group: even so, the rapid post-1917 increase in membership was mainly on the industrial side. Between 1926 and 1938 changes in the industrial/academic ratio were reversed, so that by 1938 the academic group of 384 exceeded the industrial one of 320, and was almost half of the 777 total membership. One striking feature in all three groups during these years is the great increase in overseas membership, both personal and corporate. The rapid rise in total members from 1933 onwards reflects the accelerated growth of the academic group plus a revived recruitment of industrial members. For the two decades from 1918 academic research in the U.K. universities was stimulated by support from the Department of Scientific and Industrial Research, by the University Grants Committee, and by such sponsors as Alfred Mond's in I.C.I. Typically the chemistry staff members in the Natural Science Faculty at Oxford University almost doubled, from 17 in 1920 to 30 in 1939.

The numbers of U.K. industrial members are revealing: 204 (1919), 231 (1926), 192 (1938). As the number of U.K. industrial chemists certainly did not fall between 1926 and 1938, the Society appears to have become less attractive to them, or, relatively less important in comparison with other societies serving their interests. However, a reverse trend is seen in the numbers of overseas industrial chemists: 16 (1919), 28 (1926), 121 (1938). In the list of General Discussions (Appendix C) the subjects discussed from April 1921 to January 1928 would all have been of interest to industrialists. From September 1928 to April 1939 at least nine out of the twenty-one topics could have claimed industrial interest. These figures project the considerable development of academic studies in physical chemistry, including the appearance of a chemical physics element, and a preponderating influence of academics in the Faraday Society Council. For the majority of industrial chemists the papers in the *Transactions* became too purely scientific and too scientifically pure. In this regard the *Transactions* was establishing a priority to follow and to stimulate advances in physical science which, for the years 1920 to 1939, was expanding at a very different rate from that for 1903 to 1920. One upshot was that in the earlier period, of the seven Presidents, four (Swan, Perkin, Swinburne and Hadfield) were essentially industrialists. In the period 1920–39 only one (Rintoul) was fully an industrialist, although several of the others had significant industrial connections. The Society did not make it a rule or custom, as the Deutsche Bunsen-Gesellschaft did, to have academics and industrialists alternating as Presidents. The D.B.G. has always done this; but it has a much more explicit reference in its Statutes to the need for the closest possible interaction between science and industry, or at least technology:– 'Die Deutsche Bunsen-Gesellschaft . . . strebt hier eine möglichtst ennige Wechselwirkung zwischen Wissenschaft und Technik an.' The Rules of the Faraday Society said merely that the Society 'will promote the study of' certain sciences: it was the choice of these sciences that formally determined the relation of the Society to industry. [The commitment of the German society to 'a closest possible interaction between science and technology' shows a profoundly important difference of attitude from that in the U.K. M.M.D.]

Looking back at the developments of the inter-war period we are left with a feeling of amazement that a Society which had gained so much influence in international science should have had so small a membership (895, including 61 student members in 1939) and vice versa. Matters that will be dealt with later, in Sections 3.14(c) and 3.15, may partly explain why this was.

3.12. A POT-POURRI OF ADMINISTRATION

By now we have considered several of the major problems that concerned the Society in the inter-war years. A few remain, but there are some relatively minor administrative matters which deserve attention and which might now be interpolated, being taken *seriatim*.

(a) Income Tax. The Society had battles with the Inland Revenue authorities on two fronts. One was to obtain relief from income tax for members in respect of their subscriptions. The other was to recover income tax levied on its investments. The only Minute relating to the former is of 27th July 1920. It states that a letter was received from an Inspector saying that allowance was not made in income tax assessments for subscriptions to scientific societies, and that Council endorsed the action of the President (Hadfield) in supporting the National Union of Scientific Workers in their endeavour to obtain relief for scientific workers on account of professional expenditure. This concession was eventually allowed.

The other fight started in 1924 when a Minute of 29th January states that the Chief Inspector had refused, 'for the first time', to allow a claim by the Treasurer for the return of income tax deducted at source. The Society was not registered as a Charity and so had no automatic entitlement to relief from such taxation: indeed it had no clear, legal form of existence, and this was to prove a recurring difficulty. The Minute records further that an authority was to be consulted and the

Publication, Finance, and General Administration 1918–1945

position of the Institution of Electrical Engineers discovered. Several later Minutes show the progress made. The Secretary interviewed the Claims Department of the Inland Revenue and, as a result, the Treasurer made a new application in the autumn. This met with a further refusal (Minute, 26th February, 1925), so the Secretary was instructed to consult the Royal Society with a view to the possibility of joint action in approaching the Treasury. In June of that year Council heard that, following an approach to the Treasury by a deputation organized by the British Association for the Advancement of Science 'on behalf of and in conjunction with' other scientific societies, a test case was to be instituted, but that in the meantime, because the Faraday Society had previously had taxes on investments returned, income tax that had been withheld would be returned. Bousfield raised the possibility of setting up a tax-exempt Trust; but Council decided to await the result of the test case. In October 1925 it was reported that the Treasury had, in fact, returned £13 13s 11d, for the years 1922–24. Things then went quiet until, in February 1929, the result of the High Court action was reported (but not given in the Minute). There was some discussion about the desirability of incorporating the Society, but Council decided to do nothing until difficulty was met in recovering tax. The accounts for 1929 show that £14 2s 3d was recovered for tax levied in 1926–28, and those for 1930 show £17 5s 0d recovered: so a working arrangement with the Inland Revenue seems to have been reached and there the matter appeared to rest, but a Minute of 22nd May 1930 refers to the 'trustees of the Society's funds', Mr. Robert Mond being replaced by Mr. Emile Mond, with Marlow as the other trustee.

(b) Britain and the Gold Standard. Once upon a time Britain had a stable currency. There were golden sovereigns, and up to 1914 these had kept an almost constant purchasing power for 50 years. In the 1914–18 war gold coins were replaced by paper money, by 'bank notes' which said that the Bank of England would pay the holder one pound sterling, but did not say when. By virtue of such a promise the paper currency was said to be 'on the gold standard'. But economic and social pressures on sterling, such as now are all too familiar, built up. The effort to maintain the old exchange rates with other currencies became

increasingly burdensome, and in 1931 the final link with gold was abandoned. Britain 'went off the gold standard'. The rates of exchange for sterling dropped sharply. This had effects on the Faraday Society because of the financial links that it had formed with societies in Europe and America. So, in November 1932, the Secretary told Council that the Dutch society was requesting payment on the gold basis, and he was instructed to recover the difference from those members who had joined that society.

The German, French and American societies were not making a fuss, probably because they owed monies to the Faraday Society and could now pay the sterling dues at less cost in their own currencies. A year later the Finance Committee was exercised by the alarming possibility that a German would find it cheaper to become a member of the Faraday Society and join the Deutsche Bunsen-Gesellschaft through it than *vice versa*, because he could then pay the joint, discounted subscriptions in pounds sterling which he could buy at the new, low rate of RM 13.5 = £1, leaving the Faraday Society to pay the D.B.G. a number of Reichmarks determined, under the terms of the agreement, by the old 'par' of RM 20 = £1 but which it would have to buy at the new rate. This would cost the Society 14s (70p) for every such member. In fact this did not happen. Germany seems to have imposed currency restrictions, so Germans could not purchase sterling. There was an impasse, because membership of the Deutsch Bunsen-Gesellschaft had become unduly expensive for English members, and membership of the Faraday Society for Germans had become unnecessarily dear. The difficulty was solved *ad hoc* by the Faraday Society accepting RM 24, instead of RM 36, for membership; and the D.B.G. accepting RM 24 (taken as equal to £2) from members of the Faraday Society instead of RM 28.8, according to a Minute of 28th September 1935.

(c) The Ordinary Meetings. One evolutionary change in the life of the Society deserves mention. It will be recalled that the earliest meetings were held in evenings, apparently at 8 p.m. which was intended to be after dinner, sometimes a common dinner (Section 1.1). Papers which had been circulated in proof form were read on topics that were not

necessarily related. The original intention was that most, if not all, papers should be so read and discussed before publication so that the final version would represent the best-informed view. The General Discussions, at which all the papers were related in topic, started some years later (1907) and were at first of irregular occurrence. Gradually they became more frequent and soon were more important than the Ordinary Meetings, at which attendance had already begun to flag. As early as 1906 the Council Report complained that 'attendances have often left much to be desired' and on 7th May 1913 one Ordinary Meeting had to be abandoned.

During the 1914–18 war only two Ordinary Meetings are mentioned in Council Reports; but a paper read at one in 1916, about the function of carbon in steel, so pleased President Hadfield that he gave a prize of £50 to the reader, Mr. F. C. Thompson. Vigorous attempts were made to revive them in the immediate post-war period. In 1920 Council decided to have them more frequently and to ask members to send in suitable papers. In 1921 there were no less than six such meetings: then the number settled down to between two and four *per annum*. Nevertheless, in the Council Minutes there were frequent indications of concern. The time of meeting was changed from 8 p.m. to 5.30 p.m. (October 1925) and back again (March 1926) after attendance had considerably diminished. Some of them certainly were held in the rooms of the Chemical Society e.g. in 1924; but in 1929 and 1930 attempts were made to organize some in the London Colleges at 5.30 p.m., presumably in the hope of attracting students. Unfortunately they proved difficult to arrange and only one such is clearly recorded; this was at University College, thanks to Donnan, and it was a great success.

Three meetings were held jointly with the Electroplaters' and Depositors' Technical Society. Yet, despite all these efforts, the Ordinary Meetings languished, whereas the General Discussions prospered increasingly. The trouble probably came partly from the heterogeneous fare that was offered and partly from the difficulties, or at least the discouragement that people were finding in attending evening scientific meetings, probably because of the increasing suburbanization of London and the changes in family life. Eventually, in June 1935, at the prompting of the Publications Committee whose province this had been since 1921, Council deliberated and 'having regard to the poor attendance at meetings in the past, it was decided, without committing future Councils, to concentrate the energies of the Society on holding two General Discussions each year, and on publishing papers with the utmost rapidity.'

The rule that papers must be read before being published had been rescinded in 1926, and it had been agreed that papers for General Discussions could be published before being discussed. The rule that elections to membership should be announced at an Ordinary Meeting was also rescinded. Thus, the Society was gradually forced into a new mould; but although it may have become constitutionally less like a club, those who remember the General Discussions of that time can testify that much of the old feeling remained.

3.13. ACCOMMODATION

For the first 17 years of its existence the Society's address was 82 Victoria Street, S.W. Initially, either Cowper-Coles' or Swinburne's office there was used, probably for clerical work and certainly for Council meetings. Later, Spiers used a 'works office', at an address that is not now known, and the Society shared a room at 82 Victoria Street with the British Scientific Products Exhibition. On 29th May 1919 Council was told that Spiers was giving up his office and that the shared room would not be available after July; so arrangements were made to find another office and to share this initially with the new-born Institute of Physics, as has already been reported (Sections 1.4 and 3.3).

At about the beginning of 1920 the Society moved into an office at 10 Essex Street, off the Strand. A Minute of 7th July 1924 and the addresses on the parts of *Transactions* show that there was then a further move to 90 Great Russell Street, WC1, where it was when Spiers died in May 1926. From Council Minutes of 3rd June 1926 and 22nd April 1927 it is clear that the Society agreed terms with Spiers' Executors for moving out of these premises, so Spiers had evidently bought the lease.

Although Marlow had a share of some chambers in the Temple, in King's Bench Walk, he was a member of Gray's Inn; so he was able to arrange for the Society to move there on 1st April, 1927. An architect, Mr. Charles Baker, rented some rooms at 13 South Square, one of which he sub-let to the Society. He allowed Council meetings to be held in one of his other, larger rooms. Two charming drawings by him of No. 13 appear in *Transactions* for 1953 (Vol. 49, pp. 462 and 585).

The Society had no meeting room of its own. The Chemical Society allowed some of the Ordinary Meetings and General Discussions to be held in its rooms in Burlington House. Other General Discussions, when held in London, were sometimes in Colleges of the University (especially in University College), sometimes in the rooms of the Institution of Electrical Engineers, or of the Institution of Mechanical Engineers, and twice in the Royal Society's rooms. The Presidential Lectures and the Spiers Memorial Lectures were given in the Royal Institution.

It may be recalled that in 1910 the Junior Institution of Engineers invited the Society to become tenants in a building that it was proposing to erect, but that the Society declined because of its poverty (Section 1.4). On 13th May 1924 Council considered a letter from the Institute of Physics asking if it was prepared to consider cooperating in a scheme to house the Institute and its participating societies, with their libraries, in one building. Council decided to suggest to the Institute the desirability of linking this scheme with current proposals for housing all the chemical societies together that were being considered by the Federal Council for Pure and Applied Chemistry. In June 1925 the Institute followed up by asking what office accommodation the Society required, and the Secretary replied; but in December 1926 the current working arrangement for having joint office accommodation with the Institute, at 90 Great Russell Street, was abandoned as the joint Secretary of the Institute and its participating societies, a condition which proved unacceptable to the Institute. This breakdown was not a good augury for more ambitious schemes. There is a cryptic Minute of 7th October 1929 which says that the President 'had signed the appeal' relating to a Central Building Scheme, and that Mr. Robert

L. Mond 'had agreed to act for the Society on the investigation committee.' This seems to refer to business raised earlier, but whether to the F.C.P.A.C. scheme of 1924 is not clear: the Council Report for 1929 mentions a Committee 'to consider the question of providing a Central House for Chemical and Metallurgical Societies.' and the Honorary Treasurer (Robert Mond) is named as the Society's representative, as he is again in the Report for 1930 but not in that for 1932, which does not mention the scheme. Evidently it died. Whatever the prospects may have been in 1924, by 1930 the Great Slump had developed and there was no hope of getting from industry the money which was vital for success.

On 16th April 1941 the Society had suddenly to seek other accommodation. The office was entirely destroyed in a fire-bomb attack. Most of the Society's papers and other possessions were burnt: according to a Minute of 22nd August 1941, and to the memoirs of Marlow's personal secretary, Miss Beatrice Kornitzer (of whom more later), the only things left were two filing cabinets, one being full of burnt papers, a typewriter, a card index of membership records, and a bound set of *Transactions* from Volume 1 to date. In fact, as has already been explained, the complete run of Council Minutes, the Publications Committee's Minutes and the Finance Committee's Minutes were saved. Miss Kornitzer says 'The offices themselves were completely destroyed, and it was only due to the gallant action of the invaluable charwoman, Mrs. E. Tooth, who lived in the neighbourhood and who persuaded the firemen to rescue them, that any of the above items were saved.' In the more formal language of the Council Report for 1941 'Special thanks are also due to the office cleaner, Mrs. Tooth, who by her prompt personal exertions was entirely responsible for the salvage of everything which was saved from the burning building.' But for her intervention our records of the Society prior to 1941 would indeed have been meagre. At the next Annual General Meeting, on the recommendation of the Finance Committee, it was resolved that her great services should be recognized, so a cheque for ten guineas (£10.50) was sent to her with a special letter of thanks. She replied:

15 Holsworthy Sq.,
Gray's Inn Rd.,
W.C.1

15. 11. 42

The Faraday Society.

Dear Sir,

I am in receipt of your letter of the 9th instant and acknowledge with many thanks your cheque for £10 10s 0d. This is very much appreciated, but I can assure you that the little service I rendered was not done for any award, I only wished I could have helped to recover more.

Please will you thank your committee very much for their kindness.

Yours faithfully,

E. Tooth

Apart from finding fresh accommodation it was necessary to reconstruct the records so far as possible; and with the aid of the Society's publishers, printers and bankers this was done for 1941. A gift of £25 from an anonymous member toward the cost of refurnishing did something to mitigate the financial blow. Through his connection with Gray's Inn, Marlow was able to find office space again for the Society at 6 Gray's Inn Square, in one of the rooms of Messrs. Braund and Hill, solicitors, at first on the second floor and then, after a further blitz in May, on the ground floor. Miss Beatrice Kornitzer (hereinafter referred to as 'Bea') had been invited to return to London to take over as Marlow's private secretary from her old school friend, Miss Gertrude Wakeley, who was leaving to get married. Bea came soon after the fire blitz by which time, she says, 'Trudie had got things tidied up and ready' for her coming. Marlow himself was bombed out of his chambers in the Temple and found temporary accommodation (until 1944) at Gray's Inn Square. He had taken the precious run of *Transactions* for safer keeping to his home at Sydenham. Every evening Bea took the card index of members down to the basement of No. 1 as a precaution; and indeed in the May attack, only one month after she arrived, all the buildings between No. 1 and No. 6 were destroyed. This was when the Society and the firm of solicitors had to seek asylum together on the ground floor. Consequently, Bea says, she had to share the Society's only room with one of their senior clerks. Later, Mr. Baker, the architect from 13 South Square, became the Society's sub-tenant, and she remembers how, many a time, he and she sheltered behind the filing cabinets when the bombers were overhead. Repairs to the building had been only temporary. The room was very cold. Bea says that at one time she could see the snow on the ground through the floor boards. But there were compensations. 'We looked out over the beautiful gardens of Gray's Inn, and the wonderful old catalpa trees, many centuries old, were a constant source of joy. The ruins between No. 1 and No. 6 Gray's Inn Square became rock gardens with many lovely wild flowers and seedling trees.' A more bizarre addition to the landscape was the winding gear and attendant crew for a barrage balloon, and sometimes the 'blimp' itself.

In 1942 the Society received an interesting proposal for accommodation. Council Minute 7 of 28th May of that year reads:—

'The President informed the Council that Professor Egerton [Member of the Council of the Faraday Society, 1932–35; in 1942 the Physical Secretary of the Royal Society] had indicated to him that there was a probability of premises in Burlington Gardens adjacent to Burlington House becoming available for the use of certain scientific societies. The Royal Society was anxious that the available accommodation should be utilized by these scientific societies that can properly be considered to be concerned with the advancement of science, and Professor Egerton would welcome an indication that the Faraday Society as a participating society would occupy a room there. The President considered that though the scheme was still undeveloped there would be room for meetings of the Society and of the Council as well as for conducting the business of the Society if at any time it was desired to transfer the business offices of the Society to Burlington Gardens. During the discussions which ensued it appeared likely that the scheme would not take effect until after the war, but members generally felt that the decision to cooperate in this matter with the

Publication, Finance, and General Administration 1918–1945

Royal Society was one which would affect the future of the Society for many years to come. The Secretary had intimated that it would be very difficult for him to carry on the work of the Society from Burlington Gardens, but this was a matter which could be dealt with at a later stage when the interests of the Society required. After a full discussion it was proposed by Sir Robert Robertson and seconded by Dr. Bowen and unanimously agreed that the Secretary of the Royal Society should be informed that the Faraday Society would welcome the opportunity of having rooms at Burlington Gardens.'

It was an attractive offer, and a clear mark of the standing which the Society had achieved; but like a number of later schemes to bring scientific societies into propinquity, some of them much more ambitious, it came to nothing.

After the war the Benchers of Gray's Inn decided to rebuild, and the Society then had comfortable quarters at 6 Gray's Inn Square for the rest of its time.

3.14. EXTERNAL ACTIVITIES, ESPECIALLY COOPERATION WITH OTHER SCIENTIFIC BODIES

(a) General. In the Council Minutes from 1917 onwards there are references to two bodies that were set up during or immediately after the 1914–18 war to coordinate the efforts and interests of scientific societies generally in the United Kingdom, and of chemical societies in particular. The earlier Minutes were referred to in Chapter 1 (Section 1.5), largely as evidence of the recognition that the Faraday Society had earned and of its endeavours to play a useful part in the scientific community. The bodies so mentioned were the Conjoint Board of Scientific Societies and the Federal Council for Pure and Applied Chemistry.

The former was set up in 1917: there is a Council Minute of 18th January 1917 about a letter inviting the Society to become a constituent society with representation on the Board. This was accepted, W. R. Cooper was appointed representative, and £10 was voted toward the expenses of the new body. The only action by the Board that was of potential importance to the Society and is recorded in Council

Minutes is the initiation of an attempt to persuade the Government to make money available to scientific societies for the support of publications. As we saw earlier (Section 3.4) this was not immediately successful, although later the Faraday Society was given a number of grants by the Royal Society and some of these may have come from Government support of the latter Society. In 1921 the Board asked if the Council would agree to give a 10% discount in the price for *Transactions* to members of other societies affiliated to it. Then, in March 1923 Council was informed that the Royal Society had withdrawn from the Conjoint Board on the grounds that its functions were being carried out by other bodies, including the Royal Society itself. The Faraday Society had shown waning interest, for in 1918 it reduced its contribution to £5 5s 0d and in 1922 to only £3 3s 0d. The Board was wound up in 1923.

In December 1918 the Council of the Society considered a memorandum presented by the Chemical Society which advocated the formation of a Federal Council for Pure and Applied Chemistry. Council showed interest and, again, W. R. Cooper was appointed to represent it. The new Council was set up and in 1920 the Faraday Society made a contribution of only £5 5s 0d because its interests were only partly chemical. The Council Report for 1922 explains that the Federal Council was the British Department of the Union Internationale de la Chimie Pure et Appliquée, that it made an annual contribution thereto of Fr. 4500 and was in consequence always hard up. The journal *Chemistry and Industry* had been started and was its organ. Eventually the increasing demands of the Union Internationale caused it to ask the Royal Society for help, and to be told in reply that financial assistance could only be given if it became the National Committee for Chemistry of the Royal Society. So it did this; and the Faraday Society had representation on the new Committee, the President becoming the representative *ex officio* in 1936. However, there is no record of any business being referred to the Faraday Society by either the new or the old body, so they seem to have impinged very little on it.

Another chemical body, the Chemical Council, was founded in 1935 by the three chartered bodies viz. the Chemical Society, the

Institute of Chemistry (as it was then), and the Society for Chemical Industry. Its purposes were to provide certain central services such as collecting contributions for the Chemical Society library, to organize common projects relating to the promotion and the publication of research and its application to the arts and manufactures, and to help give chemistry and chemists a common voice in affairs. For a long time the Council and the Faraday Society had little or nothing to do with each other. The latter continued to pay subscriptions for the Chemical Society directly to that society.

In 1942 the Faraday Society, with some sympathy from the Physical Society, decided to end the joint subscription scheme which it had run with the Institute of Physics, whereby members of 'participating societies' earned discounts of from 15 to $33\frac{1}{3}$% (Section 3.4). The Faraday Council had long been disenchanted with this scheme, and it decided now to give a flat discount of 10% (Council Minutes 10 of 28th May 1942 and 3(b) of 6th November 1942). This led it to offer the same privilege to Fellows and Associates of the Institute of Chemistry; but that Institute decided that with its reduced, wartime staff it could not cope with the administrative task; so the Society made a corresponding ffer to the Chemical Society. This, in turn, led the Chemical Council to consider closer cooperation in general with the Society and to invite it to send three representatives for a fuller discussion. Following this meeting the Secretary (Marlow) sent a questionnaire to members of the Society's Council. It elicited very divergent answers. In August 1943 the President summed up by saying that if the Faraday Society were a purely chemical one the proposal that it should come under the Chemical Council would have great advantages; but in fact a considerable proportion of members were physicists, and there were many overseas members. Thus, the same urge to be interdisciplinary, the sense of being special and of having a separate identity, which caused the Society to respond coolly to the proposal in 1918 that it should join the Federal Council, was evident again 25 years later. It was to prove deep and enduring. The Faraday Council appeared to decide to mark time; but a Minute of 9th November 1943 shows that it accepted the Chemical Council as the collector of contributions to the Chemical Society library, and it gave 25 guineas (£26.25), which was five times its earliest contribution. A year later that library was referred to as 'of the Chemical Council and Associated bodies.'

On 1st March 1945 the Society had to consider urgently a new development for, according to a current report 'it appeared that in future the American Chemical Society was going to become a formidable rival to the British Chemical Societies in the enrolment of members in Canada, as they proposed to offer to Canadian chemists (conditional on their enrolment *en bloc*) membership of the American Chemical Society at $1 per head.' The Chemical Council wanted to propose to the Canadians a scheme based on individual membership and a 'points' system whereby they could choose from a wide range of British chemical journals; consequently it wanted the Faraday Society's *Transactions* to be included, and it offered the bait of a possible increase of Canadian members. This meant the Faraday Society coming under the Chemical Council. Physicists already had a like privilege from joint subscriptions paid through the Institute of Physics (Section 3.4). It was agreed that the Society should have two representatives on the Chemical Council, that the *Transactions* (and membership of the Society) should be valued at 36 points, worth £1 16s 0d, i.e. a 10% reduction, and be available on this basis to members of any of the constituent bodies. In turn this required one minor change of Rules, with 'Student' Members becoming 'Junior' Members, which was effected at the next A.G.M. The Minutes of 29th August 1945 emphasize one further matter of interest. The Society had no office staff, so it did not need to come within the pension scheme for officers that might be formulated by the Chemical Council. Bea Kornitzer was employed by Marlow, not by the Society, and an assistant to her, called for at the same meeting, was likewise to be employed by him; but it was envisaged that if this arrangement were to be changed in the future then the Society might follow the Chemical Council's recommendations.

The financial arrangements that were made about 1929 to 1930 with corresponding societies in America, France, Germany and Holland have already been described (Section 3.4): they formed the only firm basis for cooperation. In the earlier negotiations with the Deutsche Bunsen-Gesellschaft it had been suggested, apparently from the German side, that there might be some unspecified collaboration in the printing of papers. The Finance Committee did not favour this but recommended that steps be taken for each Society to print short abstracts of papers appearing in the publications of the other one (Council Minutes of 25th February 1929, no. 7). Their proposal, which would have involved appreciable effort and cost, was not mentioned in the later Minute of 22nd April 1929; so it appears that the Society had second thoughts. There is an interesting Minute, of 6th December 1928, which illustrates the feeling of the Society about international relations. The President (Lowry) reported on some conversations that he had had with Professor Bodenstein about international meetings that might be arranged in concert with the French and German societies. Several members of Council doubted the wisdom of encouraging such meetings which might encroach 'upon the informal and unintentional, but nevertheless really international character of the Society's General Discussions. It was agreed that the matter should stand in abeyance.'

Another form of domestic cooperation arose from a request by a group of electro-platers that a Plating Section of the Society should be formed. This first came before the Council on 29th January 1924; but consideration was delayed 'Owing to the lateness of the hour' and because Council wanted to have the views of Professor C. H. Desch who was the most eminent scientist of the several metallurgists on Council at that time. He was not in favour of the proposal because, he said, existing societies met the needs; so Council decided to drop the matter. Nevertheless an active group continued to press their view, and they changed their tactics by proposing the founding, in London, of an independent Electro-platers' Technical Society. Council was told that the electroplating trade keenly desired a link with a scientific body, and therefore the advocates of the new society hoped that it might be described as 'Under the Patronage of the Faraday Society' or 'Associated with the F.S.'. Eventually the "Electro-platers' and Depositors' Technical Society" was founded and held its inaugural meeting in November 1925. The following March the Faraday Council agreed that it should be described as 'Associated with . . .' and regularly appointed two representatives on the Committee of the new Society, Spiers being one of the first pair. Three of the Faraday Society's Ordinary Meetings were arranged as joint meetings, the first of them in

December 1930. The interests of the new Society were obviously among those which the Faraday Society had originally been founded to encourage; but the latter Society's interests had shifted, and electroplating was now only a minor one. The new Society appears to have had no part in initiating the very successful General Discussion on 'The Structure of Metallic Coatings, Films and Surfaces' held in March 1935 (Section 2.8), though it was actively interested and asked for 550 reprints of the papers, at a favourable price.

The rather muted support given to this group contrasts with the much more substantial support given a few years later (1929) to another one, that of colloid chemists, because the opportunity for scientific advance seemed greater (Section 2.6).

In the inter-war years the Society played a full part in the life of the scientific community. There are numerous other examples of cooperation with other scientific bodies: two of them are important enough to require separate sub-sections, but the others are relatively minor and only one of them deserves mention in this miscellaneous sub-section.

In May 1922 Council received a Report which shows that a Conference had been held in 1920 between the Society, the Concrete Institution and the Portland Cement Research Association, and which set forth the subsequent developments. The Building Research Board, presumably set up by the Department of Scientific and Industrial Research, had established a laboratory, so it was likely that this body and the B.P.C.R.A. would, between them, undertake the fundamental physico-chemical research proposed by the Faraday Society. The Council then approved the holding, from time to time as necessary, informal conferences between representatives of the various bodies involved. This is a further reminder of the regard with which the Society had earned during the 1914–18 war, and its strong connection with industry at that time.

(b) *The Journal of Physical Chemistry.* As far back as 1906 there must have been some kind of approach by Spiers to Professor Wilder D. Bancroft who, in a Council Minute of 19th February of that year, was described as the Editor of the *Journal of Physical Chemistry*. He had replied to Spiers saying that the Journal 'could not for the present discuss the desirability of joint publications.'

The history of this journal is curious. Bancroft was not merely the Editor: he was the owner. He published it at a loss which, with lordly munificence, he himself carried for many years. The Faraday Society began to have a connection with it in 1922 when, on 1st August, Council met specially to consider a letter from the Chemical Society for the future control of that Journal. The proposals are not given in the Minute, but in general they were for the setting up of an Anglo-American Editorial Board. Council agreed unanimously that the Faraday Society should associate itself with this scheme, and Donnan was later appointed as its representative on the Editorial Board: he had already expressed the view that the Society ought to 'take the necessary steps for ensuring that its claims as a Society for physico-chemical questions should be properly recognised in connection with the scheme.' The issuing of the Journal under the new auspices began in January 1924; and Council, after first deciding to give it no publicity (16th October 1923) decided later (29th January 1924) to send a brief notice to members, merely informing them of the preferential terms on which they could subscribe. The tone of the Minutes suggests an ambivalent feeling, and this was indeed to be expected because internationally the *Journal* was a rival to *Transactions*. Inevitably there was some conflict of interests.

Rideal succeeded Donnan in 1927. At a meeting of the Publications Committee on 2nd July 1928, he said that the period for which the Society had agreed to collaborate was approaching an end, and that so far there had been a considerable loss on this production although the loss was guaranteed and did not fall upon the Society. He suggested that, because the Society was now publishing 'regularly' i.e. monthly, from January 1928, the Committee should at some time consider whether the Society could decently withdraw from the arrangement. He believed that British physical chemists did not now contribute papers to the American journal to any extent. In September 1929 the Publications Committee heard, in a letter from the Editor of the *Journal*, that in future the Chemical Foundation [an American body] would take over the business and financial side of the *Journal* and that this would enable it to publish everything that was good enough and to do so more frequently than hitherto. Two Council Minutes (of

February 1930 and January 1932) refer to letters about the *Journal* by Professor A. J. Allmand: the former says that Council favoured changing the name of the *Journal* to include Colloid Chemistry, in the hope that this might help to avoid the formation of a new journal. At that time the relation of the Society to colloid chemistry was greatly exercising Council (Section 2.6). Allmand is known to have had doubts about the wisdom of the continuance of the Faraday Society, for in 1931 (Council Minutes of 24th June) the Secretary reported that Allmand had expressed diffidence on this score about accepting nomination for membership of Council; so it is possible that he, like Rideal, was critical of the arrangement for the joint Editorial Board. Unfortunately, his letters are lost.

In 1932 Professor J. R. Partington was appointed to succeed Rideal as the Society's representative: at the same time the Council heard that there was to be a new publishing arrangement and that Professor S. C. Lind was to succeed Bancroft as Editor-in-Chief. They expressed the hope that the new arrangement would not eventually cause a rise in price; and they concurred in Lind's appointment [Council Minutes, 24th November 1932, 2(a)]. At the end of 1933 Council heard from the Chemical Society that it suggested ending the present arrangement in a year's time; so Council called on its representative, Partington, to make a report. They received this the following February. Later, in June 1934, they heard that the Chemical Society had decided definitely to terminate arrangements from the end of that year: they decided to follow suit, and 'directed the Secretary to convey that decision courteously to Dr. S. C. Lind.'

This caused a brisk exchange of letters. One was from Dr. C. L. Parsons, the Secretary of the American Chemical Society, who sent a copy of his letter to the Honorary Secretary of the Chemical Society, that was largely concerned with setting out the historical and current facts. There was, too, a copy of a letter from Lind which likewise was directed at clarifying the position. From these letters it appears that the *Journal* had been subsidized by the Chemical Foundation from the time when the Anglo-American Editorial Board became active, i.e. from the beginning of 1924, and not from 1929 as apparently the Publications Committee of the Faraday Society had been told in a letter from

Bancroft. The Chemical Foundation had been very generous; so subscribers had been getting the *Journal* at about one-third of its cost, but in the summer of 1932 the Foundation told Bancroft that it could no longer meet the heavy losses. Bancroft certainly could not reassume the financial burden, so this caused much perturbation. Eventually, with Parsons acting as coordinator, it was agreed that the printers should 'continue the *Journal* on a contractual basis with Professor Bancroft' which meant that they could handle the publication as a business, deciding how many pages could be printed, and handling all the business and financial matters, while the Editorial Board chose the papers independently of any constraint except for the total number of pages available. The American Chemical Society offered to take over from Bancroft, if he wished, the contract with the printers, immediately after 1st January 1933. Bancroft said that he had reached an age where he must give up the Editorship, and unless the A.C.S. did assume the burden 'he would have to abandon the *Journal of Physical Chemistry* entirely.' Thus it was that the American Chemical Society became owner of the *Journal* but that the contract with the printers, which Bancroft had signed in late 1932, continued to operate. From Lind's letter it seems that the Chemical Society believed that the A.C.S. had dissociated itself from the *Journal* and handed it over completely to the printers. It seems likely that the Faraday Society also believed this. Parsons pointed out that Lind, who was elected by the Editorial Board, received no salary or compensation of any kind. Lind said that the space limitation had not led to the questioning of any paper approved by Partington; and he hoped that a request by Partington for some editorial expenses which he had not immediately been able to meet, had not influenced the decision of the Council. There is no reason to think that this sum, of £7 18s 9d, which Council voted to Partington in December 1933, was regarded as significant. Parsons said, too, that the American Chemical Society could not make money from its new acquisition because the contract called for any profits to be used for the publication and enlargement of the *Journal*. He ended his letter on a note of pained surprise, saying that the cancellation had come suddenly and without previous adverse comment. Lind ended his letter by saying that he would have been pleased to see the association continued 'as an example of international amity and cooperation among scientists, if for no other reason.'

From this correspondence it seems that there had been misunderstanding generated, as it so often is, by inadequate communication and possibly aggravated by conflicts of interest. Marlow showed his fair-mindedness and his tact by writing to Parsons that because 'we may not have considered the matter as fully as we might if we had had before us fuller information' he would ask Council to reconsider the matter at its next meeting. He also cabled Lind. He was careful to emphasise that he was doing this on his own responsibility and that until and unless Council rescinded its decision this must be regarded as holding.

At its October 1934 meeting Council considered this correspondence, together with letters from Partington, Sidgwick and Travers, and they decided to continue their cooperation with the American Chemical Society for the production of the *Journal*. Hinshelwood was added as a second representative on the Editorial Board. These arrangements held for the remainder of the period covered by this Chapter. The Chemical Society kept to its decision and did not continue cooperation.

(c) Cooperation with the Chemical Society. The last cooperative venture to be described is an abortive attempt by the Chemical Society and the Faraday Society to produce a joint journal for physical chemistry. Although disappointing at the time, this stands as an interesting pointer to later developments. By way of background information it may be said that the general relations between the two Societies were good, although not especially close. The Faraday Society subscribed a mite toward the cost of running the Chemical Society's library, and passed on some of the foreign journals that it received in exchange arrangements. The latter Society allowed the former quite frequently to use its rooms for meetings. Nevertheless, as has previously been stressed, the Faraday Society felt itself to be a distant entity.

In 1931 the Publications Committee of the Chemical Society called together a number of physical chemists to consider the general question of the publication of physico-chemical papers. They suggested that there might possibly be cooperation with the Faraday Society. Therefore the former Society wrote to the latter telling of this meeting and saying that it had appointed a special committee to explore the position and to negotiate. As a temporary measure, the Faraday Society (Marlow) had asked Professors Allmand, Donnan and Porter, and Sir Robert Robertson to act as an informal Committee to meet and to talk with the Chemical Society's Committee. In November the Faraday Council endorsed his action and authorised further examination of the practicability of producing one journal for physical chemistry under the auspices of the two Societies, by a joint Sub-Committee on which Professors Garner and Porter, Robert Mond and Marlow would be its representatives.

The proposal went forward on the basis of joint meetings of the Finance Committees of the Chemical and Faraday Societies. The initial response was favourable but then the Chemical Society's Finance Committee sought 'a more equitable arrangement' than the original one. It also suggested delaying the operation until after 1934 when it intended to adopt a new format for all its journals. The Faraday Council unanimously decided they did not like the new format and trusted it would not be used for the new journal.

At the Annual General Meeting on 29th September 1933, the Faraday Secretary gave an admirable resumé of the situation then reached. The proposal was that members of either Society would receive the proposed journal. The General Discussions and the Reports thereon were, however, to be reserved entirely for members of the Faraday Society. The cooperation would not affect the identity of the Faraday Society and its members who belonged to the Chemical Society would benefit by a reduction in their joint subscription rates.

Reserving the Reports of the General Discussions solely for Faraday members would cause some loss in balance of the material for those only members of the Chemical Society. Also, unless the total membership of each of the Societies were to increase, they would both suffer a loss of revenue.

A detailed analysis of the probable printing costs for the two Societies served to emphasize the complexity involved in a decision on whether the enterprise would be profitable or not for each Society. Unprofitability would mean a financial loss. This would be covered only from increase of subscriptions, of membership, or from external subsidy. Adding to this the probability of increase in journal sizes enhanced the possible financial dangers. The Faraday Council seemed to regard the proposals with equanimity, even with satisfaction. The indications were that the Chemical Society gradually realized the difficulties.

In August 1933 the Faraday Council was informed that the Chemical Society had deferred further consideration until January 1934. In June 1934 this was extended until the embryonic Chemical Council, which was considering methods of cooperation between chemical organizations, had presented its report. That was the last that was heard of the journal proposals. A possible conclusion is that the abandonment of the suggestion was not so much due to a conflict of interests as to a reluctance, in a period of general price stability, to ask the consumer to pay a higher price for a better product.

Transactions undoubtedly enjoyed great scientific prosperity in the 1930s; but it is likely that some extra stimulus would have come from joint publication, so we must regret that this proved impracticable. We shall treat another, more particular aspect of joint publication in Section 3.15.

(d) The Society and Politics. A review of the Society's external activities would not be complete without some account of its relation to national and international politics. This is not because it had much political influence, indeed there is little to indicate that it had any, but because some of its reactions to events are relevant to our assessment of its general character.

Concerning national politics, little need be said. The Society played some part in recommending reforms in Patent Law, beginning in 1918 when it appointed a Committee of five members under the chairmanship of W. R. Bousfield, the K.C. and physical chemist who was an outstanding patent lawyer. No action is recorded, but in 1929 Council was informed that a Committee of the Federal Council (presumably of Pure and Applied Chemistry) had drafted a Patent Report and that the Honorary Treasurer (Dr. R. L. Mond) had been a member of it. Thereupon the Council hurriedly appointed him retrospectively as the Society's representative, so that the name of the Society could appear as one of those represented on the Committee. What influence the Report had is not known.

Publication, Finance, and General Administration 1918–1945 195

In 1921 the British Chemical Ware Manufacturers' Association asked the Council to support the Safeguarding of Industries Bill, in order to keep the chemical glassware industry in this country. Council's reply is given in its Report for 1921, and runs:–

'The Council was impressed with the necessity of investigators being able to obtain the very best optical and chemical ware, no matter what the source of supply. They therefore conveyed to the Association their opinion that in the national interest the industries of optical and chemical ware should be encouraged, but they felt that such encouragement should take the form of subsidies and not of a policy of exclusion of foreign ware except under licence.'

The Society's concern to protect the users' interests were well in line with its tradition.

Another expression of attitude came in reply to a letter from the Research Association of British Rubber Manufacturers asking the Council, in connection with the Rubber Industries Bill, to express its approval of the principle of continuing research in industry. The relevant Minute (2a, of 19th May 1933) ends:–

'There being a certain amount of opposition to compulsion in industry research schemes, no action was taken in the matter.'

The last domestic matter that deserves note is that from 1944 the Society had a representative on the Parliamentary and Scientific Committee, the first one being Dr. R. E. Slade.

Of much more interest are some of the Society's reactions to international affairs. One matter that caused heartsearching was what its attitude should be to German scientists after the emergence of the Third Reich, and then after war broke out in 1939. Council's conclusions on the first of these is best shown by quoting the whole of Minute 2 of the Council Meeting on 14th December 1938:

'The Council having considered correspondence passing between the Secretary and

(a) the Geschäftsführer of the Deutsche Bunsen-Gesellschaft between July and November, 1938, and

(b) Professors Andrade, Robinson and Wynne-Jones and Mr. E. Hatschek in November and December, 1938

affirmed:

(1) That the Faraday Society has the scientific objects set out in its

Rules and embraces members of all nationalities without regard to their political creed or racial origin, and

(2) That, whatever be the sympathies or antipathies of individual members of the Council, it is inconsistent with the objects of the Society that the Council should take cognizance of internal political action by the government of any other nation, and accordingly it was resolved that no action be taken in response to the requests contained in the correspondence (b) above that the Council should:–

(1) Pass a resolution commenting on the recent political events in Germany

(2) Bring to an end the mutual arrangements for exchange of membership between this society and the Deutsche Bunsen-Gesellschaft inaugurated in 1930; or

(3) Withdraw invitations to German nationals to attend or restrain attendance by German nationals at meetings of the Society.'

Hatschek wrote again, pointing out that Resolution 3 in Clause 2 did not deal with his question as to whether the Society proposed to continue to invite German guests. Minute 2a of 19th January 1939 gave a clear answer:

'The Council considered that their affirmation and resolution aforesaid makes it clear that, if the presence of German nationals is desirable for scientific reasons, other matters would not be considered when deciding whether to invite them.'

This response by Council is in the liberal tradition that was shown by the decision, in 1915, to leave the names of German members in a list that was about to be published (Section 1.5).

Soon after this, the United Kingdom and Germany were at war: by November the Society was having to consider the resignations caused by war conditions, which were mounting. Council directed the Secretary to keep a list of them and, when acknowledging them, to express the hope that when the war was ended membership would be resumed and, further, to say that if necessary the Rules should be amended so that such resignees on re-election should not have to pay a further entrance fee. At the same meeting the Secretary reported that he had had a letter from Professor E. Baur, of Zürich, written on behalf of the Deutsche Bunsen-Gesellschaft, to which he had replied. The President (Rideal) said that the Royal Society had as yet taken no steps to enable relations to be maintained with German scientific men, but he would notify the Secretary if any such were taken.

There were two further developments relating to foreign members. On 17th May 1940 [Minute 2(b)] Council decided, in reply to a letter from Dr. F. Gallais, a French member, that any member serving with the Allied Forces, who desired to retain his membership but not to receive the *Transactions*, should continue to be registered as a member and not have to pay the annual subscription during the war. The Council Report for 1941 re-states this policy but shows a hardening of attitude against enemy members. It runs: 'In 1939 the total number of members and student members was 895. Of these at least 80 have since become enemies and, their subscriptions remaining unpaid, their membership has ceased under Rule 15. In the exercise of their discretion under Rule 15 the Council have directed the retention on the roll of numbers, though their subscriptions remain unpaid, of members resident in territories for the time being occupied by the enemy.'

In May 1940 the Secretary informed Council that he had received the authority of the Board of Trade to import copies of the *Zeitschrift für Elektrochemie* for the use of members of the Faraday Society in this country, and to export the Society's *Transactions* to Germany, in each case through the agency of Professor Baur. The Council authorized the exchange provided that no more copies of *Transactions* should be exported than copies of the *Zeitschrift* were imported. Very characteristically, this proviso was imposed not for reasons of security but because the Council deemed it likely that at the cessation of hostilities any balance of Reichmarks due to the Society would be valueless, so it was prudent not to create one. In the same spirit the Finance Committee instructed the Secretary to collect £2 2s 0d p.a. from British members of the D.B.G. and to hold the money against a post-war settlement. At the same meeting the Secretary was instructed to 'cooperate so far as appeared possible with the Royal Society in the provision of microfilms of German scientific literature.' Such attempts to maintain some exchange seemed, however, to be blocked by the German authorities for in January 1941 the Secretary reported that although he had sent copies of the outstanding issues of *Transactions* for 1939 to Germany via Switzerland he had not received any acknowledgement of their receipt. Consequently he was holding a suitable number for post-war distribution. There are no further Minutes about exchanges of journals with Germany; and the only other matter relating to German and other enemy members is that in 1943 Council decided not to continue the membership of those whose subscriptions were in arrears, which presumably meant all of them.

A proposal was made to Council on 17th May 1940 that members of the Society should offer hospitality to children of their French, Dutch and Belgian colleagues. By the beginning of June 1940 the British troops had been evacuated from Dunkirk, and Continental Europe was sealed.

In May 1940 Council authorized an exchange of *Transactions* for the *Colloid Journal of the U.S.S.R. Society for Cultural Relations with Foreign Countries*: this and other exchanges were eventually (1943) arranged through the British Council. By August 1941 Russia was at war with Germany, so at the Council meeting that month Mr. Bury suggested that a letter of sympathy should be sent to the Russian Physical and Chemical Society. This was agreed, the President and Secretary being authorized to write it and issue it to the Press. It was published in the Minutes of the 1941 Annual General Meeting, and reads as follows:

'Mr. President,

'At their first meeting after the invasion of Russia, the Council of the Faraday Society desired us on their behalf and on that of the Society to send greetings to our colleagues in the Union of Soviet Socialist Republics. Our two countries proudly stand allied as guardians of the freedom of the world against wanton aggression. By restoring such freedom to the temporarily enslaved peoples of Europe and Asia we shall enable the work of our men of science to bless mankind. The work of Russian men of science has assuredly shewn to all the world what splendid results can be achieved.

Russia's heroic resistance against the ruthless aggressor is a source of immense pride to her ally, and will ever be remembered in history. We look forward with confidence to the day when the aggressor will be conquered and in the blessings of peace and freedom the members of our two Societies can meet in fraternal comradeship as allies in the peaceful quest of the laws of nature, just as now we are allies in the war on barbarian man.

We are, dear President and colleagues of the Physical and Chemical Society of the U.S.S.R.,

Fraternally yours,
(Signed) E. K. Rideal, *President*.
G. S. W. Marlow, *Secretary*.'

In August 1943 the Finance Committee considered a letter from the Association of Scientific Workers, signed by a number of distinguished British men of science, suggesting that the Faraday Society should make a donation to a fund for equipping the laboratory of a hospital in Stalingrad 'with a view to expressing the British people's gratitude to the defenders of that city for their great contribution to the final victory of the United Nations.' The Committee felt that it was not proper to use the Society's funds for such purposes, but suggested either that a half-page advertisement, inviting subscriptions from members, might be inserted in the *Transactions*, or that a leaflet supplied by the Joint Committee for Soviet Aid might be inserted in *Transactions*. Council did not approve of the former suggestion, which would involve expenditure of the Society's funds, but agreed to the latter.

On 10th December 1943 the Society received a long cable from the Soviet Scientist Anti-Fascist Committee. It was signed by 34 eminent Russian scientists and it described, in some detail, the mass killing of civilians (especially scientists and intellectuals), the ill-treatment of prisoners, and wanton destruction and looting by the German Army. The facts were testified to by several of the signatories who had followed the Russian Army in its westward advance that year. The cable asked the Society to participate in organizing protest meetings in all large cities against such actions. This was read to members of the Society who attended the Annual General Meeting held the following day, and they recommended that the following reply be sent:

'THE MEMBERS OF THE FARADAY SOCIETY HAVE RECEIVED YOUR TRAGIC MESSAGE WITH HORROR AND PROFOUND SYMPATHY. THE SOCIETY WILL DO ITS UTMOST TO BRING IT TO THE NOTICE OF ALL SCIENTISTS AND INTELLECTUALS OF THE UNITED NATIONS TO THE END THAT CIVILISATION SHALL BE PROTECTED AND JUSTICE BE METED OUT TO THE BARBARIANS.

(Signed) E. K. Rideal (President)
G. S. W. Marlow (Secretary)'

Members recommended further that the message and the reply be considered at the adjourned Annual General Meeting and that as much publicity as possible should be given, including publishing them in the *Transactions* and sending them to the Lord President of the Council (*Trans. Faraday Soc.*, 1943, **39**, 447–450). In January 1944 the Council directed that formal letters enclosing the Minutes of the two Annual General Meetings should be sent to the 'three senior Societies' and that they should be invited to deal with further publicity.

The following November the Publications and Finance Joint Committees recommended that the City of Petrograd (1913) and City of Moscow (1908) stock be written off next year's Balance Sheet and that the script be sent to the Russian Embassy: this was agreed at the A.G.M. which followed. Accordingly the Secretary wrote offering the abandonment of these securities, but by March 1945 he had had no reply. In August he reported that he had handed them over. The 1945 Balance Sheet states that in 1934 their total value was seven shillings and sixpence (37.5p).

The Society felt strongly about the deaths of any of its members that came from enemy action or from war service, and the President was asked to refer to them at Annual General Meetings. Fortunately there were very few. Dr. A. McKeown, a member of Council, was killed during a bombing raid on Liverpool in 1941. At the Annual General Meeting in 1945 the President referred to 'the gallant death of Professor Leif Trønstad in Norway after the accomplishment of his mission in April' and Donnan spoke of 'the murder by the enemy of the Society's distinguished member Professor Ernst Cohen.' (*Trans. Faraday Soc.*, 1945, **41**, 790). One tragic loss, mentioned in the Minutes of the 1941 Annual General Meeting, was that of Oliver Gatty. He came up to

Publication, Finance, and General Administration 1918–1945

Balliol College, Oxford, in 1926 as a Williams Exhibitioner; so he was a pupil of Sir Harold Hartley, F.R.S., and a contemporary of Professor Paul Ubbelohde, F.R.S. (President, 1963–64). He was briefly Tutor in Natural Science at Balliol; but his interests changed, so in 1934 he went to work with Professor Gray in the Department of Zoology at Cambridge. Later, he joined Rideal's group. In 1940 he was helping with an experiment involving the passing of superheated steam into oil, presumably as a means of generating a smoke cloud. The equipment overturned: he was very badly burnt, and he died (D. D. Eley, *Biog. Mem. Fellows R. Soc. London*, 1976, 386). The writer remembers him as having a dreamy manner but obviously being highly intelligent. He would probably have gone far. In the same Minutes there is mention of the death of Professor Sinhoven, in an air-raid on Helsinki in 1939. He was the first of the continental scientists to adopt Langmuir's concept of monolayers on surfaces as the seat of chemical action; and he applied this very successfully to elucidating the mechanism of the oxidation of carbon.

These instances, taken together, remind us of the very intense emotional stresses of that time, and show that the Society was not always unaffected by them. Nevertheless, for the most part the Society acted in a calm and balanced manner, and with a keen sense of self-preservation.

3.15. *TRANSACTIONS* AND CHANGE

In the first Section of this chapter it was said that one of the tasks ahead was to consider 'how well, by its publications, the Society served the field that it professed to cover, and especially how it reacted to the changes and developments therein.' Therefore we will first review the ways in which *Transactions* actually changed and then ask if these were the right ways, given the changes in the field to be covered.

The great improvement in the scientific quality of the papers, resulting from the advances in understanding, has already been stressed (Sections 3.10 and 3.11) and this is clearly a matter for satisfaction. If we were correct in suspecting that the greater emphasis on science and the reduced interest in industrial practice and economics made *Transactions* less palatable to industrialists then we must feel some regret; but with the pressure of new knowledge and ideas, and with the constraints within which the Society had to work forcing it to make a choice – either consciously or unconsciously – this was probably the way things had to go. We can only hope that the industrialists' needs were met by other journals.

The decision to give more attention to colloid chemistry, especially to its biological implications and the consequences thereof, has also been stressed and discussed at length in Section 2.06, where its wisdom was questioned. In one field where the Society once had good standing, namely metal science, a marked fall was noted in the number of General Discussions, and therefore in papers. The fact that the Hume–Rothery rules, one of the major clarifications of the era in the understanding of alloys, did not appear in *Transactions* but in a metallurgical journal supports the suggestion that this fall was coupled with a greater readiness by these journals to accept theoretical papers (Section 2.12); so the metal scientists' needs were met. There was also a fall of interest in the physico-chemical problems of agriculture: this interest had been evident in the early 1920s (Sections 2.1 and 2.3) but appears to have faded by 1934 (Council Minute 2(c) of 28th February 1934). It had not become of major importance to the Society.

We have already described the rise of a new field of activity which became known as 'chemical physics', and we have discussed the amount of attention given to it by the Society through its General Discussions (Section 2.12). Because of the eventual importance of this field, it is worth while to inquire how popular *Transactions* became as a medium for the publication of chemical physics papers. Before embarking on the more serious side of this task we may note that after the Society decided in 1919 to say that it supported 'physical chemistry' rather than the 'chemical physics' of 1903, the title page setting for *Transactions* was changed accordingly but the plate for the covers of the monthly parts was not changed, and for another thirty years it said 'chemical physics'.

One way of proceeding is to ask how successful *Transactions* was in attracting papers from typical workers in this field, between 1933 and 1940. In quantum theory, three of the most influential authors, namely

J. E. Lennard-Jones, N. F. Mott and W. G. Penney, published substantial papers, but these were mostly contributions to General Discussions. *Transactions* was certainly not their primary journal, but one to which they contributed more chemically oriented or more general papers. They published their major theoretical papers either in the *Proceedings of the Royal Society (A)* or in other physical journals. They published little in the *Journal of Chemical Physics*.

Of two infrared spectroscopists, one, G. B. B. M. Sutherland, who was trained as a physicist, published three papers in *Transactions* but most elsewhere, while H. W. Thompson, who was trained as a chemist, switched more to *Transactions* as his interest in spectroscopy increased, finding ready acceptance for his papers and willing help from Marlow. Of two workers who used electron diffraction for different purposes, G. I. Finch for the study of surfaces and the writer for the structure of small molecules, Finch published almost half of his papers in *Transactions* and the writer published all of his electron diffraction papers therein.

Of those who applied wave mechanics to reaction kinetics we may take M. Polanyi, M. G. Evans, and R. P. Bell as representative. After he settled in England Polanyi published more than half of his papers in *Transactions*, while Evans published about two thirds of his therein; so it was their prime journal. On the other hand, Bell published only about a quarter (11) of his papers in *Transactions* compared with almost a half in the *Journal of the Chemical Society*. It is interesting that C. N. Hinshelwood, who was a rather more classical kineticist, published only about one seventh (10) of his papers for that period in *Transactions*: he published twice as many in the *Proceedings of the Royal Society (A)* and still more in the *Journal of the Chemical Society*. W. E. Garner, who worked mainly on reactions of solids, but had interests in adsorption and in some gas reactions, like Bell published almost a quarter in *Transactions* and a half in the *Journal of the Chemical Society*. Evidently, therefore, much depended on individual loyalties.

The senior British X-ray crystallographers, the Braggs, published very little in *Transactions*. Of their pupils, W. T. Astbury published only two papers, both for General Discussions. Bernal published seven substantial papers, including several on general problems of the structure and properties of condensed phases; but mostly he published elsewhere, about a half as letters to *Nature*. Of the next generation, typified by E. G. Cox, D. M. Crowfoot Hodgkin, H. M. Powell and J. M. Robertson, only Dorothy Hodgkin published at all in *Transactions* and that was a joint paper with Bernal. They published mainly in the *Journal of the Chemical Society*, by letters to *Nature*, and also (especially J. M. Robertson) in the *Proceedings of the Royal Society (A)*.

We have already dealt with the change in the relation of the Society to metal science.

If we judge from the proportion of available chemical physics papers that was published in *Transactions* we must conclude that it was not the prime British journal for that field: indeed it was a good deal less important than the *Proceedings of the Royal Society (A)* for theoretical papers or the *Journal of the Chemical Society* or *Nature* for the publication of experimental results. Nevertheless we may claim that the Society's influence in this field was greater than the proportion of papers would indicate, because many of the papers were contributions to General Discussions and therefore, apart from their being of very high quality, they gained greatly in their effect by being grouped together. It is reasonable to claim this for the General Discussions on e.g. 'Molecular Spectra and Molecular Structure' (1929), on 'Dipole Moments' (1934), on 'The Structure of Metallic Coatings, Films and Surfaces' (1935), on 'Structure and Molecular Forces in (a) Pure Liquids, (b) Solutions' (1936) and on 'Reaction Kinetics' (1937). It will be noticed that these meetings came close together in time.

From these considerations we are bound to ask if it would have been better for the development of chemical physics in the United Kingdom if there had been a single, major primary journal for this field? No certain answer can be given but if we can judge from the stimulus that appeared to be given to American workers by the founding of the *Journal of Chemical Physics* in 1933 the answer would be 'Yes.' Furthermore, if we are right in claiming that the grouping together of papers in General Discussions increased their total effect, we must argue that similar benefit would have come from a wider bringing together of papers in one journal. If this is the answer then we must ask if it would have been possible to create such a British journal. From our sampling it would seem that *Transactions* alone would not have been large enough; and the only available option for increasing size would have been for the Chemical Society and the Faraday Society to have published jointly; but we have seen that the attempt to do this failed, probably because of practical difficulties, especially financial ones [Section 3.14(c)]. Even if this joint publication had gone forward it would not necessarily have helped chemical physics enough: a receptive attitude by the two Societies would have been necessary. The proposed joint journal might have promoted the chemical side but under-emphasised the physical one: it will be remembered that the *Journal of Chemical Physics* was founded by a physical society.

Historically, the Faraday Society ought to have had the right attitude, but it should be noted that on 25th February 1938 the Publications Committee agreed that 'papers containing lengthy theoretical discussions are to be discouraged more than hitherto'; although whether this restriction was on grounds of scientific particularism or was due merely to an urge to economize is not stated. It was a partial retreat from the policy enunciated in the euphoric days just after the receipt of the Bourke bequest (Section 3.8) which was that 'The Council anticipate that as a result of this legacy and of the economies and increased charges indicated above, the Society will in future be in a position to print all papers submitted which come up to its high standard of merit.' (Council Report, 1933). Therefore it is possible to find some consolation from the actual publishing facilities. Monolithic systems have great advantages if the ideas behind them are correct, but this does not always happen, so it may be better to have a multiple system which, although somewhat untidy and unable to muster such united resources, can at least prevent the mistaken policy of a single body having too adverse an effect. It depends upon circumstances, and these can change with time. We shall take up this story again in later chapters.

3.16. PEOPLE

It seems proper to conclude this Chapter in the same manner as Chapter 1, with a section essentially about people. First there is a mixed bag, viz. comments on a benefaction, on some individual benefactors, on the non-existence of honorary members, and on a celebration. Then, the brief sketches of those who served as Officers or as Ordinary Members of Council, given in Chapter 2, are supplemented for a few individuals, and in addition there are brief profiles of others who served the Society.

(a) The Spiers Memorial Lectures. The earliest corporate known benefaction to the Society was a fund, raised by friends, to commemorate Mr. F. S. Spiers, the first Secretary and Editor, who died in 1926 and whose vital work for the Society has been described in the two previous Chapters (see, especially, Section 2.4). In the Council Report for 1927 it was stated that £250 had been raised, and that as interest from this fund accumulated it would be used to provide a memorial lecture from time to time: it was hoped, every three years. The first of these Lectures was given by Sir Oliver Lodge, F.R.S. on the occasion of the 25th Anniversary Celebrations of the Society in 1928, of which more later. The second was given by the President of that time (1931), Sir Robert Mond, F.R.S., who spoke about Michael Faraday (Section 2.6); and the third by Sir William Bragg, F.R.S., in 1934, who talked about 'Molecular Planning' by which he meant the determination of the spatial distribution of electron density in a crystal. In June 1935 Council heard that Sherard Cowper-Coles wanted to publish a paper on 'Some Practical Applications of Faraday's Discoveries'. It may be remembered that he was one of the friends who founded the Society in 1903 (Section 1.1), that he resigned in 1906, apparently in a huff, rejoined in 1911, and then seems to have lapsed. Council deemed the material more suitable for a general lecture and felt that with some modification it might be given as a Spiers Memorial Lecture. It asked C. H. Desch to give an opinion: he approved, so Cowper-Coles was encouraged to go ahead. He joined the Society for the third time in October 1935. Unfortunately, he died in September 1936 but, since he had practically completed his Lecture, Council agreed to pay £20 as an honorarium to his widow. In October 1936 Council decided to look for a young, Jewish refugee from Germany as the fourth Lecturer, but eventually chose W. T. Astbury who was not Jewish, nor a refugee, nor especially young (39) although certainly he was very lively. He lectured

Publication, Finance, and General Administration 1918–1945   207

on 'X-Ray Adventures among the Proteins', in 1937. In May 1938 the Secretary commented in a Council meeting that these Lectures 'had never attracted audiences which were adequate either to the distinction of the Lecturer, or to perpetuate the memory of the late Mr. F. S. Spiers.' Therefore, Council eventually decided (September, 1938) that the Introductory Lecture at such General Discussions of the Society in London as Council thought fit should be made a Spiers Memorial Lecture, and an honorarium paid from the income of the Fund.'

(b) Sir Robert Hadfield, F.R.S. Of personal benefactors the Society had several. One of the earliest was Sir Robert Hadfield, F.R.S. He was a generous benefactor while he was President (1913–20) by his hospitality and his support of the exchange scheme with the American Electrochemical Society. As already related, in 1915, when the American society ran into financial difficulties, he offered to make an unspecified contribution to it if this would help maintain the exchange; and in 1916, when a new exchange scheme was being considered, he offered a £50 indemnity to the Faraday Society to cover the loss that it might incur therefrom (Sections 1.5 and 3.6). The Society lost £125 11s 0d in 1917. In 1927 he expressed concern to the Secretary (Marlow) that the British optical industry had failed to take advantage of the work which the Society had done on microscope design through its 1920 Discussion on this and related matters. He offered a prize of £50 (evidently his standard unit of giving, and worth about £1000 in 1980) 'to forward the project of rendering the country independent in the matter of Optical Instruments.' Marlow brought this to the attention of the Publication Committee because of the implied suggestion that another meeting should be held; but that body was inclined to think that the problem was more economic than scientific. This view was supported by subsequent events; indeed this particular problem was an early example of what has become recognized as a basic weakness of the British economy.

In 1933 Hadfield offered £50 'to meet in some measure the depletion of the Society's reserves' which was gratefully accepted by Council. Finally, in 1943, after his death in 1940 at the age of 82, the Finance Committee recorded the receipt of a legacy of another £50 with accrued interest.

(c) Sir Robert Mond, F.R.S. Sir Robert Mond, F.R.S., also was notable for his generous entertainment of the overseas guests of the Society at General Discussions, during his time as Treasurer and then as President (1917–32). Several such occasions are specifically mentioned in Council Minutes or elsewhere, but his obituary notice implies that this was his usual custom (*Trans. Faraday Soc.*, 1938, **34**, 1369). He also gave the Society £100 so that it could give 100 guineas (£105) to the Michael Faraday Centenary Celebrations in 1930; and in the same year, when he relinquished the Treasurership, he gave £348 to its capital fund to bring this up to £1000.

In 1933 he gave a complete set of *Transactions* to the University of Jerusalem, as it then was, and subscribed to maintain it. After his death, the Council looked around for another benefactor to take over the subscription but resolved, in 1939, that if none were found the Society would give a subscription as a memorial to him.

One last, rather odd matter, is that because he died in Paris in 1938 it was impossible in 1942 to get a death certificate for him: this was needed for the transfer of some stock inscribed in his name and Marlow's. Eventually a way round the difficulty was found.

(d) Lieutenant-Colonel J. J. Bourke, I.M.S., C.I.E. The most surprising and most generous benefactor of all was Lt. Col. John Joseph Bourke (Indian Medical Service, Companion of the Most Eminent Order of the Indian Empire). An obituary notice for him appears in *Transactions* (1934, **30**, as a preface) and this can be summarized as follows. He was an Irishman, born in 1865 in Kilkea, which is a small village about 37 miles to the south-west of Dublin, in County Kildare. He first studied experimental physics at what was then the Royal University of Ireland; later, less the Queen's University in Belfast, it became the National University. After he graduated as M.A. in 1886 he read medicine: in his final year (1890–91) he held a Final Exhibition at the then Catholic University of Ireland and studied at The Coombe which, in those days, was the maternity hospital serving a poor area of Dublin. Two years later he went to India as a Surgeon Lieutenant in the I.M.S. In 1897 he joined the Chemical Department of the I.M.S. and went to Madras. In 1898–99 he was acting Chemical Adviser to the Government of Bombay. Clearly, this aroused his interest in chemistry because he applied for two years' leave, in which he was willing to include a year's

furlough, to study the subject further; but this request was refused, so he had to be content with studying during his year's furlough, and he did this at the Royal College of Science. After a year in China, during the Boxer rebellion, he returned to India as Probationary Assay Master at the Bombay Mint, and a year later became Deputy Assay Master. It was probably his scientific interest in metals that led him to join the Faraday Society when it was formed about a year later, in 1903, for he wrote a joint paper on the sampling of coins and another one on 'The Amount of Kinetic Energy transformed at the moment of impact in a Screw-Coinage Press.' Later, in 1911, he became Assay Master at the Calcutta Mint from which post he retired, aged 48, in 1913. He started to fit out a laboratory with physical and chemical apparatus, but on 1st August 1914, three days before war broke out, he was directed to hold himself ready for immediate return to India, and he sailed ten days later to take up his old post at Calcutta. When he finally retired, in 1919, he had become Master. A letter to him from the Viceroy (Lord Chelmsford) thanks him warmly for his 'unflagging efforts when our currency difficulties seemed insurmountable.' In 1919 he was appointed C.I.E.

In his final retirement he went to live at Hertford. He seems to have been a lonely man. He occupied himself with experimental work in two rooms well equipped for physical work (mainly optical and electrical) and for chemical work, respectively. He had a number of 'excellent' X-ray tubes. He kept notebooks but he wrote largely in a form of shorthand, so they were not easy to read, and what he did in his experiments is not known. Also, he was a very knowledgeable collector of books – especially French books and scientific text books – and of Japanese sword guards and bronzes.

He met M. W. Travers (President, 1936–38) in 1907 when Travers was living, as Bourke was, in the Bombay Yacht Club Chambers: Travers was stricken with malaria, and Bourke treated him 'very kindly'. It has been suggested that it was this acquaintance that may have led to his making the Society his residuary legatee; but, while this may have helped, it should be remembered that he was already a Founder Member of the Society. The announcement of the legacy, at the beginning of Volume 34, concludes:

'The Society had no knowledge whatsoever of the intention of its generous benefactor, and Members will feel deep regret that the Society was not able during his lifetime to express thanks to its late member.'

The final value of Bourke's legacy was £15 490 12s 7d after his house, books and other effects had been sold. The securities, of total value £13 725 2s 9d, included two lots of Russian stock viz. 300 City of Moscow 1908, valued at 7/6, and 1000 City of Petrograd 1913, value nil [See Section 3.14(d)]. The annual income from this legacy was estimated to be £480, i.e. 3.1%. Bourke made this bequest 'with the request and in the confidence (but without imposing any legal obligation) that the Society will use the same for such purposes as the Society may consider best fitted to advance human knowledge.' Initially the Council decided that the best way of meeting his wishes was to use the money for the ordinary purposes of the Society, which meant mainly to subsidize *Transactions* (Council Report, 1934, p. 7 and Section 3.9).

In May 1938 Council considered generally 'proposals for perpetuating the memory of benefactors of the Society.' but with particular reference to that of Bourke. Various suggestions, some of them made previously, were reviewed, including the provision of a medal, making the introductory lecture at General Discussions a Bourke lecture or having a special Bourke lecture, or providing travelling expenses and entertainment for invited foreign guests. Difficulties were seen in the awarding of a medal. Experience with the Spiers Memorial Lecture was against providing further special lectures. Eventually, in July 1938, Council decided to adopt the third of these proposals and to set up a Bourke Fund with a minor part (ca. £3500) of the Bourke legacy for this purpose. In September, realizing that they had omitted to do anything about the Spiers Memorial Lecture, they decided (as already noted) to have this occasionally as the introductory lecture to a General Discussion.

(e) Michael Faraday. Michael Faraday might be regarded as a benefactor of the Society, albeit an involuntary one, for since he died 36 years before the Society was founded he could neither agree nor object when it was suggested, probably by Sherard Cowper-Coles, that his name should be used as a means of gaining respect (Section 1.1). In

Publication, Finance, and General Administration 1918–1945

1903 it may have been presumptuous to do this but, fortunately, the mixture of hope and cheek shown by the founding friends proved to be fully justified. It is possible that in the early days the Royal Institution may have had reservations about this upstart society, but it was represented at the Twenty-Fifth Anniversary Celebrations in 1928 by the Director, Sir William Bragg, O.M., F.R.S., who had contributed a far-sighted paper to the X-Ray Discussion in 1919 (Section 2.1). Certainly by the time Dr. Robert Mond became President (1930) relations were good: he gave his Spiers Memorial Lecture in the Royal Institution lecture theatre, and a General Discussion was held there in 1933. Sir Robert Robertson, one of the staunchest of Faraday men, was Treasurer of the Royal Institution in 1934, when Bragg gave his Spiers Memorial Lecture and gave it on his home ground.

For no known reason, J. C. Richardson, a Founder Member, wrote to the Council in October 1925 saying that there seemed to be no public memorial to Faraday, so Sir Robert Robertson agreed to find out from the Managers of the Royal Institution if this was so. He quickly learned that there was a memorial and statue at the Royal Institution, paid for by public subscription, and that the family had not wished a memorial to be erected in Westminster Abbey. By coincidence, at the same meeting, Council was given the addresses of two surviving grand-nieces of Faraday: they were living at Orleton, near Ludlow in Herefordshire.

Faraday was born in the Borough of Southwark, and when that Borough decided in 1927 to commemorate its illustrious son the Society gave it the current *Transactions* for a special collection that it was forming.

There is one further reference to be mentioned. It tells that in 1921 Lady Cecilia Kennedy gave the Society a portrait of Faraday, at the suggestion of Dr. Robert Knox who was eminent in the Röntgen Society. It was an engraving made by her father, George Richmond, R.A., in 1852 (Council Report, 1921). This seems to have been lost in the flames of 1941.

(f) Honorary Membership. In February, 1936, Council considered a form of recognition for those from whose outstanding services or great distinction the Society had benefitted, by creating a class of Honorary

Members. It also considered the possibility of remitting the annual subscriptions of those who were Founder Members and who, therefore, had been members for almost 33 years, or those who had paid not less than a certain number of subscriptions. Both suggestions were referred to the Finance Committee for its opinion. That Committee advised against changes of Rules to give effect to these proposals. It pointed out that the Council already had power to determine any sum, not exceeding £30, whereby a member might compound for Life Membership; and that if members of very long standing expressed an intention of resigning Council could, if it wished, accept their current subscription as a Life Composition Fee. In April 1936 Council accepted this advice, so the class of Honorary Members was not instituted and none is so designated in the 1938 list. Several senior members were forthwith offered Life Membership, including C. R. Darling, M. W. Travers and A. W. Porter. Darling and Travers accepted immediately; but there is no record that Porter ever did, and he does not appear as a member in the 1938 list. As the senior Vice-President of the Past-Presidents he had to retire from that Office when Sir Robert Mond joined the group in 1932, and he resigned as Chairman of the Publications Committee in December 1933; so he seems to have severed his connection with the Society which he had served so well for 20 years.

(g) *The Twenty-Fifth Anniversary.* Loosely related to benefaction is celebration, and something should be said about the 25th Anniversary Celebrations in 1928. Although the first meeting of the Council was on 4th February 1903 and the first meeting of the Society was on 30th June of that year the Celebration was not held until 9th November 1928. The Council did not start thinking about it until 16th May 1928. Naturally the Celebration included a meal which was a luncheon at the Hotel Jules in Jermyn Street. This was followed by the First Spiers Memorial Lecture given by Sir Oliver Lodge [Section 3.16(a)]. Originally it had been intended that this should be followed that evening by a 'Chemical Dinner', but this idea seems to have been dropped, though whether from organizational or gastronomic difficulties we do not know. There was a plethora, or at least a plenitude of speeches, for there were five at the luncheon and then seven at the

Lecture – plus the Lecture. Sir Robert Hadfield was the Chairman for the Lecture and he gave not merely an introduction of the Lecturer but an Address on 'The Faraday Society: Its Past, Present and Future Work.' which was printed and which gives a lively, idiosyncratic account of the Society's activities. T. M. Lowry was the President of the time.

The Society published a brief, 11-page history of the Society that was well balanced and full of useful facts: it was probably Marlow's work.

The representatives of learned societies were an impressive group. Among them may be mentioned Professor Ernst Cohen of Utrecht, who had lectured to the Society in 1911, Sir Richard Glazebrook, F.R.S., who represented the Royal Society on behalf of its President who could not come, and who had been a President of the Faraday Society 1911–13, Dr. Robert L. Mond, the Honorary Treasurer, representing the Royal Society of Edinburgh, Sir William Bragg, F.R.S., representing the Royal Institution, Professor Mauguin from the Société de Chimie Physique, Professor Kasimir Fajans representing the Deutsche Bunsen-Gesellschaft, Professor K. Moltkenhausen for the American Electrochemical Society, and Senator G. Marconi representing the National Research Council of Italy. Obviously it was a great occasion for rejoicing.

(h) *Dr. Henry Borns.* In the previous chapter (Section 2.8) we saw that several eminent members of the Society served it over long periods, in one capacity or another, sometimes for 20 or even 30 years with only short interruptions. It was one of the characteristics of the Society that it generated such loyalties: there were 'Faraday men' and, as we have seen, for better or worse they had great influence over its development.

Not all those who served long and faithfully were eminent in the worlds of research or affairs: there was one who was humbler and less grand, yet whom we may remember with gratitude and whom we may select as representative. He was Henry Borns, who has already been briefly mentioned (Section 1.2). Born in Germany, near Stettin, he came to this country some time in the late nineteenth century and joined the Society on its foundation in 1903, when he was 49 years old.

He never made a name in research: he earned a living as a scientific translator and journalist. He soon made an impression in the Society by urging, in 1906, that the Patent Abstracts which then were part of *Proceedings* should be dropped. Council had already, in October 1905, decided to discontinue the Science Abstracts provided for a fee by the I.E.E., so it sought members' views and consequently decided to continue with the Patent Abstracts. But Borns returned to the attack in 1909, when he was invited to attend at a Council meeting, evidently to argue further the case that he had made at the previous Annual General Meeting. This time he won and was, moreover, forthwith made a member of the Publications Committee. He went off that Committee when he became an ordinary member of Council in 1911, but returned in 1912 and remained on it for the rest of his life. As a professional writer he was an especially useful member.

During the 1914–18 war, in succession to Dr. Wilsmore, he represented the Society on the International Electrochemical Commission. In 1916 he offered to resign, no doubt because of the popular feeling against Germans then current, but Council asked him to continue. He translated German papers for the Nitrogen Committee (Section 1.6).

In 1919 Borns returned to the Council. Borns and Partington seemed to form an alliance. In 1921 the former moved and the latter seconded first that no reductions in the joint subscriptions under the Institute of Physics scheme should be allowed if paid after the end of March, and second that the need for the utmost brevity be emphasized in all papers for publication. Council agreed to both proposals.

There are two more references to Borns. In 1928 he became the Society's representative on the 'Technical Committee of Nomenclature and Symbols of the British Engineering Standards Association' – an odd job, suitable for a man of 74. Finally, in the 1930 Council Report, the Council recorded his death with deep regret.

(i) *Dr. Roland Edgar Slade.* Dr. R. E. Slade succeeded Emile Mond as Honorary Treasurer: he was an interesting man. As already stated in Section 2.4 he was a Manchester graduate: he held a Lectureship at Liverpool University for the years 1909–13 and so would have been appointed by Donnan who was Director of the Muspratt Laboratory for Physical Chemistry from 1906–13. He evidently came to University College, London, with Donnan and was nominally a Lecturer there from 1914–18; but for part of that time he served with the Royal Engineers, attaining the rank of Captain and being awarded a Military Cross. In 1920 he joined Brunner, Mond and Company, and so was absorbed into Imperial Chemical Industries Ltd. when the new Company was formed in 1926. A year later he was made Managing Director of the subsidiary company, Synthetic Ammonia and Nitrates Ltd., which was the fruit of the Society's wartime efforts (Section 1.6). In 1935 he was appointed Controller of Research for the Company and he held the post until 1945.

Slade had joined the Society in 1913; but it was not until 1926, when he was 40, that he became an ordinary member of Council and served for the usual stint of three years: this was probably F. A. Freeth's doing. He did not serve the Society again until he became a Vice-President at the Annual General Meeting held in September 1938. In the following January he was elected to fill the vacancy caused by Emile Mond's death. It seems unlikely that Slade was nominated in July with that succession clearly in mind, for Mond was well and attended the Council meeting: it was a few months later that he was known to be seriously ill.

There is an apparent anomaly in Slade's nomination as Honorary Treasurer and his acceptance of the duties, for by 1939 he must have been regarded by some as an iconoclast in the temple of Academe. This can be seen from the very interesting account of that era given by W. J. Reader in his book *Imperial Chemical Industries, Ltd., A History*, Vol. 2 (Oxford University Press, 1975).

There it can be found that Alfred Mond, the younger brother of Robert Mond and a principal founder of Messrs. I.C.I. Ltd. set up a Research Council in 1927 aimed to develop links between industry and the universities. [In the 1990s it will seem remarkable to some that a *British industrialist* initiated the funding of research in our universities: but it must be pointed out that few British industrialists have been born here only four years after their fathers left Germany. M.M.D.] This Council had funds from industry to support research in universities, which it did to an immediate total of £14 000. Slade greatly resented the influence in this I.C.I. Research Council of those he referred to as "the professors". When he became Controller of Research for I.C.I. in 1935, one of his first actions was to neuter the Council by depriving it of spending powers and so making it a purely advisory body. The Council seems to have faded away within a year or two.

In his key position relative to the scientific interests of Messrs. I.C.I., Slade was a very busy man, yet he found time to be Honorary Treasurer, thereby paying the Faraday Society a very handsome compliment.

(j) *George Stanley Withers Marlow*. In her recollections Bea Kornitzer says that in 1926, when Spiers died (Section 2.4), Marlow was Assistant Secretary at the Royal Institute of Chemistry. Through being an expert witness he had become interested in the Bar and started to read for it when he was nearing forty. He needed money so when Donnan, acting on a suggestion by Robertson, invited him to consider the post of Secretary and Editor to the Society and Secretary to the Institute of Physics presumably at an increased stipend, he was interested. As was related in Section 2.4 he was summoned by telephone for interview with the Council and appointed with a minimum of formality. Either shortly before this, or soon after, he was called to the Bar, for in January 1927 he reported that he had started to practise. In time he became very successful.

During the 1939–45 war Marlow was bombed out of the Temple and simultaneously the Society was out of 13 South Square in Gray's Inn (Section 3.13). He found temporary refuge at No. 1, Gray's Inn Square, and for the Society at No. 6 where, after a further blitz, it had to share space with a refugee firm of solicitors. During this phase Bea Kornitzer had to share an office with the solicitors' head clerk, and while that situation held Marlow could not visit her because the ethics of the Bar prohibited him from entering the office of a solicitor. Later, the Society became the main tenant of a ground floor set and had a room to itself, so Marlow could come over daily for a cup of tea and a talk.

Marlow's influence on the Society was much more than that of a highly competent, articulate, clear-minded, efficient Secretary and Editor, invaluable as that was. He had a geniality and an obvious,

Publication, Finance, and General Administration 1918–1945

genuine desire to be helpful to authors of papers and also to participants in the General Discussions which won him and the Society hosts of friends. He loved to have people dining and discussing together, so at dinner during Discussion Meetings he would encourage parties to form at High Table or to join in informal talk after dinner. He played a major part in generating the happy, family atmosphere of these meetings. He believed in the solvent properties of moderate doses of alcohol. During the war his resourcefulness was tested severely. Even just after it, in September 1945, concerning a reception for overseas guests during a Discussion Meeting, he wrote to Donnan '. . . and we are trying to collect alcohol. I've promised 1 gin (1 rum and 3 Algerian to make some hectic concoction) and, if I can get it, 1 whisky. Garner 2 Burgundys, and if I can get anything else so much the better.' Marlow had had to bear several severe bereavements. His father died in middle age. His young brother, who had enlisted illegally in the Army at the age of 16 during the first world war, was killed at Gallipoli. His only grandson died from a seemingly trivial accident; and his only son, Lieutenant Howard Marlow, was killed in action with the First Army in North Africa in 1943. Bea can remember his head clerk coming over to break the news. Marlow had, in rare degree, the power of transforming personal grief into an urge to help others, although Bea thinks that he never recovered completely from the loss of his son.

Only once is Marlow known to have lapsed from his customary efficiency. The Council Minutes of 30th March 1935 report that the Westminster Bank, which held certain securities for the Society, had pointed out that the Secretary should be re-elected annually under the Society's Rule no. 22. Marlow had not been re-elected for years! He had omitted to include this business in the agenda. Needless to say the omission was immediately remedied and thereafter he was re-elected almost every year.

(k) *The Assistant Secretaries.* The first assistant secretary of whom we have any record is Miss Parsons. She is mentioned in the Minutes of the newly constituted Finance Committee of 18th January 1922, as receiving one quarter of an annual salary of £91. F. S. Spiers at that time was receiving £150 p.a. The Council Minutes for 10th July 1922 show that Miss Parsons attended the meeting on behalf of Spiers, possibly because he was beginning to have trouble with his health. After Marlow began to practise at the Bar in 1927 he visited the Society's office morning and afternoon during Law Terms, and Miss Parsons could make appointments for him if necessary.

It would be interesting to know how salaries changed from 1922 onwards, but after 1923 neither the Council Reports nor the Finance Committee Minutes give any detail: only the totals of editorial, secretarial and office expenses are given. These increased from about £240–290 p.a. in 1918 to about £425 p.a. by the early 1920s, to £600 p.a. by 1930, and then more slowly to about £740 p.a. by 1938. The last mention of a salary for the Secretary was in 1934 when it was increased to £650 p.a., this sum to include the salary of an assistant secretary.

Sometime in 1928 Miss Gertrude Wakeley succeeded Miss Parsons. In turn she resigned in April 1941, in order to marry: she was entertained to luncheon by Council on 18th April and given a cheque as a wedding present. Her friend and successor, Miss Beatrice (Bea) Kornitzer had heard many stories about the Society from her, but Bea's own experience of arriving to find the office just recovering from a fire bomb attack, as narrated in Section 3.13, went far beyond any of them. In such unpropitious but challenging circumstances Bea began her 30-year stint of service with the Society.

As has been explained already [Section 3.14(a)] Bea was Marlow's employee. He paid her and was reimbursed by the Society. She recalls, somewhat ruefully, that 'he at first offered me £4 a week, which was a little more than I had been earning, but after a month he increased my salary to £5 which was the same as my predecessor had been getting after 13 years' service. (You must remember that compared with the astronomical salaries girls receive today, even well trained secretaries were an exploited section of the business and professional world.) Yet, even compared with other friends, I could not say that I was well paid for what I did.'[1]

---

[1] [Qualifications to the assessment of both L.E.S. and Bea should be considered. 'Astronomical' is overstated for the salaries of well trained secretaries who contribute so largely to the functioning of others often paid ten times their secretary's earnings. And £5 per week in 1941 was more than a 1st Class Honours scientist would earn in a secondary school teaching up to Form VI at scholarship level. M.M.D.]

During the war, because Marlow's clerks were away on war service, Bea had to deal with his legal work as well as coping with the complex business of the Society, including dealing with subscriptions. After the daily discussion with Marlow over tea she was left pretty much on her own 'but on the whole it suited me very well. I liked the responsibility and enjoyed the work until it got a bit too much and I needed an assistant. I remember, Mr. Marlow saying reluctantly – "What shall I tell Council?" So I left him to cope and went away on holiday leaving the office to him and Mrs. Marlow who had been pressed into service. In no time at all I had a junior.' A Council Minute for 29th August, 1945, records that a junior assistant was needed for Miss Kornitzer and that a salary of up to £150 p.a. was authorised. Bea adds 'there were never more than two people apart from myself at any one time: we really ran the office on a shoe-string.'

A consequence of Bea's being employed by Marlow and not by the Society was that when the Society became associated with the Chemical Council in 1945 it did not need to come within the pension scheme for officers that the Council might formulate [Section 3.14(a)], and in fact no pension provision at all was made for her until much later. One can but feel that the Society's concern to do things on a shoe-string occasionally led to some insensitivity. Yet, so far as Marlow was concerned, he asked no more than he was himself willing to give; and so he had Bea's great respect – 'I liked working for Mr. Marlow. Who could not? He was so very human and kind.'

Bea's feelings about the Society are best expressed in her own words:
'If I were to write a history of the Faraday Society from my own standpoint, that is of an employee, seen from the close familiarity of 30 years, I would say that what impressed me most was that when any major decision had to be made Council almost unanimously considered the scientific reputation and integrity of the Society before its financial profit. It seemed to me that they had a right sense of values and I was proud to work for them.

Although at times it might have seemed, I dare say, to some of the members that Council lacked an influx of new blood and the society was ruled by the "old guard", ringing the changes, in fact the needs of the Society in preserving a continuity were always the first consideration. Two other things impressed me: the warm, informal atmosphere of the General Discussions which encouraged scientists, whether members or not, to meet together and discuss their problems with a great lack of regimentation and protocol. I always thought of them as "parties" with a serious purpose, and look back with great pleasure on the many friends I made through them, both at home and abroad. This happy atmosphere of the General Discussions was in a great measure due to Mr. Marlow . . . who in spite of his informal approach was always completely in charge of the situation.

The other impression that remains with me is the truly international flavour of the Society and its meetings. From the first, reading through the early Minutes, . . . it seems the Society sought links with like-minded bodies overseas and was sought by them: . . . At the time of amalgamation in 1971 the overseas membership was 1233 as against 1137 home members.'.

In a postscript to a letter written on 26th July 1982, Bea says 'And I must add that it was a very great pleasure to work not only for Council and the Officers of the Society but also the membership. I have some *very* happy memories – and no unpleasant ones. You were all so nice to me.'.

Bea retired in March, 1972. She died on the second day of December, 1987.

PART II

1945–1971

MANSEL DAVIES

I
Accommodation

THANKS TO THE Marlow connection, the Faraday Society office and meetings were located at 6 Gray's Inn Road in 1945, when the premises were still leased in the name of the Honorary Secretary at £100 p.a. Marlow died in March 1948. F. C. Tompkins was already in place as Editor, and at the Council meeting of June 1948 he was appointed the new Secretary. The first sign of possible changes appears as a minute for 13th July 1949 when: 'The Secretary reported that the Society having outgrown its present accommodation, an application had been made to the Society of Chemical Industry for rooms in its new building at 61 Green Street, Park Lane, W1.' In the meantime, the lease with Gray's Inn was held in The Honorary Treasurer's name (R. E. Slade). At the next Council meeting, 21st September 1949, Slade undertook to sign a new agreement with Gray's Inn at the increased rental of £215 p.a. At the same meeting it was agreed that from 1st October 1949, Wansborough-Jones would take over as the new Honorary Treasurer.

No further reference in this context to the Society of Chemical Industry occurs: rather, there are mentions of a Science Centre in 1950 and 1951 which came to nothing (*Supplement to the Record of the Royal Society, 1940–1989*, pp. 8–9), and in November 1958, the renewal of the Gray's Inn lease at the increased but still 'very reasonable' figure of £285 p.a. was agreed.

In June 1962 the Biochemical Society had written *à propos* the aim of the British Science Foundation to help smaller Societies with accommodation and publishing. At the same time, the Secretary reported that The Royal Institution in Albemarle Street might now be able to offer new accommodation for the Faraday Society. It was decided to 'consider with sympathy' the former possibility and also to explore the latter.

Discussion with The Royal Institution on the surplus accommodation they anticipated at 19 Albemarle Street is recorded for November 1962 and March 1964. The Faraday Society emphasised its inability to pay a strictly commercial rent for larger premises in Albemarle Street. An even more transient interest occurs as a mention, in a message from the President of the Royal Society, of a building – The Hammersmith Heights Scheme – being planned by City Hall Properties.

At the same March 1964 Council meeting it was revealed that with the move of The Royal Society to Carlton House Terrace, accommodation for other scientific societies could become available at Burlington House. This stimulated considerable discussion. The President (Ubbelohde) emphasized the essential activities of the Faraday Society as being the organisation of the General Discussions and the publication of the *Transactions*. An application on that basis should be submitted to the Ministry of Public Buildings and Works. The Chemical Society would be informed of this independent submission.

By June 1964 the Office of Works had replied that the needs of the Societies already at Burlington House would leave no room for others. Accordingly, further consideration of The Royal Institution's possibility led to the definition of the needs of the Faraday Society. There were listed:

1. Security of tenure.
2. A rent, of say, £600 p.a.
3. Office space of 2000 sq. ft. minimum.
4. Complete independence.

A panel of the President, Treasurer and Rideal was appointed for further negotiation.

There were no developments, for it was reported in March 1966, that the Royal Institution was unlikely to commence any rebuilding in less than three years. In June 1966 the Gray's Inn lease was renewed for seven years at £525 p.a. However, the situation was inherently unstable. With an increase in staff arising from activities on additional Discussions being planned, it appeared that Council meetings would have to be held outside 6 Gray's Inn Square; the office would have taken over the Council Room.

In March 1967 a minute explains that larger premises were available at 19/20 Albemarle Street at three times the current rental: but there was no urgency even for some years.

Accommodation

At last, in November 1967, the President (Bawn) and Treasurer (Musgrave) were able to report their inspection of accommodation at The Royal Institution, where substantially more space would be available to the Society, on a lease for up to 14 years at three or four times the current rent (£525). At the same Council meeting Porter, who was then Director of The Royal Institution, was able to extend their information. No. 18 Albemarle Street was also to be incorporated with Nos. 19/20 into the main building, and the Faraday Society would have to decide fairly quickly on its future, and a rental of around £2750 p.a.

All this is detailed in the minute having the heading Future of the Society, and is followed by the President reporting a confidential talk with (Sir Harry) Melville. As President of The Chemical Society (he was a Past President of the Faraday Society) he had put forward some ideas on the federation of the major U.K. Chemical Societies. The Royal Institute of Chemistry had already agreed, in principle, on amalgamation with The Chemical Society. This moved the discussion of the future of the Faraday Society to a higher or more fundamental plane than the question of accommodation. The Society, as such, was to maintain its office at 6 Gray's Inn Square, but the Council Room having been taken over, the Council met from March 1967 to June 1968 at The Royal Institution, where an appropriate room was made available. On 14th November 1968 The Faraday Council met 'in the rooms of The Chemical Society, Burlington House'. This continued up to and after the amalgamation in January 1972.

2

Early Post-war Years

IT WAS AT the Council meeting on 29th August 1945 that signs of post-war developments became obvious. The Secretary (Marlow) reported that the increase in membership which was taking place, and additional activties, necessitated the appointment of a junior assistant for Beatrice Kornitzer. At this date she was the only full-time staff member of the Faraday Society: Marlow was the executive officer responsible for the continuity of activities, which he achieved by dropping into the office, most usually at tea-time. It was agreed that a sum, up to £150 p.a., should be added to the Secretary's honorarium to pay for the new assistant.

This *modus operandi* serves to show that the Faraday Society minimised its own profile. It existed largely in the person of its Secretary. That this was not an uncommon concept can be illustrated. Some years previously the author watched a member (then Dr. Delia Simpson, later Agar) writing her cheque for the annual subscription to The Marlow Society before correcting it for The Faraday Society. A Miss Irene Coles was appointed as an assistant in the office at £2 10s (£2.50) per week.

The replacement of Rideal as President needed a nomination to be made at the next A.G.M. Three names were proposed and seconded within the Council by the eight members, with the President and the Secretary, who were present. Those nominated were Garner, Hinshelwood and Hartley: of these, the first two received four votes each, and Hartley two. On a second vote, Garner had six and Hinshelwood four.

This event was noteworthy. Usually the officers had held consultations, probably including Past Presidents who remained Vice-Presidents, and arrived at the Council meeting with their nominations. Other names could, of course, be put forward by members of Council, and if seconded, would lead to a vote. Rarely did this occur. In 1945 the choice of Garner probably reflected his more active participation in the Society's affairs; he was in any case a Vice-President. It will be noted that both President and Secretary voted on the nominations.

Early Post-war Years

At this meeting also the Secretary reported that he had handed over to the U.S.S.R. Embassy and the Moscow Novodny Bank Ltd, pre-1917 Russian securities which the Society was prepared to abandon. In this way it was recognising the disappearance of the Imperial Russian regime and the worthlessness of its undertakings.

What must now appear as an extraordinary proposal was presented to the Council in February 1946. Professor M. G. Evans had been in The Netherlands and reported on suggestions being considered there for a reorganisation of West European chemical publications. Firstly, a Western European Journal of Physical Chemistry was being envisaged to incorporate the contributions previously appearing in *Zeitschrift für physikalische Chemie.*, *Kolloid Zeitschrift*, *Zeitscrift für Elektrochemie*, etc. Even more inclusive a proposal from Dr. de Boer (of Messrs. Rosbaud Phillips, Eindhoven) and Dr. Rosbaud involved bringing in three major national journals, essentially of organic chemistry: *Berichte der Deutsche Chemische Gesellschaft* (German), *Recueil des Travaux de Chimie des Pay Bas* (Belgian–Dutch) and *Helvetica Chimica Acta* (Swiss). Dr. de Boer was specific: 'that the Faraday Society, as the only already international organisation [having one full-time secretary and a typist!] should take over everything in the way of physical chemistry, chemical physics and colloids'. Further, 'The Secretary (Marlow) favoured such a suggestion, provided that no diminution in the standard of the *Transactions of the Faraday Society* followed, and he had discussed the whole problem with the President (Garner) and Professor Rideal, who thought that it should be given consideration by Council.'

Many readers may find this proposition to the Faraday Society almost hilariously optimistic. It reflects in dramatic fashion the great status accorded in Europe to the U.K. and all its institutions during the early post-war years. *Facilis descensus in averno.*

After considerable discussion, Council decided the matter was of sufficient interest to warrant sympathetic consideration by a sub-committee. To this were appointed The President, Slade, Rideal, M. G. Evans, Allmand, Polanyi and the Secretary: with power to co-opt any Dutch, Danish, Scandinavian and Belgian members in this country. It is a matter for comment that no French members were envisaged: nor was the *Bulletin de la Société de Chimie de France* mentioned in the first place. This probably arose because the French chemical journals were expected to continue as in 1939. It is a remarkable comment on the status of the Faraday Society that so many other European groups thought of it as the best nucleus for the formation of a Western European Chemical Journal.

No further mention of the sub-committee nor of the inflated proposal they were to consider sympathetically appears in the minutes. In the same direction, however, is a comment from Marlow (14th January 1947). There were distinct signs of the *Transactions* developing into an international journal of physical chemistry with the accretion of papers which would earlier have gone to German scientific journals. This led to the attachment to papers in the *Transactions* of abstracts in French and German, a practice, however, almost immediately abandoned at the end of 1947.

Senior Faraday Society members in Europe had been questioned. They are reported as saying they saw no need for their national groups to be represented on the Editorial Board of the *Transactions*, but arrangements to publish papers in French and German would be welcomed. This larger suggestion was never pursued. It is easy to see the problems it would have given in the Gray's Inn Office. The same mainland contacts had indicated they thought that French and German summaries were of little practical value. This was probably not a representative opinion. It almost certainly came from senior colleagues whose knowledge of English was excellent.

The war had disrupted circulation of the *Transactions* to members and institutes on mainland Europe. Appeals came for the provision of back-issues: currency difficulties meant that these should, if possible, be gifts. Accordingly, information was sent of this need to U.K. members. To this appeal the response was pathetic: one incomplete run of the *Transactions* arrived (16th July 1947). However, Council itself had made one specific arrangement (23rd April 1947). Thanks in part to Allmand's representation, two sets of the *Transactions*, 'so far as was available', would be sent to Warsaw: one to Professor Zawadski in person, payment for which would be made when it could conveniently be transmitted, and one to the University of Warsaw, where Zawadski was located. This seems the more remarkable when the same minute reveals that the Polytechnic School [University] of Warsaw had requested the Society's help to re-establish its library. Be it noted that *á propos* the *Transactions*, Polish universities for the 1939–45 years were in no very different a position from those of many other European countries, which included the universities of Budapest, Prague and Vienna, not to mention those of Germany itself. Poland, of course, had been an ally: but there is no mention of concern for the universities in Norway, Denmark, the Netherlands, France or Italy. Possibly this arose because these countries had adequate funds available in London to ensure their chosen universities, at least, receiving the *Transactions*.

In the July 1947 meeting of Council it was decided to accept membership of ex-enemy nationals when their country signed a peace treaty with Britain. At the same time an exchange of publications, including the *Transactions*, was approved to take place via the Control Commission in Germany. Already in 1946, arrangements were suggested for exchanges between the Faraday Society and the Academy of Sciences of the U.S.S.R. Quasi-membership exchange was possible for those British members interested in acquiring Russian journals. They could choose the latter to the approximate value of the then Faraday Society annual subscription: £1 16s (£1.80). There is no indication how this suggestion developed.

Post-war adjustments were not the only important changes in 1945–47. Marlow indicated (1st August 1946) that he found he could no longer devote sufficient time to the Society's business. One aspect of this he revealed when he proposed the Society should consider appointing an assistant editor to help him, and that this appointee might eventually take over the editorial duties and become the 'papers' secretary'. To consider this matter and the work in the Society's office, a sub-committee consisting of the President (Garner), the Treasurer (Slade), the chairman of the Publications Committee (Allmand) and the Secretary (Marlow) was appointed. As a measure of the Society's expansion at this time, the minutes of the same meeting reported 125 applications for membership during the preceding three months: the overseas names included Dervichian, Magat and Böttcher, with one from Finland and three from Italy, these four latter applications to be held over until the countries had concluded peace treaties with Britain.

Three months later (23rd October 1946), a report from the sub-committee had been circulated to members of Council who then appointed: as Honorary Secretary and Editor, Mr. G. S. W. Marlow: as Assistant Editor, Dr. F. C. Tompkins, at a salary of £400 p.a.: as Assistant Secretary, Miss B. E. Kornitzer at a salary of £300 p.a.; to rise by annual increments of £25 to £400 p.a. Miss Kornitzer was also authorised to engage a suitable typist–clerk at a salary of £4 per week, rising by annual increments of 5s per week to £5 per week.

The minutes offer no comment on Dr. Tompkins and his suitability for the assistant editorship. Undoubtedly the sub-committee had gone into this and other questions carefully and details were probably present in their report, which is not appended to the minutes.

What appears above was not the end of the changes. Firstly, at the next Council meeting (14th January 1947) Marlow reported an error. Bea's salary had not been increased as the sub-committee had intended: it should have run from £400 in four steps to £500. This was generous, being half-way up a university lecturer's salary scale; but she was a key element in the Society's activities.

The Assistant Editor was incorporated as a (non-voting?) member of Council: he appears as such in the minutes for 23rd April 1947. However, what is remarkable about this procedure is that for the meeting during which these appointments were made, the minutes open with: 'The Chairman (Findlay, later replaced by a late-arriving President Garner) welcomed the new member of Council and Dr. F. C. Tompkins the Assistant Editor'. This implies that the officers accepted as a foregone conclusion that Council would approve of their suggestions. Even those familiar with the Faraday Society being run from on high will be surprised by this. Then, without any recorded comment, he appears at the next meeting (16th July 1947) as Editor. Tompkins was clearly running into place successfully, as at the July meeting it is reported that he had been asked by the Publications Committee to approach the Paper Controller at Reading personally on the difficulty with regard to paper supplies for the *Transactions*.

Early Post-war Years

THE LOSS OF MARLOW

Only one Council meeting (31st October 1947) occurred with the new regime fully installed before that on 18th March 1948, from the minutes of which the first item must be quoted in full:

'The President in reporting the death of the Honorary Secretary, (on 3rd March 1948), said that the Council of the Faraday Society had learned with profound regret of the untimely death of its Honorary Secretary, Mr. G. S. W. Marlow. For twenty years, during which the membership increased from about 350 to over 1400, Mr. Marlow carried out almost single-handed the manifold duties of Secretary and Editor with never-failing zeal and effectiveness, as also with a tact and a personal touch which was peculiar to him'.

Members of the Society and visitors at its scientific meetings would best remember him by the part he played in the organisation and conduct of its General Discussions. Their actual form in later years was largely due to his initiative, and in the planning and carrying through of the details he found full scope for the exercise of those outstanding social qualities which made him with everyone so deservedly popular a figure. In all that Mr. Marlow did, he displayed an almost eager personal interest in the affairs of the Society and of its members. In return he received, not only popularity, but also, to a unique degree, the personal regard and affection of its members. The Council of the Society offered to Mrs. Marlow and to other members of his family its deepest sympathy, and instructed the Assistant Secretary to send a copy of the above minute to Mrs. Marlow'.

As the President's statement makes clear, the loss of Marlow was a shattering blow. Without repeating what is in Garner's tribute or what L.E.S. has already written (Section 3.16) perhaps one specific point can be made. Marlow is rightly credited with 'the organisation and conduct of the General Discussions.' This was the hallmark of the Faraday Society. Members naturally came to accept, and even to expect, their effectiveness. Overseas visitors were invariably impressed, and when views and comments on the Society were sought, this has been the most recurrent feature on which praise has been lavished. Their success was a unique memorial to Marlow.

Rideal reported that he had sent an obituary to *The Times*, and F. I. G. Rawlins would prepare one for *Nature*. Donnan was to write an appreciation for the *Transactions*. Rather than quote from these notices, brief extracts are given from letters to be found in the Faraday Society's archives.

Firstly, a letter to Beatrice Kornitzer from Marlow (17th February 1948):

'Dear Miss Kornitzer
Please apologize [to the Council and others].
I've been off colour for a week and tomorrow I'm going to King's [Medical School Hospital] so that my innards may be explored: Probably nothing serious but an obvious precaution – although a confounded nuisance. Best of luck and by this time next week I hope to be *mens* though perhaps not quite *corporis sannis*.'

From that most distinguished of physical chemists, Professor Michael Polanyi: (12th March 1948):

'My dear Allmand,
...I was very much moved by the passing of a friend and a lovable public figure. He was the first man I talked to on my first visit to England twenty years ago, and every time I met him since I felt the better for the contact with his wonderful good nature and purity of purpose.'

From Sir John Lennard-Jones (12th March 1948), a professor of few words:

'It has been a privilege to know Marlow and to enjoy his warm and cheerful outlook. His tact and personality have been of inestimable value to the Faraday Society and much of its growth and success are due to his personal qualities. No man could be more popular with his colleagues or with the general body of members than Marlow. He was the life and soul of the Faraday Discussions, and he contributed greatly to their success by his presence and generous hospitality.'

Early Post-war Years

From Professor Michel Magat, the leader of French physical chemists:

'J'ai eu le plaisir et l'honneur d'assister depuis 1936 à de nombreux Congrès de la Faraday Society, et j'ai pu pleinement apprécier le rôle que jouait M. Marlow dans la vie de votre Société et, je dirai même, dans la vie scientifique européenne: ainsi que le caractére exceptionnel de cet homme.'

A committee was immediately appointed to consider matters arising from Marlow's death, including the appointment of his successor, the future accommodation (the Society was located in Gray's Inn because Marlow as a member held the rooms on its behalf), and the form of a memorial. This important committee had Slade as chairman: Garner, Goodeve, Rideal, A. E. Alexander and Guggenheim as members.

The committee reported its findings at the next Council meeting (3rd June 1948). After some discussion (in the presence of the Editor and of the Assistant Secretary) Council decided

(i) To appoint Dr. F. C. Tompkins to be (Honorary) Secretary as well as paid Editor and to allow him up to £250 per annum for a secretary to handle his Faraday Society work, in addition to £400 p.a. as Editor. The (Honorary) was removed in a later correction of the minutes.
(ii) To rent from Gray's Inn (in its own name) the rooms then occupied at a rental of £100 per annum. One of the rooms was already being used by Mr. C. W. Baker and it would be sublet to him.
(iii) To thank Miss Kornitzer for the excellent work she had done for the Society and to award her a bonus of £25.

As a memorial to Marlow, the first proposal was the purchase of loving cups to be used at the dinners which were part of the General Discussions. An appeal went to the Society membership: 233 responded with a total of £111.35. Later it was resolved that one good cup be purchased and that it be placed before the President when he proposed the toast to the memory of Marlow at Guest Night Dinners. As reported in June 1950 a cup costing £105 was purchased. This purchase shows that the Council did not use any of the Society's funds to add to the total collected. The disappointing collection total could have been because the members at large were not enthusiastic about a silver cup they would, at best, rarely see.

THE APPOINTMENT OF TOMPKINS

It is relevant to emphasize that what has appeared above is all that the Faraday Council Minutes record of the appointment of Tompkins to the most influential position in the Society: for the remainder of its life a great deal depended upon the Secretary, his views and his *modi operandorum*.

Bare facts on Tompkins are available from *Who's Who*: Frederick Clifford Tompkins born 29th August 1910: educated at Yeovil (Somerset) School and Bristol University; 1934–37, lecturer King's College, London; 1937–46 lecturer/senior lecturer Natal University, S.A.; 1947 reader in physical chemistry and 1959–77 professor at Imperial College, London; 1950–77 editor and secretary of the Faraday Society; F.R.S. in 1955.

This very acceptable academic record offers no insight into his special status for the Faraday Society appointment. But the revealing clue is 'Bristol'. There he graduated and did research with W. E. Garner, and when Marlow died, Garner was President of the Faraday Society. No advertisement nor public announcement of the position was deemed necessary. The President must have persuaded several senior colleagues on Council that he had a suitable appointee: no questions, explanation or discussion were recorded.

An important memory came relating to Tompkins' first introduction to office with the Faraday Society. Geoffrey Gee, who was already a member of Council at the relevant time, wrote: '... my strong impression is that Council had for some time been considering the situation which would arise when Marlow retired. He had for so long given a continuity which had immensely strengthened the Society, and greatly helped successive Presidents. No one could replace his highly personal contribution, but the sub-committee thought they had identified a possible successor in Tompkins. They therefore proposed that he be appointed assistant editor without any committment to subsequent promotion. I do not know how much of this is clear from earlier minutes, but I am confident it is correct.'

There is no doubt but that Tompkins operated in a very efficient way. The activities of the Faraday Society were dealt with each Monday: the remainder of the week was spent at Imperial College. There he led a successful research group studying adsorption and interactions, including catalysis, at inorganic solid surfaces.

To Faraday members he was 'Tompkins': to his close friends and research associates he was 'Tommy'. He was a dry rather than affable character, with a distinct, if limited sense of humour: quite approachable and always helpful on Faraday Society matters. From 1962 his home was at Portsmouth (where he was a very keen gardener), but he retained a flat in central London where he usually spent from Monday to Friday. Some of his heavy editorial duties could be done during the London–Portsmouth train journeys. That he was meticulously careful and inculcated care in his research students is suggested by the fact that he did not allow them to wear laboratory coats at the work bench: they were unlikely to cultivate messy habits.

3
Publications and Finance

THE EARLIER CHAPTERS have made clear that these two aspects of the Faraday Society's activities were so closely interdependent that they must be considered jointly. This is a consequence of the simple fact that, for most of its existence, the principal business of the Society was that of publishing both the papers approved for the *Transactions* and those contributed at the Society's Discussions. The costs of printing made it inevitable that most of the Society's income would be expended in covering them.

This inter-relation was all along clear in the Council minutes. Whilst there were separate sub-committees dealing with Publications – their control via referees, their organization and printing, etc. – and Finance, which examined the income from the membership and from such capital reserves as were invested, and the expenses incurred in all aspects of the Society's activities, there were not infrequent joint meetings of the two when current problems were considered and suitably resolved. On this relation L.E.S. had made many notes. There are some pages on *A theory of printing costs* in which this key item is related to the number of members and the number of papers and pages in the *Transactions*. The present account makes no attempt to present such an analysis: only what appear to be the main facts will be presented. Unquestionably the matter was of central importance, as the level of activity and its limitation in new directions, was at least greatly influenced by the income–expenses balance.

The surge of activity requiring physical chemists during the 1939–45 war led to increases in membership even from 1943. At the same time, the print run of the *Transactions* was adjusted with the intent of providing issues later for those many overseas members who could not be supplied during the war. This purpose was only partly achieved because of the severe paper rationing which continued at least into 1947: it was in part alleviated by the marked reduction in the number of papers being published during the war years.

The increased membership without increased level of publications led to a significant cash surplus accumulating. For 1943, there was a net income/expenditure credit of £1152: the total investments of £15 541 led to an overall favourable balance of £2319. Already in 1944 printing costs were increasing, principally because of the enlarged print runs resulting from the increased membership.

Shortage of labour in the printing trade must have applied at Messrs. Oliver and Boyd, who published the *Transactions*. Delays in publication occurred, and a backlog on the issue of both the *Transactions* and the reports on General Discussions developed. A new departure was the printing, separate from the *Transactions*, of the General Discussions held in 1946: these appeared in 1947 as two separate volumes, 42A and 42B, each with separate pagination, contents table and author index. This step initiated the publication of the *General Discussions* as a separate series carrying their own volume numbers.

The rapid increase in post-war membership must be noted: for 1945–48 the totals were 1081:1413:1640:1783. This led to demands with which the publishers found it difficult to cope. Accordingly, in 1948, the *General Discussions* were sent to the printers, Messrs. Hazell, Watson and Viney. This arrangement applied from the volume *The Interaction of Water and Porous Materials*, Vol. 1 of the General Discussion series, to Vol. 5 *Crystal Growth*, of April 1949.

The consequence of this increase in printing and its much larger cost in London was a sharp deterioration of the Society's financial position. In 1946 there was still a surplus of income over expenditure of £551: in 1947 this had become a deficit of £77. In 1948 the current year deficit was £4600, which more than absorbed the Society's own reserve of £4524, leaving an overall deficit of £76. (The income from the Bourke Fund capital contributed to the Society's reserve.) The printing costs had been exceptionally high owing to the appearance of four of the General Discussion volumes previously in arrears (13th March 1949). A further contributing feature was the appearance in the *Transactions* of research reports pertaining to studies conducted in the war years. Their release for publication had been anticipated. The Treasurer (Slade) had reported that the Society was currently running at a loss of £1000 to £1500 per annum (15th July 1948).

Again, in 1949, the working year gave a deficit of a further £1792, reduced, however, by 28% thanks to a grant of £500 from the Chemical Council. Although it had earlier been resisted, all this led to an increase in the membership subscription from £2 to £3 p.a., which was accepted by the A.G.M. in 1949. At that meeting it was pointed out that members had been receiving for £2 publications which had cost £2 10s (£2.50) to produce.

The typical features of the interaction between membership numbers, publication costs and subscription fees are illustrated in the foregoing paragraphs. Now the central role of publication activities receives further emphasis.

A request from Messrs. Butterworths for permission and terms for their reprinting of Vols. 1–20 of the *Transactions* was agreed in principle in June 1948. The request, of course, showed the status of the *Transactions* and the market available for their further dissemination. At a later Council (27th October 1948) the Secretary, who had been asked to seek advice on a suitable contract for the republishing operation, reported a difficulty: *the Society had no legal status*, that is, it had no existence in law. That question has its own separate account. In discussions with Messrs Butterworths a further question had been raised; the terms under which they would take over as the Society's publishers. They were prepared to do so but wished for a five-year contract which Council (13th July 1949) felt should be reduced to three years. It was at this juncture that the President, Lennard-Jones, asked whether the Society could better be its own publisher.

There was a rapid response (21st October 1949). Enquiries with a number of publishers had established that the sales promotions effected by them counted for less in the case of a science journal than non-trade aspects such as its widespread reference in other journals, its availability in academic libraries, and its quotation by science reviewers. Were the Society its own publisher in 1949, paying all printing and distribution costs, it would gain £500–600 relative to its then current arrangement. Much of this gain would come from the sale of the *Transactions* to non-members, on which there was a 20% profit margin. It was immediately agreed that from January 1951 the Society would become its own publisher: the Aberdeen University Press to be its printer and distributor.

The question of reprinting the *Transactions* had led to a substantial change. It was obviously profitable, too, to Messrs. Butterworths, who later sought and obtained permission to reprint Vols. 21–25 of the *Transactions* (7th June 1951).

In October 1949 Slade's period as Treasurer ended and Wansborough-Jones took over. There were immediate consequences. At the Council meeting of 2nd March 1950 the Finance Committee's report generated a notable discussion. It started with Travers objecting to 'the appointment of Messrs. Lazard Bros. as the Society's investment advisers' on the basis that the Finance Committee had no power to make any appointments. The Treasurer pointed out that he wished to have their advice, which they were giving without charge. He had obtained confirmation that investment in industrial securities was legally allowable. The imminence of a General Election had made it desirable to act quickly. (A new Parliament with a Labour majority assembled on 1st March 1950.) He had informed members of the Finance Committee and a representative number of Council members, including the President (Lennard-Jones). All were in complete agreement for immediate action except Travers, who thought a Finance Committee meeting should have been called: he remained the sole objector at the Council meeting.

A comment may be in place. In a technical sense Travers had a point: only Council could appoint the Society's officers and advisers. However, the non-rigid stance, perhaps one should say, the empirical approach of Council, was true to the character of the Society; both this appointment and the freedom the Treasurer had found to invest in equities were possible, thanks to the absence of a legal constitution precluding such flexibility.

The transactions which involved Lazard's advice were far larger than any normally reported: some £10 370 (nominally), essentially Government stock, part of the Bourke Fund, being transferred into £2000 (nominally) distributed between Messrs. Unilever, ICI and Babcock and Wilcox, and a similar sum into National War Bonds (1954/56).

What becomes very clear is that with Wansborough-Jones as Treasurer, there was a firm and enlightened control of the Society's finances. It was accepted that the Society should seek to keep its expenditure within the anticipated income for each year. Partly in response to enquiries and suggestions in a letter from Travers (21st September 1949), the Treasurer now, with each annual report to Council, presented a budget for the anticipated income/expenditure for the following year. (The reports were often made in June so that the year was half through before its budget appears: earlier comments were frequently offered.) This immediately promoted stability, if only because it projected a figure available for publication costs. The effect of this was to suggest what could reasonably be expended on the publication of the General Discussion volumes: this, in turn, led to guidance on the number of papers which the Discussion organizers should envisage. Council indicated that in such Discussions review material should be kept to a basic minimum (24th November 1949). Only very rarely did review accounts appear in the *Transactions* (1st March 1945).

On 24th May 1950 there was discussion at Council on the terms of reference under which the Publications Committee and the Finance Committee operated. The memorandum on the regulations agreed was attached to later minutes (23rd November 1950). Some details may be summarized: The Publications Committee to operate as follows:

Members (of unspecified number) are appointed for one years, which can be extended;
The chairman to be appointed annually from amongst members of Council;
To accept (with qualifications on quality, suitability, style, editing) papers for publication;
To refuse publication, with right of appeal to Council;
The Organizing Committee of Discussion Meetings shall be responsible for the standard of the papers accepted but the Publications Committee shall decide what number of papers can be allowed;
To be responsible for the standard of the *Transactions* but not to allow expenditure to exceed that authorized by the Finance Committee.

The Finance Committee:

To consist of the President, Treasurer (who shall be chairman) and no more than six other present or past members of Council;

Members are appointed for five years and are re-electable;
A quorum to consist of the President (or Treasurer) and two others;
To be responsible for the management of all financial aspects of the Society;
To recommend annually to Council the sums to be spent on the Transactions, the Discussion Volumes and any other publications;

The Publications Committee being responsible for the best use of such funds, there shall be joint meetings of the Finance and Publications Committees at least once a year.

THE PUBLICATIONS COMMITTEE 1947–1951

Allmand was chairman of the Publications Committee until 1947, when he became President of the Society. In October that year Guggenheim took his place.

The record shows that the Faraday Society was run by a small number of senior British physical chemists who usually knew one another very well. Their choice of individuals for senior office in the Society almost invariably showed notable discrimination: even by the highest standards their performance was rarely other than excellent. In that context, and with some direct evidence, one can question whether Guggenheim was an appropriate choice to head the Publications Committee. He was a theoretician, especially able in the field of statistical mechanics. The leading British exponent and contributor to this subject was R. H. Fowler. Guggenheim joined him in writing an account oriented towards molecular problems, which became a standard work. Parenthetically, it is surprising in retrospect, that whilst Fowler himself had been a member of the Faraday Society Council, and Lennard-Jones became a most effective President, the *Transactions* only rarely, before the 1970s, contained purely theoretical papers. In the 1940s, 50s and 60s very few papers in the *Transactions* were not based on new experimental data.

One reason for mentioning Guggenheim's chairmanship of the Publications Committee is that only during his period (and a few instances after 1968) were there *details* of several papers being rejected.

It must be emphasized that this happened only when at least two referees reported unfavourably, and then the Editor and Chairman took the final decision. But early in Guggenheim's office the Publications Committee 'agreed to have further meetings when a sufficient number of difficult papers appeared.' These included a series of three papers on intermolecular energy terms being deduced from experimental data for liquids, and one paper whose sole author was a future Nobel prizewinner. No such entries are found before 1947 or after 1951. In the Publications Committee itself (23rd November 1950) Lennard-Jones asked that 'members should be given a letter of instructions' and, in the same vein, Schulman 'wished guidance to be more specific'. There was a sequel – at least in time. Six months later at a Council meeting (7th June 1951): 'The Secretary read a letter from Professor E. A. Guggenheim resigning from the chairmanship of the Publications Committee'. However, although the letter was as read, the minute was amended (1st November 1951): 'The Secretary read a letter stating he would not be seeking re-election for the coming session.' Guggenheim was replaced by Bell, and the minutes of the Publications Committee revert to the record of papers received, rejected, in press, etc. Throughout these changes the rejections varied little from the 25% mark.

During the whole of its existence, or certainly after 1920, the *Faraday Transactions* kept a very high standard in its refereeing of papers. Each was assessed by two individuals experienced in the relevant field, one often being a member of the Publications Committee. A full listing of its members is not given but as a brief indication, and for the years now under consideration, the Publications Committee for 1950 can be given.

Chairman: E. A. Guggenheim

N. K. Adam	C. K. Ingold
W. T. Astbury	A. S. C. Lawrence
C. E. H. Bawn	H. W. Melville
C. A. Coulson	J. T. Randall
C. W. Davies	E. K. Rideal
H. J. Emeleus	D. W. G. Style

M. G. Evans
W. E. Garner
G. Gee
C. N. Hinshelwood
A. R. Ubbelohde
W. A. Waters
W. F. K. Wynne-Jones

These 20 members represent the widest range of interests of the Faraday Society and, in 1950, it would have been very difficult in Britain to replace even three of them without decreasing the overall strength of the respresentation. They were the living proof of the status of the Faraday Society and its *Transactions*.

THE FINANCIAL SITUATION 1951–1959

The reorganised publishing arrangements were successful. The Transactions were to be increased by 10% per annum (11th March 1951). Some relevant figures were: for non-members the *Transactions* subscription was increased from £4.50 to £6.00 p.a. in 1952: individual issues went from 5s (25p) to 6s (30p): printing costs for the *Transactions* and *Discussions* volumes were £10 000: offprints cost £1500, and advance proofs (for the *Discussions*) £300. Specially nominated committees of three or four individuals, not necessarily members of Council, organized the General Discussions. Although with Wansborough-Jones in charge they were told what was the limit on the printing costs they could incur, this expenditure, which for long had been a wild item in the annual accounts, still produced difficulties. Specifically, the intrusion of reviews was limiting the number of original papers which could be accepted (5th June 1952). It was agreed that review material must be kept to a minimum and to promote better control the chairman of the Publications Committee was made, ex-officio, a member of all Discussion Committees.

By 1953, the draft budget indicated a substantial surplus of income over expenditure. Despite the increase imposed on prices, outside sales were increasing. Discussion in Council led to significant consequences. The *Transactions* could be increased by 20% 'since this could be done without lowering the standard', and the publication delay, currently about six months, would be reduced to the practical minimum of four months. The hope was that this policy would eventually lead to the flow of more and better papers (5th March 1953). It was the improved income from the Bourke Fund which allowed Dainton to suggest what became established as the very significant Bourke Lectures, and the question was raised of printing a list of the membership of the Faraday Society (4th June 1953). The cost estimated by The Royal Society led to agreement to publish the list (3rd June 1954).

Quotation from the minutes of the Council meeting of 4th March 1954 is appropriate: 'The Treasurer said that the annual report on investments sent to him by Lazard Brothers showed that the Society was in a stronger financial position than it had ever been before... The total assets available at the end of last year amounted to £23 000; taking into account all the funds, the general capital reserve and the Bourke and Spiers Funds, the annual income from investments was £900'. This led Longuet-Higgins to propose an instant vote of thanks to the Treasurer (Wansborough-Jones) 'which was warmly endorsed'. The reports from the Finance Committee presented to Council changes in investments to which they had agreed, but which always needed ratification by Council. These details are omitted in this account.

Through 1955 there is scarcely a mention of these interests in the Council minutes. Then, the March meeting in 1956 heard a report from the Finance Committee: largely due to a significant loss of income from non-members of the Society and increased printing costs, a total loss of £3300 was to be anticipated on the year's working. This was the endemic Faraday situation and led to a typical, one can say recurrent, response in three parts. Referees should be informed of the policy 'to improve the quality of papers by conciseness and the exclusion of irrelevant matter'. (Readers may recall that when it was possible to expand the *Transactions* by 20%, this also was without harm to the quality.) Secondly, non-membership subscriptions should be raised from £6 to £8. Thirdly, the number of papers in the Discussions should be controlled.

Much further consideration (7th June 1956) led to only one new proposal: that recruitment to the Society could be enhanced by local representatives, but none was appointed.

That the financial situation needed careful attention is evidenced by further meetings of the Finance Committee. In March 1957 the results for 1956 showed a loss of essentially £4000 on running costs: £5310 had been anticipated. This had been reduced to £500 by grants, not to be further counted on, from the Royal Society and the Chemical Council. The pattern of the financial situation can usefully be given in terms of the budget forecast for 1957.

Expenditure		Income	
Printing	£16500	Subscription & Entrance fees	£6300
Postage	700		
Administration	3800	Non-membership subscriptions	11000
Advertising	70		
Distribution[a]	2100	Transactions[b]	650
Meetings	200	Discussions[b]	1500
Donations	400	Reprints	1000
		Advance proofs	250
		Advertisments	20
		Interest and dividends	1600
	23770		22320
Estimated loss £1450			

[a] Paid to Aberdeen University Press for their service.
[b] Income from sales to non-subscribers.

Without pursuing details, the figures show that Faraday Society members were receiving the *Transactions* and *Discussion* volumes for less than the cost of printing, and that the administration costs were 16% of the total expenditure.

In this context the offer of the Pergamon Press to distribute the publications of the Society without charge, not surprisingly, 'was discussed at length and in detail'. What transpired (and the minutes of the Finance Committee repeatedly establish this) was that the Society had such extremely good treatment from their printer and distributor, the Aberdeen University Press, that when Norrish proposed and

Dainton seconded, Council decided it would 'not proceed with negotiations with the Pergamon Press'.

Again, without pursuing details, the impact of the financial situation on publishing was spelt out in Council on 6th June 1956. Bell, as President, spoke from his experience as chairman of the Publications Committee. Higher rejection rates for papers followed from budgetary restrictions: contributing authors were dissuaded and certain types of papers were rejected as unsuitable if they 'contained too much pure biology or pure physics for the *Transactions*'. This led to a thorough examination of the question. It was established that the rejection rate (ca. 22%) had not increased: the refereeing was not too severe, unless standards were to be lowered drastically: that Council should not finance an increase in the number of papers published (7th November 1957).

Grants of £1250 from the Chemical Council and £500 from the Royal Society came for the 1957 year. In allocating the latter, 'the Royal Society had given notice that no further grant could be expected from them whilst the Society's membership subscription remained so low in relation to the value of the publications received'. There immediately resulted an increased subscription for members, from £4 to £4 10s (£4.50), and for Junior Members from £2 to £2 10s (£2.50). At the same time the age limit for Junior Members was raised, in yearly intervals, from 25 to 28.

Through all this period the Finance Committee discussed and decided on changes in the invested capital. This was always done on the basis of observations from Messrs. Lazards. Even then, Council had to approve all such changes. As these details are reckoned to be of scarcely more relevance than the rainfall in the corresponding year, they are routinely omitted here. A sample can be given (5th November 1958): Council approved: (i) Sale of 1000 North British and Mercantile Insurance Co. Ltd., 5s Ordinary shares and reinvestment in 205 Prudential Assurance Co. Ltd. 4s 'A' shares. (ii) Sale of 100 US Steel Corporation $16.67 Common Shares at $65.375 and reinvestment in 400 shares of the Lazard Fund Inc. at $15 per share, the balance in 7 shares of B. F. Goodrich Co.

In 1959 subscriptions from members and non-members were

estimated to increase by about £3000. This would allow the Society to make £1750, or possibly more, available for travel grants to the Canadian meeting. Messrs. I.C.I. Ltd. and the Royal Society had already promised a total of £750 towards that sum. In this well provided state, Council proposed to refund £500 of the grant from the Royal Society for publication costs. This was not accepted, as it had become available owing to printing delays. The surplus was invested in the general funds.

A NEW TREASURER

In June 1959 having previously indicated his wish to retire after ten years in the Treasureship, Wansborough-Jones reported that Professor Ubbelohde had declined to accept nomination. He had, accordingly, written to Dr. Traill who was shortly to retire from I.C.I. However, his intention of living in Edinburgh would cause difficulties, and Dr. Forrest Musgrave of Messrs. Albright and Wilson (of which company Wansborough-Jones was a director) was to be approached. His services were so highly appreciated that Council had elected Wansborough-Jones to Honorary Life Membership when accepting his wish to retire. Musgrave accepted the Treasurership, to commence from January 1960.

Travel grants to members attending the Kingston, Ontario, General Discussion amounted to £1425. As the Canadians had raised funds towards this, a total of £1726 had become available towards the cost of publishing the Discussion. No backlog in printing the *Transactions* existed.

At the end of his term as President, Melville expressed concern that the membership (834 in 1939, 2251 in 1958) had become stationary, despite a rapid increase in chemistry graduates. Reference is made to local representatives at the universities and of the need for a recruitment drive (3rd November 1960). It was proposed to have pamphlets and posters for display to help in that effort, and the fact that the Chemical Society arranged meetings of their local representatives was mentioned. Rather astonishing was the revelation (2nd March 1961) that 'any University, Society, Library, Partnership, Firm or Company' were all being elected as normal members of the Society and so they paid the minimum possible charge for the *Transactions* and other rights of membership. This was to be rescinded: such institutions would become non-member subscribers. It is difficult to understand how this unnecessary detriment of income persisted through years when ordinary members and junior members were having their subscriptions increased for want of funds. The reader will encounter other instances of blind spots in Faraday Society affairs.

PUBLICATION AND FINANCE AFTER 1960

A change in the format and cover of the monthly *Transactions* and the use of a slightly larger type, all to increase printing costs by £1200 per annum, was approved in June 1961.

On his assumption of the chairmanship of the Publications Committee, Ubbelohde suggested that its members who had served for ten years or more be thanked for their services and replaced by others. This was done. It indicated that there had been no policy for length of membership of the Committee. Later it was restricted to three years.

The financial tide was again turning. During 1961 and 1962, the *Transactions* had each year published 600 pages, or approximately a quarter of the total, by non-members who contributed in no other way to the Society. Musgrave, as Treasurer, saw no reason why page-charges should not be introduced. In the Publications Committee, Twigg insisted that his opposition to page charges for members be minuted. Council decided in March 1963 to impose a £3 per page charge when non-members were the authors. This action was later praised as it had led to the elimination of some papers and an increase in the membership. However, the situation was frequently considered, as a long discussion reported in the Council minutes for 4th November 1965 illustrates. The only positive result was an approach to the Science Research Council, through its Secretary, Sir Harry Melville, to see if it could help fund the Society's publications, but two interesting figures were produced. The cost to the Society of the publications acquired by each member (12 issues of the *Transactions* and 2 *Discussion* volumes) was £2 14s (£2.70) p.a. The cost per member of the hard-pressed administration was £4 8s (£4.40), making a total cost per member of

£7.10. This compared with the subscription of £6.00 p.a. From January 1967 the non-membership subscription (applicable to Libraries, etc.) was raised to £15.00 p.a.

In 1896 Bancroft, on his own initiative, had set up the (American) *Journal of Physical Chemistry*, and, the Faraday Society maintained, from its members, a panel of referees who usually assessed all the papers submitted from Europe. This involved a substantial amount of secretarial work, correspondence, and postage costs. 'Recently', that is, to 1967, the Society had been allowed to advertize three times per annum in the *Journal of Physical Chemistry*. As the papers handled for the journal represented 20% of the total passing through the Faraday Society office, the Secretary reasonably estimated that on the annual overheads for the Secretariat of £7000, the overseas journal should contribute £1350 or $5000 p.a. To this suggestion, the reply from Baltimore was that they could not provide such a payment and so, from January 1968, all European papers would be handled there.

Hindsight comes readily: but one must ask how could this item have been contributed *gratis* by the Faraday Society for so many years to what had been a privately run journal? It could scarcely have been consciously intended when the Faraday Society itself was, almost invariably, short of funds.

In 1968 the future separate existence of the Faraday Society was already under consideration. A backlog of papers had been cleared in 1966 (thanks to a grant from the Royal Society), the report for 1967 showed a substantial balance, publication costs had been reduced, and a substantial increase in non-members' subscriptions imposed. Also there was an increase of £10000 in the book value of the Society's investments. However, a further charge on publications arose when the *Symposia* volumes appeared: the first in 1968.

Already in 1969 the Publications Committee were discussing changes which could be involved in the amalgamation. They were strongly in favour of retaining the refereeing procedure as operated in the Society: they were opposed to the form of double-column pages favoured by the Chemical Society: and they wished the title *Transactions of the Faraday Society* to be retained. They were successful with their first two intentions, but in the last, seeing the Faraday Society as such was due to disappear, their lack of realism was corrected by fact.

Many, in truth most, implications of the merger with the Chemical Society were considered by the Finance and Publications Committees. However, it is thought appropriate to present these under the subject of Amalgamation.

Ample evidence has appeared of the impact of the publications upon the financial position of the Society: less crucial, but equally long-term, was the influence of the Publications Committee on the presentation of Physical Chemistry by the Society. This becomes explicit only in the minutes of the Publications Committee, and will be illustrated by two instances.

The discriminatory policy, that is, the choice or exclusion of topics to appear in the *Transactions*, was essentially the sum of the weighted views of the many, and always experienced, members of the Publications Committee. Here 'weight' means influence and strength of expressed opinion. The gradual elimination of electrotechnical, metallurgical and applied science had taken place in the 20s and 30s, perhaps without any explicitly recorded decision. As late as April 1966, Dr. Peter Gray asked what was the leading question (since 1903) for the Publications Committee. 'Was there any policy with regard to the subject matter of papers accepted for publication?' The Secretary and Editor (Tompkins) replied: 'There was no discrimination except against papers largely devoted to experimental techniques and theoretical papers with a large mathematical content, although those concerned with physicochemical problems were acceptable'. He went on to say – presumably describing rather than defending – 'that there had been a reduction with respect to papers on many aspects of pure spectroscopy, the physical properties of polymers and biophysicochemical topics.'

This statement could be followed by a long and debatable commentary. The Biophysical Society arose in 1960 and, about the same time, various polymer journals also appeared. As to papers on 'aspects of pure spectroscopy', the obvious comment is that these had appeared in the *Transactions* only less rarely than a near zero number on diffraction methods, or other electro-optical techniques for studying molecular structure. There were very large areas in the more physical aspects of molecular science for which one would not send students to the

Publications and Finance 251

Transactions of the Faraday Society.

Death-bed revelations have constantly adorned romantic literature. The last meeting of the Faraday Society's Publications Committee took place on 6th July 1971: it was about to become the Publications Committee of the Faraday Division, reporting to the Primary Journals Committee in the amalgamated Society. Discussing the guidelines to be advocated, one senior Faraday officer, perhaps unwittingly, revealed aspects of the unexpressed policy of the Society. He is reported as saying that papers with purely structural chemistry should not be included: and in regard to reaction kinetics in solution, the *Dalton* (i.e., the Inorganic Chemistry Division of the amalgamated Chemical Society) *Transactions* would be more suitable. Many thermodynamic papers would be more appropriately placed in other journals or other sections of the *Journal* of the (new) *Chemical Society*.

We all make unconsidered assumptions. This applied also to members of the Faraday Society's Publications Committee; their decisions, more than anything else, defined the character of the publications and so the image in the science field of the Society. It must be remembered that so small a group, with such limited resources, could not possibly cover the whole of the field between chemistry and physics. Selection was inevitable, and it possibly took the form of a consensus of unexpressed personal assumptions as to what constituted the engaging aspects of physical chemistry.

It will not help to consider the whole width of the physics–chemistry interface. However, it borders on the impossible to exclude structural aspects from physical chemistry. How is one to discuss the behaviour of methane or any other molecule with the disadvantage of not pursuing its structure? When it comes to the silicates, almost all important physicochemical aspects are determined by structure: in the zeolites, even refined details, such as precise channel-widths, are critical to the behaviour. Nevertheless, the study of such structures was scarcely ever represented in the *Transactions*. Similar remarks apply to molecular geometry in general.

Reaction kinetics in solutions more suitably belongs to Inorganic Chemistry? Well, since the first chemical reaction rate was measured (Wilhelmy, 1850) it has overwhelmingly been concerned with organic molecules. Could the rate of inversion of sucrose be a study in inorganic chemistry? Here, the Faraday Society record is not so dim. To mention only perhaps the most significant such studies between 1930 and 1960, many of Moelwyn-Hughes's papers, all organic molecular reactions in solution, appeared in the *Transactions*. Between 1945 and 1972, a third of one General Discussion and two halves of other Discussions considered special cases of reactions in solution.

'Many thermodynamic papers' did not indicate what proportion of the total this exclusion might be. Thermochemistry, which provides much of the experimental data for chemical thermodynamics, including lattice energies, heats of reaction, heats of solution, etc., had been poorly represented in the *Transactions*, and did not appear as a subject for a General Discussion. It could be argued that free-energies are at the foundations of chemical science, as they determine the direction and extent of chemical reactions. Instances of their evaluation in the *Transactions* do not come to mind.

Some further aspects of Finance and Publications will appear in considering the amalgamation process.

4

The Colloid Committee, the Colloid and Biophysics Committee and the Biophysical Society

A COMMITTEE TO promote the discussion of colloid studies under the *aegis* of the Faraday Society had been set up in 1929. Its promotions grew and were little dampened during the 1939–45 war years. Then, for several years, the Colloid Committee came to play an ever increasing role in the Faraday Society. This reached a peak when, for a period, it was accepted that one Discussion meeting should be organized each year by the Colloid Committee: this at a time when two General Discussions per annum constituted normal practice. Much of this exceptional status was due to Rideal and his prominence in the Society's affairs. That colloid interests might occupy one half of physical chemistry was a modest claim in comparison with the quip popular in the Cambridge Colloid Science Laboratory: 'Science can be divided into two parts, colloid science and astronomy'.

The frequent reports on discussions in the Faraday Council of the proposals, contributions and activities of the Colloid Committee are here suitably reduced. Their positive outcome was two-fold: the organization of many successful discussions and the eventual formation of a new society.

In March 1945 the Secretary (Marlow) explained that he was approaching the Societies represented in the formation of the Colloid Committee. The need had arisen because the records, including the constitution of the committee, had been 'destroyed by enemy actions'. The Societies which could identify representatives on the Colloid Committee were:
The Royal Society; The Chemical Society; The Physical Society; The Physiological Society; The Society of Chemical Industry; The Oil and Colour Chemists Association; whilst neither the representation of the Biochemical Society nor of the Society for Experimental Biology was established. The Faraday Society had acted as the parent body for the

Colloid Committee, and so in August 1945 Council appointed three members to the Committee: J. H. Schulman (of the Cambridge Colloid Science Laboratory to act as convenor), N. K. Adam and T. R. Bolam.

The reconstituted Colloid Committee arranged a *three-day* General Discussion on *Swelling and Shrinking*. This was obviously a thoroughly successful meeting as can be seen from its proceedings printed in the Transactions, Vol. 42B.

Schulman played a leading role in the next major meeting. This was one jointly between the Faraday Society and the Société de Chimie Physique on *Surface Phenomena*, a suitably broad topic to honour Professor Devaux, at Bordeaux on October 5–9th 1947. This major event was a notable success. It served to revive relations with French colleagues on a working basis. Sixteen Faraday members attended and, remembering the leaness of the war years, the Bordeaux meeting was an unforgettable event for them: *verb. sap.* The papers presented were not published by the Faraday Society as it had not organized the meeting. Initially the Cambridge University Press was to handle the volume. However, after failure to secure coverage of possible losses from the Faraday Society, it eventually appeared from Messrs. Butterworth.

In quick succession there was a one-day meeting at U.C.L. on 19th December 1947. *Colloidal Crystallites and Micelles* were discussed on the basis of papers which appeared in the *Transactions* (Vol. 32. pp. 811–830). For March 31st, April 1st and 2nd (three days) a General Discussion was staged at Southampton on *Interaction of Water and Porous Materials*. At the Council meeting when Marlow's death was announced (18th March 1948) J. T. Randall and J. A. V. Butler (who were not members of the Council) clearly on the basis of their prior consideration, suggested in correspondence the formation by, or within, the Society of a new group 'to foster the study of biophysics and the physical chemistry of living processes'. This proposal so obviously raised major questions as to the future direction of the Society in the post-war years that a committee was appointed to assess the situation. In view of the generality of the future developments involved, it could be thought that the committee did not have a total of impressive experience: E. K. Rideal, A. E. Alexander and G. S. Hartley. All three, as physical chemists, could be described as 'strongly surface oriented'. Only little insight is needed to appreciate that the biophysical area was certain to expand rapidly and in several directions. The Faraday Society had to decide on the scale of its involvement and to discriminate on its directions.

In this context, the committee's recommendations seem reasonable, if, apparently, modest in scope. Council accepted that the Colloid Committee should be broadened and renamed the Colloid and Biophysics Committee. Representation on it should remain as before but could be added to, the individuals being elected for three years, and the chairman would be a Vice President or member of Council of the Faraday Society. The newly named committee 'Shall organize one General Discussion of the Faraday Society each year'. In view of the wide interests covered by Colloids and Biophysics, one meeting per annum was in no way unreasonable, but in relation to the Faraday Society's other concerns, not surprisingly, it created difficulties.

Rideal was immediately elected chairman of the Colloid and Biophysics Committee (later C.B.C.), taking over from Schulman's chairmanship of the Colloid Committee, but he (Schulman) became one of the Faraday Society's representatives on it. An Appendix to the minutes (3rd June 1948) gave further details of the new committee's organization. Suggestions which came back to the Faraday Council from the Committee led to Council asserting its control: e.g. 'The chairman of the C.B.C. to be appointed by the Council and approved by the Committee'. Informal meetings organized by the committee 'shall be subject to the approval of the Council' and it shall 'in addition propose to the Council subjects suitable for one General Discussion each year' (15th December 1948). Conflict between the intentions of the Faraday Council and those of the C.B.C. soon arose. The former had planned a General Discussion on *Spectroscopy and Molecular Structure*: the latter had envisaged a meeting on *Optical Methods of Investigation of Cell Structure*. The clash led to the unusual calling of a special Council meeting in December 1949. Several pages of Council minutes are concerned with this problem which led (how effectively?) to a *four-day* General Discussion in Cambridge embracing both

Molecular Spectroscopy, with a bridging session on large molecules, followed by *Optical Methods of Investigation of Cell Structure*.

The next problem quickly appeared (2nd March 1950). The C.B.C. had suggested the Discussion title: *Applications of Radioactive Tracers to Chemical and Biological Problems*. Emeleus as a member of Council and of much experience in this area, concluded the title 'was too diffuse for a good meeting'. A much better title would be 'Chemical and Biological Effects of Ionizing and Nuclear Radiations'. Macaulay's schoolboy could not fail to detect that this was not merely a change of title but a change of subject. Some would be quick to point out that 'the title is the subject'. The Faraday Council thus appointed Emeleus as the chairman and Dainton as their other member of the organizing committee. One is inclined to suggest here that that should have fixed it.

A quotation from the same minutes must be given: 'Finally in view of the above, it was decided that *Nucleoproteins* be deferred for further consideration as a topic for 1952.' However the C.B.C. did, independently, organize an informal meeting in May 1950 on *Physico-chemical Properties and Behaviour of the Nucleic Acids*. The anticipated 'fixing' suggested earlier did not occur. A clash with an international meeting of similar content meant that even the new topic, accepted by the C.B.C., could not be fitted in for the autumn of 1951. Discussion of the possibilities in two meetings reported in over 1000 words in the minutes, showed much unhappiness in the Faraday Council. At one time Council had to be reminded that it had itself agreed that the C.B.C. had the right to organize one General Discussion each year (28th June 1950). That was when Council anticipated having five General Discussions in two years, that is, three in every second year. The small Faraday Society secretariat could not cope with that third Discussion in one year.

There was one immediate conclusion. The C.B.C. accepted a spring 1952 meeting under the Emeleus title and were to consider a 1951 meeting on *Size and Shape Factors in Colloid Systems*. In November 1950 this was reported to the Faraday Council as to be held at Ashorne Hill, Leamington Spa from 18th–20th July 1951. This gave only eight months from the acceptance of the subject to the actual conference: the

The Colloid Committee

Faraday Council normally planned a Discussion two years ahead. The desirability for such an interval is illustrated by the next Council's meeting, in March 1951 considering topics for General Discussions during the next *five years* – those for the next two years having been decided. Later (6th March 1952) the C.B.C. was allowed to have preprinted summaries of papers to be presented at their informal meetings.

Perhaps it is advisable to explain to the reader that almost the only details of the C.B.C. coming for consideration to the Faraday Council were their annual General Discussions. This resulted from the Faraday Society's acceptance of such Discussions as part of its own programme. The meeting so organized on *The Physical Chemistry of Proteins* on 6–8th August 1952 at Cambridge was one of the largest of its kind: about 400 were present, of whom 90 were from overseas. Informal meetings arranged by the C.B.C., whilst often reported ahead, needed far less attention, as they did not impinge on the Society's own programme and normally required no financial consideration. Two such meetings were: *Biological After-Effects of Ultra-Violet and Ionizing Radiation*, held at Cambridge, 12th–13th September 1952: and *The Nature and Structure of Collagens*, held at King's College London on 26th–27th March 1953. The latter was handled by Professor J. T. Randall and came very close to the Crick–Watson announcement of the DNA structure (*Nature*, 25th April 1953) which also related to work in his laboratories. The collagen meeting had a quality which led to a full report being published by Messrs Butterworths: again arranged by Randall. At the Council meeting of 4th June 1953, at which accounts of these informal meetings were given, Professor J. A. V. Butler reported that a meeting was being organized at the premises of the London School of Hygiene and Tropical Medicine and Hygiene on 30th–31st October 1953: the subject, *The Structure and Properties of Nucleic Acids and Nucleo-Proteins*. This clearly would incorporate discussion of the Watson–Crick, the Rosalind Franklin and the Maurice Wilkins studies, the last two being contributions from Randall's King's College, London, Physics Department.

Despite the high professional standards usually attained in organizing General Discussions, there were occasional lapses. The meeting at

Leeds on *The Physical Chemistry of Dyeing and Tanning*, 8th–10th September 1953 was not a successful one: the subjects, as could have been foreseen, were principally of interest to those in the industries concerned and they had not then generated significant activity in physicochemical research. For a meeting planned at Sheffield in September 1954 on *Coagulation and Flocculation*, the chairman of the organizing committee had to be replaced by Dr. D. D. Eley, with the result that 'a very successful meeting' was held. What was unusual at the time, was that two Russian scientists were part of the very good attendance. Contributors from the U.S.S.R. were not infrequently invited: sometimes they expressed the intention of attending, but rarely appeared.

At the end of 1954 criticism arose because Schulman asked whether the C.B.C. General Discussion on *The Physical Chemistry of Enzymes* for Oxford in August 1955, could be extended 'by a day or a day and a half'. Council agreed it might be given a half-day extension but expressed dissatisfaction on the way the Discussion was being organized. Professor F. W. Roughton of the Cambridge Colloid Laboratory was the convenor: it is reasonable to suggest he had small experience of Faraday Society procedures.

Remembering that the C.B.C. constitution and its share of Faraday General Discussions had been thoroughly assessed only a few years previously, it was surprising that the President, R.G.W. Norrish, 'raised the question of the Colloid and Biophysics Committee, and said he thought too many of the Society's discussions were being offered to the Committee'. It was a brave suggestion to make at a Council meeting when both Rideal and Schulman were present. It will be seen that it had consequences.

DEVELOPMENT AFTER 1954

The C.B.C. on (7th December 1954) discussed the concern expressed by the President of the Faraday Society and agreed 'that should the occasion arise they would consider sympthetically foregoing a Discussion'. Norrish was not satisfied (3rd March 1955) and it was decided to consider what action was necessary at the next Faraday Council. The Council accepted Norrish's wish that the C.B.C.

consider an allocation of one in three of the General Discussions. This they, in turn, accepted.

However, pressure built up in the C.B.C. At a Faraday Council meeting in November 1957 with R. P. Bell as President, Professor J. A. V. Butler reported the wishes of C.B.C. members, characterized as of biophysico-chemical and biophysical interests, to hold an increased number of Informal Meetings. The Faraday Society had sponsored such meetings by advertizing them and providing small grants towards travelling expenses, apart from publishing summaries or abstracts and helping with organization. Butler 'would like to see many more meetings of all kinds held under the aegis of the (Faraday) Society'. The C.B.C. was composed of representatives of a large number of Societies and Associations whose interests were neccessarily much wider than those in physical chemistry of the Faraday Society. The C.B.C. would be acting within its powers by increasing its number of meetings. The Faraday Council got no further than to express its concern at the time and money which might be spent on these meetings, which it welcomed in principle, but might have to refuse commitment to further expenditure. It is not recorded that whilst colloidal and surface phenomena came within the Society's interests, biological aspects were an additional feature. Note that the C.B.C. were careful always to avoid any suggestion of biochemical interests.

The next step was reported by Butler. A sub-committee on Biophysics and Biophysical Chemistry had been set up by the C.B.C., whose chairman would be that of the C.B.C., and who could report directly to the Faraday Council as he would be one of its members. The C.B.C. sub-committee asked Council on 5th November 1958 to reconsider its decision (taken by the Secretary and Treasurer?) not to print abstracts of Informal Meetings.

More discussion ensued at a Council in April 1959 (incidentally, when both L.E.S. and M.M.D. were present) when Professor Butler reported that the sub-committee of C.B.C. wished to be promoted to a full and independent committee of the Faraday Society. This clearly led to consideration of the interests to which the Society was committed. 'Professor Norrish said that the interest of the Society should be physical chemistry, and if it were to be reoriented towards biology then there should be 'A' and 'B' publications.' Such a designation would copy Royal Society publications. The fact that Faraday himself engaged in minimal, if any, biological studies was, perhaps, not relevant.

The situation was further clarified when the Faraday Council was informed in June 1959 of a strong movement for the formation of a Biophysical Society. 'Should one section of the membership wish to organize itself differently they should be encouraged to do so. Members of Council considered that the Society had done good work in encouraging interest in biophysics and biophysical chemistry. They would not stand in the way of any new Society.' At the next Council there was extended consideration but limited decision, as the aim clearly was that partition should come about not only without confrontation but with good feelings on both sides. A working group – J. A. V. Butler (chairman of C.B.C. and its sub-committee), G. B. B. M. Sutherland and J. T. Randall – was asked to recommend how to proceed. They had co-opted eight colleagues (Eley, Gray, Huxley, Kendrew, Pantin, Preston, Roughton and Swann) and in reporting on 3rd March 1960 announced the birth of the new British Biophysical Society. For the steering committee setting up the new Society, the group had chosen officers from amongst themselves, but no names were reported.

The Faraday Finance and Publications Committees proposed to make a grant of £500 to help float the new Society, but in a Council attended by eleven members and the Secretary (2nd June 1960) it was voted by five to four to increase the grant to two instalments each of £500. When the British Biophysical Society sought free advertising space in the *Transactions* (2nd March 1961), Council granted it through 1961, but ruled that subsequently normal advertising charges would apply.

The Colloid and Biophysics Committee (C.B.C.) had ceased to exist with the formation of the British Biophysical Society: it could be said that the tail that had tended to wag the dog had been amputated. However, regrowth was established when a Colloid and Surface Chemistry Discussion Group asked for recognition by the Faraday Council. They wished to organise regular one-day meetings on the lines of the Informal Discussions: this led to a more defined pattern for Informal Meetings. When sponsored by the Faraday Society, preprints would not be issued but papers would be subject to an informal refereeing. To appear in the *Transactions* the normal assessment would apply and papers arising from the same meeting would be grouped together.

There is subsequently no further mention of the Colloid and Surface Chemistry Group except when its Informal Discussions are recorded. And thus, in 1960, there terminated the special preoccupation of the Faraday Society with the field of colloid science. The reader may well conclude that excessive attention has been given this one interest. In fact, and in relation to the time taken in Council and space occupied in the minutes, the C.B.C. is dealt with as briefly as might be adequate. If the British Biophysical Society which sprang from it can be claimed as resulting from Faraday Society activities – its early character was certainly sponsored more by the Faraday Society than by any other of the many participating societies – then this is notably to the credit of the Faraday Society. Much more open is the question whether, by committing itself to foster a special section on colloid science, the Faraday Society neglected to develop other topics which could (some would say 'should') have had at least equal attention.

This question also inevitably arises in the context of the scope of the General Discussions. The absence of notable consideration for valence theory, for the techniques and results of structural studies in the gaseous and solid states, for the development of irreversible thermodynamics, for statistical and computational modelling of molecular behaviour – in fact, of those aspects of science 'between chemistry and physics' which are embraced in Chemical Physics – was this, in part, due to programme-filling by colloid interests?

On how colloids came to play so large a part in the Faraday Society there is no doubt. It was Rideal's influence: it is a fact that Rideal himself felt frustrated in the Cambridge Physical Chemistry Department until he acquired his own Laboratory of Colloid Science. That he made little distinction between colloid science and the totality of physical chemistry is seen by such facts as that the former included the infrared spectroscopy of simple molecules, and by his verbally challenging Irving Langmuir to suggest the limits of colloid science. Not until Norrish, the then Cambridge Professor of Physical Chemistry, became President did the Faraday Council consider limiting the scope of 'Colloid Science'.

5

The Legal Status of the Faraday Society

REPEATED COMMENT HAS been made that over the period of his Secretaryship, the Faraday Society existed, in practice, in the person of G. S. W. Marlow. His successor, Tompkins, had clearly grasped this, and in Council (15th December 1948) raised the question of the legal status of the Society. He had been advised that the Society should be registered as a limited company. The Finance Committee were asked to proceed on this matter, the resolution of which was to take a significant time.

The initial decision, taken in February 1949, was that the Society should petition the Privy Council for a Royal Charter. This aim would, *inter alia*, require the Society to have the reaction of allied interests, *e.g.* the Royal Society, the Chemical Society *et al.* to its application. A loss-showing balance sheet in 1949 indicated the advisability of postponing the application. Later, in November 1950, in the same context, Council accepted a definition of the aims of the Society:

'To promote the advancement of those fields of science lying between chemistry, physics and biology'.

This can hardly be seen as a very meaningful or grammatical statement. It could imply that biochemistry and biophysics were as important parts of its field as physical chemistry and chemical physics. More appropriate, perhaps, was one definition not chosen by Council: 'to promote the advancement of chemical physics, physical chemistry and biophysical chemistry'.

Discussion led to postponement of the Royal Charter application. It was seen to restrict the flexibility and the informal character of the Society. Alternatives were to be considered. Legally approved trustees could be appointed or a Limited Liability Company could be created. Probably as a result of the consultations with the Chemical Society which had been necessary, there was the significant agreement of

Council 'that closer collaboration with the Chemical Society was desirable'. Specifically, the Faraday Society would send advanced notice of its General Discussions to the Chemical Society. These observations all came in 1950. At this time also, Council decided its meetings would normally be on the first Thursdays in March, June and November. In such a relevant detail, it could be that one sees the hand of the new Secretary.

The next position adopted on the Society's status was to pursue incorporation as the Faraday Society Limited (with permission to drop the 'Limited' from the title) which would bring it under the Companies Act. The Society's solicitors did, in fact, present Articles of Association appropriate for the formation of a Faraday Society Ltd.: consideration of these Articles led Council to conclude, *inter alia*, (5th June 1952):

(a) That the Society should aim to retain the maximum flexibility:
(b) That the governing body should consist of Members of Council and Officers, the Officers to be defined as the President, the immediate Past-President, Treasurer, and Chairman of the Publications Committee.

No further reference appears before 1955 on this major question of the Society's legal status, first raised in specific form in 1948.

In November 1956 the Treasurer, Wansborough-Jones, reported on the two main alternatives which a sub-committee had investigated. The first was to obtain a Royal Charter which would convey considerable status but was only available if all other interested parties, which would include all the other chemical societies in the U.K., raised no objection. One veto would debar the granting of a Charter. The second possibility was the incorporation of the Society as a limited company. This would be a speedy matter and legal costs would be about £500. The disadvantage was that under various Companies' Acts, much flexibility and informality, part of the Society's greatest assets, would be lost. As its capital assets were deemed adequate for any likely contingency, he, the Treasurer, was in favour of retaining the status quo.

The Council agreed with Norrish, who expressed concern and advocated that the situation be kept under consideration, at least on an annual basis. Despite this clear indication, it was five years later before the question was next considered. In the meantime Wansborough-Jones had been replaced by Musgrave as Treasurer. He had asked the Society's solicitors (Messrs. Bristows, Cooke and Carpmael) to draw up Articles of Association. A committee – the President (Hinshelwood), the Treasurer, the Secretary, Norrish, Style, Topley and Ubbelohde – was appointed to report on the Articles to the next Council meeting.

This was the first step in a lengthy process. Council members were each provided with a copy of The Third Draft of the Articles: Topley asked whether a Royal Charter could be reconsidered: the President established that the Royal Society would have no objection to a Royal Charter application, but Sir Alexander Todd had informed him personally that the Officers and Council of the Chemical Society would deplore it – thus finally killing a very lame duck. Members of the Society were told by letter of a form of resolution to be put to a Special Meeting, and that 'reference to the proposed incorporation' would be made at the next A.G.M.: they would be able to approve or disapprove within a limited time: the Solicitors would submit the draft memorandum and Articles of Association to the Board of Trade, the Charity Commissioners and the Inland Revenue.

The Charity Commissioners proved unexpectedly difficult, and the Board of Trade raised some difficulties. Much detail is thereby covered but not reported here. In November 1964, Council minutes report that a Special General Meeting, held at the Royal Geographical Society's premises, had approved a legally drafted resolution of some 330 words appointing a sub-committee of three – Ubbelohde, Dainton and Musgrave – to deal with the Incorporation of the Society.

At the Council meeting on 11th March 1965 the Assistant Secretary reported that the Society had now become incorporated on 8th January 1965 'under No. 833636 of the Board of Trade' as a Company Limited by Guarantee and not having a share capital: that all members would be circulated and asked to apply for their new membership of the Incorporated Society; that Messrs. Musgrave and Ubbelohde be authorized to make an application on its behalf to the Charity Commissioners.

The Legal Status of the Faraday Society

And so, sixty-two years after it was founded, seventeen years after it first sought recognition of its being, and six years before it passed out of reality, this very English institution, the Faraday Society, came into legal existence. It was nine years after incorporation had been reported to the Council as 'a speedy matter'.

Its title remained what it had been, the Limited being omitted by Licence from the Board of Trade. As a charity there were small gains, such as that members could pay their subscriptions by covenant. The officers and *modus operandi* of the Society were virtually unchanged as recorded in the detail of the first meeting of Council after incorporation. This was for recording the location of the Registered Office, the Common Seal, the Auditors and the Bank. It was held at the Hyde Park Hotel with merely six members of Council present: that is apart from the Secretary, Assistant Secretary and one of the Company's Solicitors. It now became important for each A.G.M. to have a quorum – ten being its modest number. And the Secretary, whilst able to attend all meetings, would no longer have a vote in Council.

6

The Fiftieth Year Jubilee

NINETEEN FIFTY-THREE was the fiftieth anniversary of the founding of the Faraday Society. Already at the November meeting in 1951 the President, Goodeve, reminded Council of this anniversary. Of the seven individuals most actively involved in the birth of the Society, two still survived in the persons of Sir James Swinburne and Professor F. G. Donnan. It seemed appropriate to associate the celebrations with the General Discussion on Non-Electrolytes planned for the spring of 1953, and to hope that this could be located in the Royal Institution with its Davy–Faraday Laboratories. A further point was made: members of Council should give thought to the next choice of President, as he would be in office in the Jubilee year.

The following Council meeting on 6th March 1952 opened with a statement by Goodeve on the question of his successor as President. In informal discussions with a considerable number of members, 'five or more people had been suggested'. He mentioned only one name in proposing that they ask Professor H. S. Taylor of Princeton to be President for one year in 1953. This possibility had clearly already been explored with Taylor as the President was able to indicate that he (Taylor) would probably be allowed leave of absence to spend much of the year in the U.K. As they would both be at the Toronto General Discussion in the autumn of 1952, Taylor could be received into office there. Council agreed unanimously that this suggestion be recommended to the A.G.M. the following July.

Although Taylor may not have been the most distinguished physical chemist the Council could have chosen, there were substantial reasons – not mentioned in the minutes – for the choice. It was a symbolic confirmation of the international character of the Faraday Society: it would appeal to the substantial number of members in North America and could be expected to increase that number: Hugh Stott Taylor had graduated in Liverpool and his close collaboration with one of the best regarded of Past Presidents, Eric Rideal, was well known. Also, he had retained his British nationality.

The Fiftieth Year Jubilee

Other aspects of the Jubilee were put in the hands of a worthy committee: R. P. Bell, Donnan, M. G. Evans, Goodeve, Hinshelwood, Lessing and Taylor. Suggestions for consideration included a commemorative volume and special lectures, all aimed 'to show what the Faraday Society has meant in physical chemistry'. Presumably it was a reflection of the careful control which he maintained on the Society's finances that the Treasurer proposed allotting £300–400 for the Jubilee items. The committee later raised this figure to £750.

The recommendations accepted by Council in November 1952 were that at the General Discussion, now fixed for the Royal Institution for 16–18th April 1953, an opening session would take the form of short talks. In particular, Sir Harold Hartley would remind the meeting of the physicochemical work of Faraday himself, at the Royal Institution, and the two surviving founding members would each speak. Then Professor J. H. Hildebrand (Berkeley, California) would give the 7th Spiers Memorial Lecture. For the general interests of Faraday Society members no-one could better survey the situation on non-electrolyte solutions than Hildebrand, and the outstanding role of the Society's first secretary would be appropriately commemorated.

There would be a banquet (at the Dorchester Hotel) with minimal formality: evening dress would be worn, ladies would be welcome, and the principal guest should be an outstanding personality. By this, it transpired, what was meant was a leading figure in the Establishment or Government circles. The Lord President of the Council (responsible, it is true, for Government science policy) was first invited but was unable to accept. Lord Samuel had accepted to be the principal guest of honour. He was the notable Liberal politician–statesman–philosopher, certainly of a distinction not to be seen in the 1990s. He had twice been Home Secretary, and a High Commissioner for Palestine, where his successor to the title became a well known professor of chemistry. The choice of Lord Samuel, however, was scarcely original: he had been the principal guest when the Chemical Society celebrated its centenary (six years late) in 1947. In his speech Lord Samuel remarked 'how glad he was that the Society was shortly to discuss 'Free Radicals''. On the second day of the meeting, I.C.I. would hold a reception for the Society.

At the first Council meeting for which he took the chair (4th June 1953), Professor H. S. Taylor was able to report that the Jubilee celebrations had been a great success. The volume *The First Fifty Years*, in which authors had given accounts of Faraday Society contributions in physical chemistry areas to which they themselves were major contributors, had been widely appreciated. The May issue of the *Transactions* contained these historical essays, as well as a full account of the celebrations.

The June meeting of the Council had opened by Rideal expressing the great pleasure of all members at the honour conferred on their new President by the Queen in the Coronation Honours List, where his knighthood was announced. It was undoubtedly a recognition the Society could rightly share as it was doubtful, despite his distinction as a physical chemist, whether it would have been conferred were he not the President in its Jubilee Year.

It is important that the reader be made aware of the contents of *The First Fifty Years*. It comprises (xv + 82) pages: a black and white reproduction of a painting of Faraday lecturing at the Royal Institution: excellent small photos of Sir Hugh Taylor; Sir Joseph Swan; Lord Kelvin; Sir William Perkin; Sir Oliver Lodge; Sir James Swinburne; Sir Richard Glazebrook; Sir Robert Hadfield; Sir Robert Robertson; Professor A. W. Porter; Professor F. G. Donnan; Professor C. H. Desch; Professor T. M. Lowry; Sir Robert Mond; Professor N. V. Sidgwick; W. Rintoul; Professor M. W. Travers; Sir Eric Rideal; Professor W. E. Garner; Professor A. J. Allmand; Sir John Lennard-Jones; Sir Charles Goodeve; F. S. Spiers; G. S. W. Marlow; Professor F. C. Tompkins. There are also reproductions of two sketches showing 13 South Square, Gray's Inn and its entrance.

The membership of the Faraday Council is given for 1903–04, with its two O.M.s who were Vice-Presidents, Lord Kelvin and Lord Rayleigh; and also for the Council 1952–53; the Presidents for 1903–53; the Honorary Life Members from 1950–53; and the three Secretaries 1903–53. As a part of its Jubilee Celebrations the Society had elected Bjerrum, Garner, Hartley, Hildebrand, Rideal and Tizard as Honorary Life Members.

There is an Editor's Preface by F. I. G. Rawlins, a Foreword by Sir Hugh Taylor and an item which must be reproduced.

SOME PERSONAL REMINISCENCES

F. G. DONNAN

HAVING been given the honour of writing a short personal account of the life of the Faraday Society, I propose to relate my recollections of many happy days and remarkable events. The first of these was a small meeting held at the London office of the consulting engineer, James Swinburne (now Sir James Swinburne, Bart., F.R.S., and still alive and well, I am glad to say). The date was either late 1902 or early 1903. The meeting was called, I think, at the instance of Sherard Cowper-Coles, who took an important part in the proceedings. Several other people were present besides myself, but I cannot recall their names. The object was to start a new scientific Journal and a new scientific Society, which should be concerned with such subjects as electrochemistry, electro-metallurgy, and physical chemistry in general. I remember that Swinburne pointed to a pile of 'Science Abstracts' (of which, I think, he was the originator), and remarked how much attention was now paid to these subjects. A letter from Sir William Ramsay was read, in which he objected to the appearance of a new scientific journal, saying that he had not time enough to read the existing ones!

The actual foundation of the new Society, to be called 'The Faraday Society – to promote the study of Electrochemistry, Electro-metallurgy, Chemical Physics, Metallography, and Kindred Subjects' took place in London on 30th June 1903. It is interesting to note that in April, 1951, the new description 'To promote the Study of Sciences lying between Chemistry, Physics, and Biology' was adopted. The importance immediately attached to the new Society is evident from the names of the earlier Presidents: Sir Joseph Swan, Lord Kelvin, Sir William Perkin, Sir Oliver Lodge, Mr. James Swinburne, Sir R. T. Glazebrook, Sir R. A. Hadfield (period 1903–20). The first list of Vice-Presidents included the names of Professor Crum Brown, Dr. Ludwig Mond, Lord Rayleigh, and Alexander Siemens. So the Faraday Society received early help and encouragement from many eminent scientists. Under theses happy auspices I do not include the fact that I was a member of the first Council. The honour was mine, and greatly appreciated. But I must assuredly include the name of Frederick Solomon Spiers, the first Secretary of the Society, who acted as its Secretary until his death in 1926. He had a deep interest in physical chemistry and possessed high organizing ability. The initiation of the Faraday Society and its successful development during the years 1903–26 were largely due to his knowledge, zeal, and energy.

I now come to tell of the greatest and most famous part of the Society's work. Unlike other Societies such as the Royal and the Chemical, it has never held regular meetings at which papers on various 'unrelated' topics are presented, though such papers are constantly received and published in the

Transactions. The greatest feature of the Society's activity, the greatest source indeed of its fame and international influence, has been the holding of 'General Discussions', at which papers are presented and discussed which deal with the various aspects of some main, important, and pre-selected subject, and which are printed and distributed *before* the meeting. A number of eminent foreign scientists are invited who present papers, and take part in the discussions. Such meetings have been held in many of the great Universities of Great Britain (and recently at Toronto University). They are great *social* events; ladies are present, banquets are held, speeches made, toasts drunk. The discovery, in fact, of *scientia humana*, a discovery which we share with our friends of the German Bunsen Society.

Ever since 1913, [at least] two such General Discussions have usually been held annually. Occasionally there have been three or more. It is an astonishing fact that more than ninety such meetings have been held. I wish I had space to give the famous list of the subjects discussed–a record *sans pareil* in the history of science in this country.

Perhaps I may be allowed to make a few personal remarks. I remember the Discussion on 'Osmotic Pressure' in 1917. I have had a great interest in that subject ever since I had the temerity to publish a new method of deducing van't Hoff's osmotic pressure law. Also in 1917 I took part in the Discussion of the 'Training and Work of the Chemical Engineer', and in the Discussion on 'Pyrometers and Pyrometry'. I shall never forget the exciting Discussion in 1921 on 'Catalysis with Special Reference to Newer Theories of Chemical Action'. The trouble arose in connection with the origin of the necessary increment of activation energy required for *unimolecular* chemical reactions at ordinary temperatures. Jean Perrin and W. C. McC. Lewis read papers in which this origin was attributed to the infra-red *radiation* present. But Irving Langmuir and F. A. Lindemann (now Lord Cherwell) got up and demolished this theory with immense gusto! At the Discussion in 1930 on 'Colloid Science applied to Biology', A. V. Hill sat on my vain attempt to introduce 'ionic membrane equilibria' into living organisms. It was, I think, at the meeting in 1932 concerned with the adsorption on solid surfaces that H. S. Taylor (now our new President) explained his new theory of 'activated' adsorption (chemisorption). Colloid Electrolytes were discussed in 1934, on which occasion I fear I inflicted on the meeting a long paper dealing with the osmotic pressure and some other properties of solutions containing electrolytes of high molecular weight.

On 9th November, 1928, the Society celebrated its Silver Jubilee. An anniversary luncheon was held at the Hotel Jules, Jermyn Street, London. Many representatives of learned Societies were present, including Professor E. C. S. Biilman, Mr. Emile S. Mond, Senator Marconi, and Professor Ernst Cohen (representing the International Union of Pure and Applied Chemistry), and Professor Mauguin (Société de Chimie Physique), Professor K. Fajans (Bun-

The Fiftieth Year Jubilee

sen Gesellschaft), and Professor K. Molktenhausen (American Electrochemical Society). The President, Professor T. Martin Lowry, was in the Chair. The first Spiers Memorial Lecture was delivered by Sir Oliver Lodge at the Royal Institution, his subject being 'Some Debatable Problems in Physics'.

I cannot conclude this brief sketch of the Society's life without paying a sincere tribute to the memory of Mr. G. S. W. Marlow, who acted as our Secretary and Editor from 1926 until his death in 1948. Nothing could exceed the skill, energy, devotion, and high good humour with which he conducted our affairs. He was indeed the heart and soul of the Faraday Society for many years. I am proud to say that it was my good fortune to get such a brilliant young barrister to come and help us.

Let me also pay a sincere tribute to the excellent services of Miss Wakeley, who was Assistant Secretary during Marlow's period. I need scarcely tell present-day members of the valuable work of Marlow's successor, Dr. F. C. Tompkins, and of Miss B. E. Kornitzer, Assistant Secretary.

Permit me now, Gentle Reader, to depart in a blaze of reflected glory. The three Faraday Presidents, Professor W. E. Garner, the late Professor A. J. Allmand and Sir Charles Goodeve, worked as research students in my laboratory in their earlier years. And our new President, Dean Hugh S. Taylor, had to listen to my lectures on Physical Chemistry some forty odd years ago.

The other contents are brief essays on the interests fostered by the Faraday Society over its first fifty years. Brief though they are, these accounts admirably encapsulate significant developments in the physical chemistry cultivated by the Society.

CONTENTS

	PAGE
Some Personal Reminiscences. *F. G. Donnan*	1
Metals, Alloys, Fluxes and Slags. *A. H. Cottrell*	5
Soil. *R. K. Schofield*	13
Solutions. *C. W. Davies*	21
Electrochemistry and Corrosion. *J. N. Agar*	25
Crystal Chemistry. *E. G. Cox*	33
Structure, Spectra, and Dielectric Properties. *L. E. Sutton*	37
Photochemistry and Kinetics. *D. W. G. Style*	47
Free Radicals. *C. E. H. Bawn*	53
High Polymers. *H. W. Melville*	61
Colloids and Colloidal Electrolytes. *D. C. Henry*	67
Biological Applications. *J. A. V. Butler*	71
The Colloid and Biophysics Committee. *Sir Eric K. Rideal*	75
Surface Chemistry. *F. I. G. Rawlins*	79
General Discussions (a list, 1907–52).	83

7

The Faraday Society General Discussions

GENERAL CHARACTER OF THE DISCUSSIONS

THE FARADAY SOCIETY had two main functions: publishing a physical chemistry journal of international repute, and organising discussions which, from the 1920s, were widely recognised, even outside the U.K., to be uniquely good. Accordingly, this is one of the central sections of the Faraday Society History.

The evidence for it is in three forms: The Council minutes report the selection of topics, the choice of convenors, the time and the place: in the printed Council records of the meetings there are usually perfunctory expressions of success and satisfaction. Secondly, from 1945 the substance of the meeting, that is a full printout of the papers contributed and a report, provided by the individual speakers, records the main items raised and answered in the verbal discussion. The reader should be reminded that those attending the meeting would have received, some ten days or more before it opened, galley-form preprints of the papers to be presented. This had two important consequences. Firstly, the one contributor chosen to present the paper was asked to assume that the participants had read the paper, and he/she was allotted no more than five minutes to explain and illustrate its main points.

Then an interval, usually 15 minutes, was available for questions, comments and assessment. This arrangement meant that each paper occupied, say 20 minutes. It was the chairperson's important duty to ensure that each paper and its presenter received an equal share of the time: the last paper listed in any session was to be given the same treatment as the first. Rare exceptions to this uniformity arose when a paper attracted very little discussion. The time so gained was shared between later papers, but such lack of interest, when it occurred, was one of the signs of a contribution of limited significance. The more stimulating papers invariably led to much informal exchange between its author(s) and those colleagues who wished further to pursue relevant aspects. The controlled time for verbal comments meant that papers of unusual interest would attract written questions. These were sent via the Faraday Society Assistant Secretary to the presenter, and both questions and replies could then be printed in the Discussion volume.

It is clear that the Discussion volumes usually carry a completely adequate account of the scientific content of the meeting. The third, and only other source of information could be the memories of the participants. But 1994 is more than 20 years after the last Faraday Society General Discussion.* It is not surprising that sample enquiries directed to that minority of senior participants now extant have drawn little substance.

However, several replies assured the writer of the exceptional stimulus to research interests which came from these meetings. This, of course, is a common, if not universal experience from such discussions. It was maximized in Faraday Discussions thanks to the sharp focus ensured by the preprinting of the papers. Several commentators indicated that Faraday Discussions had stimulated not only interest and appreciation but also enhanced activity on a topic, certainly in the UK, and not infrequently, in a broader scene. This was particularly true of a number of the early post-1945 Discussions.

Obviously, the war years saw a reduction, almost a cessation, of communications and publications of new work and new developments. Accordingly, when the General Discussions resumed in 1945, there was a head of steam in many research topics, including some new ones. Their exposition to an immediately critical audience did much to disseminate widely the stimulus for new work in the post-war years. One obvious example was the meeting on Dielectrics at Bristol in April 1946, and another that on Crystal Growth, also at Bristol, in April 1949.

However, that mere discussion does not generate advances in a topic can perhaps be seen in the subject of Liquid Crystals. There was a General Discussion in 1933 which served to reveal the special features involved. The meeting in 1958 at Leeds was followed much later by a short discussion at The Royal Institution in 1971. There, de Gennes' contribution was outstanding but no great progress or enlightenment

* Although they have continued, with sustained vigour, since the Society's amalgamation with the Royal Society of Chemistry. Ed.

had obviously been achieved. It took several more years and the stimulus of liquid-crystal displays in watches, computers, etc. to raise liquid-crystal studies to a major world-wide activity and, incidentally, to form the basis of a Nobel Prize in Physics: one of the first in Physics for purely molecular science, that is, de Gennes' in 1991.

One other qualification: senior physical chemists have often credited a Faraday Discussion with revelatory powers. This can have been partly because of the impact of such a meeting on anyone just entering its field. Senior overseas contributors often, and rightly, made notable impressions. Especially could this be true for those of them not usually publishing in English-language journals. Reading competence in foreign languages has never had much emphasis in the training of U.K. scientists. It is well known that unless they had published in English (as has become increasingly likely in post-1960 years), European scientists have been astonished by the ignorance of their British counterparts in not knowing what may have been published years earlier. Until recently, this could still apply *a propos* work in eastern Europe and the now defunct U.S.S.R. It must still apply in some degree to Chinese and even Japanese work.

CONSIDERATION OF THE DISCUSSIONS IN COUNCIL

In order to approach an adequate, if brief, account of the Discussions, we must look at the Faraday Council minutes where these activities were considered, often in much detail, so that a proposed meeting could give rise to a 500 word entry. The minutes of one Council meeting has three typed foolscap sheets (ca. 1400 words) devoted to General Discussions, and a total of seventeen possible topics were once listed for consideration (5th November 1953).

As the General Discussions that occurred are detailed later, here we shall favour comments on the less fortunate items, as it is not irrelevant to see what topics were put aside. The member or members of Council responsible for suggestions are identified in parentheses.

Early on, *Distillation Principles and Practice* (E. K. Rideal) was turned down (1948), and, more significantly, *The Nature of the Chemical Bond* (M. G. Evans) and an alternative, *The Strengths of Chemical Bonds* (M. G. Evans and E. A. Guggenheim): both disappeared. In 1950 a proposal to arrange three General Discussions in one year showed the accumulation of topics, many of which had been in abeyance and others which had newly come forward during the war years. The proposal was rejected, specifically by the Finance Committee, who concluded the office staff could not handle the work involved. There were two consequences. The first was the merger of two topics into one meeting, when two quite disparate interests were combined [see Vols. 9A and 9B (1950) later]. The second consequence was the new appearance in the minutes of the heading Informal Meeting (28th June 1950). D. C. Henry was organizing such a one-day meeting on *The Electric Double Layer*, at Manchester, for the following September.

In 1951, new titles which appeared included *The Physical Chemistry of Geochemical Reactions*. Barrer eventually reported 'considerable interest and sufficient material', but the topic was rejected as most of the work was being done overseas and there would be too few U.K. contributors. At the same time there was the reappearance of *Aerosols*. The Colloid and Biophysical Committee (C.B.C.) immediately pressed forward to suggest eight titles for Informal Meetings.

The Institute of Physics approached the Faraday Society to cooperate in a meeting to discuss *Luminescence*. Perhaps for the first, certainly not for the last time, this invitation to collaborate was not accepted and a rather tart comment made: the Faraday Society would encourage its members to attend, on the presumption that they could do so 'on the same terms as I.O.P. members'.

Of the Informal Meetings in 1953, one was on a subject much repeated by the C.B.C.: *The structure and properties of nucleic acids and nucleo-proteins*. There was an attendance of 133, including 12 from overseas. Regular meetings every two years on this theme were advocated (J. A. V. Butler). One topic never taken up was *The Thermodynamics of Irreversible Processes and Hysteresis*, suggested by Everett in 1954 and repeatedly suggested subsequently.

In 1954, Lessing argued the time was ripe for consideration of *The Physical Chemistry of Air Pollution*. He was told it could not be allotted a General Discussion, but a one-day Informal Meeting could serve. On this he reported his failure to raise organizers. Council regretted this and 'agreed it was of national importance but it was, after all, a government problem' (11th November 1954). Did this mean that

Council thought the Government should be more far-sighted and capable of greater public concern than themselves? The subject would not go away: Dr. C. N. Davies wrote to advocate a General Discussion, but Council were not prepared to think other than of a one-day meeting.

Norrish was frequently critical and constructive in his comments on Faraday Society activities. When Informal Meetings, one on *Experimental Thermochemistry* at Manchester for April 1956 (H. A. Skinner), were being approved by Council, he regretted that no details were presented for their programmes (3rd November 1955).

In March 1956 Garner drew the Council's attention to the fact that the Chemical Society was organizing a research discussion on *Catalysis*. For this they had acquired support from industry. No concern is recorded in the minutes, rather that 'there was no need for action'. It was, of course, an indication that the Chemical Society was to build up its own Physical Chemistry activities: 15 years later it led to the inevitable fusion of the Faraday and the Chemical Societies. A topic of some importance, *Valence Problems of Inorganic Chemistry*, was suggested (3rd November 1960) for the General Discussion in the spring of 1962. Linnett and Longuet-Higgins were asked to proceed, with powers to co-opt members to the organizing committee. Nothing more appears on this subject, which was being very actively developed at the time. The Faraday Society had not discussed valency since 1923. The April 1962 meeting was on *Inelastic Collisions of Atoms and Simple Molecules*, the ninth General Discussion since 1945 on molecular dynamics (see later).

An informal discussion on *Reversible Photochemical Processes* was planned by the U.S. Army Research Office, to take place at Duke University, North Carolina in the spring of 1962. The Faraday Council regretted (2nd March 1961) they could not collaborate as 'the programme for 1962 was too full'.

INFORMAL DISCUSSIONS AND THE FORMATION OF A
STANDING COMMITTEE ON CONFERENCES

Almost immediately after the Colloid and Biophysical Committee had ceased to exist in 1960 (on the formation of the British Biophysical

The Faraday Society General Discussions 277

Society), more than 40 colloid/surface chemists in the Faraday Society proposed to Council the formation of their own Discussion Group. They envisaged regularly arranging one-day Informal Meetings. A group of members interested in catalysis wished for similar meetings. The two groups had independently thought out a pattern of procedure which would minimize, if not eliminate, financial dependence on the Faraday Society. Council appointed a panel to consider these requests: Eley, Longuet-Higgins, Norrish, Pople, Ubbelohde and Tompkins. The result was probably not what the petitioning groups intended.

The panel recommended: (i) the Society should continue to sponsor Informal Meetings; (ii) preprints would not be issued; (iii) papers to be presented would be subject to informal refereeing; (iv) publications would follow after subjection to the usual Transactions procedure; (v) suggestions for topics would be submitted to Council 'who would control the number and, in part, the subject matter of the meetings held' (2nd November 1961). In the context of the antecedent situation, the members of the groups would perhaps be justified in saying 'thank you for nothing'. This upshot would be interpreted by some as Council's inability to foster the formation and growth of subject groups which greatly appealed to younger members. General Discussions on their particular interests would be infrequent, and at too high a level for them to have a large part. The flourishing of numerous subject groups in the Chemical Society, both before and after 1971, is the proof of this conclusion. It was later (11th March 1965) that the Secretary (Tompkins), who had attended more meetings than any other member of Council, insisted that despite the controls listed above, attendance at Informal Meetings was not restricted and they often 'engendered more enthusiasm than the formal (General Discussion) meetings'.

The interest in Informal Meetings was growing. There were five in 1962. To illustrate the type of topics they embraced we can list their titles: Chemical reactions of graphite; Gaseous Free Radical Reactions; Fibrous Proteins; Forces in Colloidal Systems; Chemisorption and Catalysis. Two subjects were rejected: Nuclear Chemistry; The Liquid State. Also rejected as General Discussion topics were: The Thermodynamics of Irreversible Processes (again); The Evaluation of the Thermodynamic Parameters of Molecules (sic: The writer is unaware of the thermodynamic parameters of a molecule or of molecules, *tout court*.); Energy Conversion.

A major question was raised by Dainton in March 1965. He had heard criticism of the Faraday Society's General Discussions and wondered 'whether they were outmoded by other types of conferences?' He had presented a paper on the Gordon Conferences to the British National Committee for Chemistry and a meeting organized by the Royal Society had helped set up Eurochem Conferences, of which four had already taken place. There was an immediate reaction against restricting attendance at Faraday Discussions: they would become private discussions inaccessible to the majority of members if they became like Gordon Conferences. However, a small panel was established to report on the possibilities: Dainton, Agar, Whiffen and Tompkins.

There are minutes for two meetings of this advisory panel. Those for the first (15th October 1965) opens with an extended statement by its chairman, the President (Dainton). Of several significant observations a number must be summarized: (i) The Chemical Society was now in competition with the Faraday General Discussions. With funding up to £1000 it had recently staged a conference attracting young chemists some of whom indicated they would publish their (physical chemistry) work in the *Journal of the Chemical Society*. (ii) 'Already some areas of physical chemistry had been lost to the *Transactions*, such as biophysics, molecular spectroscopy, etc.' (iii) A criticism of the General Discussions was that they were becoming too highly specialized and so unattractive to younger people. (iv) The two-year interval between choosing and staging a General Discussion had several disadvantages, including the delayed publication of work which may have been completed many months even before its presentation at the meeting.

A second meeting of the panel, on 28th May 1966, formalized proposals to the Faraday Council. These amounted to the creation of a Standing Committee on Conferences (S.C.C.) to be responsible for the overall arrangements for Discussions of all kinds. The pattern recommended was: (i) General Discussions as already established: (ii) Informal Meetings, ditto: (iii) Special Discussions, or, as they later became known, Faraday Symposia: these were to differ from General

Discussions in usually being one-day events on more limited (and therefore more specialized) topics. The chairman of the S.C.C. was to be appointed by Council and would be re-electable for a second three-year term. The chairman of any group arranging a Discussion would become a member of the S.C.C., finally presenting a report on his meeting before withdrawing.

The first meeting of the S.C.C., held on 30th September 1966, embraced virtually all the top brass of the Faraday Society: Dainton (President), Bawn, Bell, Eley, Kemball, Porter, Rowlinson, Whiffen, and the Secretary (Tompkins). This Committee spelt out the details of its own composition. In addition to the chairmen of all General and Special Discussions, it would have three members elected by Council, and also the President, the chairman of the Publications Committee, and the Secretary, these three *ex officio*. The second meeting of the S.C.C. (12th June 1967) had an attendance of three: Dainton, G. J. Hills, and Tompkins.

Before the formation in 1966 of the S.C.C. all Discussions were considered at Faraday Council meetings. It became the aim to decide on a topic, if possible, two years before the meeting. Often three or four subjects would have been proposed and their probable contents considered. A rolling list, subject to additions and deletions, sometimes having more than a dozen titles, would be before Council. Inevitably the topics chosen were those most fully and convincingly advocated by some member or members of Council and, not unnaturally, such a subject would often be one of its advocate's research interests.

With a topic in mind, the Council nominated a member of the Society, maybe from Council, to form a committee of three or four colleagues and to act as its chairperson, which committee sought and selected the contributions to be presented. Future Discussions were announced in the *Transactions*, and often in *Nature*. The quality of the meeting depended very largely on the organizing committee: its composition has usually, but not always, been found in the Council Minutes and the chairman, when known, it given first in the listing of the Discussions later.

After 1966 all such details on conferences are only to be found in the minutes of the S.C.C. whose recommendations were acted upon by

Council. Accordingly it is appropriate to offer brief notes from the S.C.C. minutes.

At its second meeting, already mentioned, two of the topics 'selected for consideration' were Far-infra-red Spectroscopy and Molecular Quantum Mechanics. The latter became the subject of a Symposium in December 1968, by which time its title had become Molecular Wave Functions.

On the formation of the S.C.C. one recommendation accepted by Faraday Council in November 1966 was that the Committee's minutes should go to 'the local representatives to stimulate interest and to invite suggestions of topics for meetings.' This must be one of the first mentions of 'local representatives'. As far as Faraday Council minutes reveal, they came into existence spontaneously at some date unrecorded. There is another surprising, although welcome, aspect to the enlistment of local representatives' interest. The reader will be aware that the Faraday Council was an effective, self-perpetuating body, having minimum interaction with the membership at large even at A.G.M.s. It so happens that the writer (not at the time on Council) had written in 1965 making a number of suggestions, including the use of local representatives *inter alia*, to stimulate enhancement of the membership. Of these suggestions 'Council took note', but added, 'it was thought that the expense that this would entail would probably not be justified'. There was thus a helpful change of attitude within twelve months: see above (3rd November 1966). Within another year the Council minutes recorded 'many useful suggestions for future topics' coming from the local representatives (2nd November 1967).

This acceptable interest of the membership shows up in the S.C.C. minutes of June 1969: there twenty-six titles from seven sources are listed. Amongst those not further considered were: Aspects of Vibrational and Electronic Energy Transfer: Clathrates and Zeolites: Conformation of Polymer Molecules including Networks. These topics subsequently became of major research interest. Simultaneously 13 subjects were listed for Symposia, including Nuclear Quadrupole Resonance Spectroscopy and Photoelectron Spectroscopy. Surprisingly, the latter eventually turned up in 1972 as a General Discussion although it had been preceded by a Chemical Society meeting on

The Faraday Society General Discussions

Photoionization Phenomena and Photoelectron Spectroscopy in 1970.

The topics chosen by the S.C.C. for Discussions and Symposia are revealed as the items so listed as coming after 1968. The overlap into the Faraday Division is illustrated by the S.C.C. deciding in 1969 the subjects for the General Discussions in 1972.

LISTING THE DISCUSSIONS AND THE GROUPS OF THEIR TOPICS

The first General Discussion, that of September 1945, is reported from its appearance in the *Transactions of the Faraday Society* Volume 42. Two others come in separate volumes, 42A and 42B. All later ones are in individual volumes numbered 1 to 52. So the record, now presented, in each case gives the volume number and its year: followed by the title for the meeting, its place and date: The convenors for the Discussion are given (when recorded) as the Committee, the chairman coming first.

The chairman actually taking charge of the General Discussion was usually the Faraday Society President. His name is given, followed by the overall attendance at the meeting and the number of participants from overseas, most of whom will have been named in a brief note which precedes the contents sheet for the printed Discussion. Then the number of a sub-class is given: participants from overseas who actually contributed either by presenting a paper or by participating in the discussion of the papers. In some cases there is uncertainty in this lesser figure of perhaps 25%, but it offers some differentiation from overseas visitors who only attended the meeting. A small number of the senior scientists from outside the U.K. are named: usually, but not always, they were contributors at the meeting. Cognoscenti will then often be immediately able to make their assessment of the quality of the Discussion.

The extent and content of the Discussion volumes are given, with such general comment as may serve further to describe the aspects of the subject presented, or to suggest the character of the meeting in the context of the Faraday Society. The earlier meetings get the more extended comment. For them, also, is recorded what later never occurred: there were a few contributors who presented (or were part-authors of) more than two papers at the same Discussion. Their names are followed by that number in brackets.

L.E.S. has emphasised that the Discussion topics show the pattern of interests cultivated by the Faraday Society. This can be most simply expressed by assembling the Discussion titles under general headings which are themselves major areas in Physical Chemistry. This classification is only an approximation aimed to delineate roughly the areas of chemistry represented in the Discussions. As readers may wish to follow the sequences of topics in one particular area, the groups suggested and the titles they contain are given with the volume and year of the Discussion:

Reaction Kinetics and Molecular Reaction Dynamics:
    Vol. 42, 1945 (T.F.S.) Oxidation
    Vol. 2, 1947 The Labile Molecule
    Vol. 8, 1950 Heterogeneous Catalysis
    Vol. 10, 1951 Hydrocarbons
    Vol. 14, 1953 The Reactivity of Free Radicals
    Vol. 17, 1954 The Study of Fast Reactions
    Vol. 22, 1956 The Physical Chemistry of Processes at High Pressures
    Vol. 29, 1960 Oxidation–Reduction Reactions in Ionising Solvents
    Vol. 33, 1962 Inelastic Collisions of Atoms and Simple Molecules
    Vol. 37, 1964 Chemical Reactions in the Atmosphere
    Vol. 39, 1965 The Kinetics of Proton Transfer Reactions
    Vol. 41, 1966 The Role of the Adsorbed State in Heterogeneous Catalysis
    Vol. 44, 1967 Molecular Dynamics of the Chemical Reactions of Gases
    Vol. 46, 1968 Homogeneous Catalysis with Special Reference to Hydrogenation and Oxidation

Colloidal, Biological and Related Systems
    Vol. 42B, 1946 Swelling and Shrinking
    Vol. 3, 1948 Interactions of Water and Porous Materials

Vol. 6,   1949 Lipo-proteins
Vol. 9B,   1950 Optical Methods of Investigating Cell Structure
Vol. 11,   1951 The Size and Shape Factor in Colloidal Systems
Vol. 13,   1953 The Physical Chemistry of Proteins
Vol. 16,   1954 The Physical Chemistry of Dyeing and Tanning
Vol. 18,   1954 Coagulation and Flocculation
Vol. 20,   1955 The Physical Chemistry of Enzymes
Vol. 27,   1959 Energy Transfer with Special Reference to Biological Systems
Vol. 30,   1960 The Physical Chemistry of Aerosols
Vol. 42,   1966 Colloid Stability in Aqueous and Non-aqueous Media

Crystals and the Solid State
Vol. 5,   1949 Crystal Growth
Vol. 23,   1957 Molecular Mechanism of Rate Processes in Solids
Vol. 28,   1959 Crystal Imperfections and Chemical Reactivity of Solids
Vol. 38,   1964 Dislocations in Solids
Vol. 48,   1969 Motions in Molecular Crystals
Vol. 51,   1971 Electrical Conductivity in Organic Solids
Vol. 52,   1971 Surface Chemistry of Oxides

Molecular Spectroscopy and Structure
Vol. 42A, 1946 Dielectrics
Vol. 9A,   1950 Spectroscopy and Molecular Structure
Vol. 19,   1955 Microwave and Radio-frequency Spectroscopy
Vol. 26,   1958 Ions of the Transition Elements
Vol. 34,   1962 High Resolution Nuclear Magnetic Resonance
Vol. 35,   1963 The Structure of Electronically Excited Species in the Gas Phase
Vol. 47,   1969 Bonding in Metallo-organic Compounds

Systems at Equilibrium
Vol. 15,   1953 The Equilibrium Properties of Solutions of Non-electrolytes

Vol. 21, 1956 Membrane Phenomena
Vol. 25, 1958 Configurations and Interactions of Macromolecules and Liquid Crystals
Vol. 32, 1961 The Structure and Properties of Ionic Melts
Vol. 40, 1965 Intermolecular Forces
Vol. 43, 1967 The Structure and Properties of Liquids
Vol. 49, 1970 Polymer Solutions

Electrochemistry
Vol. 1, 1947 Electrode Processes
Vol. 24, 1957 Interactions in Ionic Solutions
Vol. 32, 1961 The Structure and Properties of Ionic Melts
Vol. 45, 1968 Electrode Reactions of Organic Molecules

Radiation Chemistry
Vol. 12, 1952 Radiation Chemistry
Vol. 31, 1961 Radiation Effects in Inorganic Solids
Vol. 36, 1963 Fundamental Processes in Radiation Chemistry
Vol. 37, 1964 Chemical Reactions in the Atmosphere

The two most obviously questionable placings in these lists are Vol. 27, 1959: Energy Transfer with Special Reference to Biological Systems; and Vol. 21, 1956: Membrane Phenomena. In the former, many papers dealt with simple atomic states and only half the programme mentioned the colloidal–biological structures. In membranes, the ultimate interest is in the dynamics of molecular transfers. And there are two General Discussions not classified: Vol. 4, 1948: The Physical Chemistry of Process Metallurgy; Vol. 7, 1949: Chromatographic Analysis. The reader may arrive at a different classification.

THE RECORD OF THE GENERAL DISCUSSIONS

These summaries are presented in the chronological order of the meetings:

Vol. 42, 1945. (*Trans. Faraday Soc.*) Oxidation   University College, London, 27th–28th September.

The Faraday Society General Discussions 285

Committee: *M. G. Evans*, C. N. Hinshelwood, H. W. Melville.
Chair: E. K. Rideal (President).
Total attending ca. 350. Participants from overseas 15+; contributing 4 papers; including A. E. van Arkel (Leiden), J. H. de Boer (Netherlands); G. Coumoulos (Athens); L. Farkas (Palestine); F.O. W. Gledrye (Poland); E. N. Gorter (Netherlands); H. R. Kruyt (Utrecht); M. Magat (Paris); M. Mathieu (Paris); H. Theorell (Stockholm); E. J. W. Verwey (Eindhoven); H. Vichnievsky (Bellevue).
Main headings: (I) Electron Transfer Reactions; (II) Oxidation of Hydrocarbons.
Papers 33; discussion pages 50; total pages of text 297.
This was the first Faraday Society Discussion after the cessation of hostilities. In his General Introduction the President said '. . .On behalf of the Society I extend a most hearty welcome to all those visitors who have come from overseas. Our Discussions have always attempted to be international in character and the large attendance of distinguished overseas visitors from Europe and from across the Atlantic present here today augurs well both for future Discussions of the Faraday Society and for the international cooperation of scientists. I only regret that it has proved impossible for our Russian colleagues to attend. . . The Society, however, has survived (although, as our foreign visitors may like to know, all that survived of Faraday Society records when Miss Kornitzer took over the work of our office was half a typewriter buried in a wall which once formed part of the Society's office) and indeed has entered its post-war with a larger membership and a more active body of members than ever before. . . . Presidents and Councils may come and go but the Society goes on. In other words, our General Secretary, Mr. G. S. W. Marlow, is the man around whom the physico-chemists of this country coagulate or crystallise. I take this opportunity of expressing my indebtedness. . .especially to our Secretary, Mr. Marlow.'

Fortunately there is a good photo, taken at lunchtime on the Thursday, in which as many as **226** of those attending are recorded. It contains major figures of the Faraday Society, including the elderly F. G. Donnan, one of the founder members, Mr. G. S. W. Marlow, other senior officers and members of staff. See the penultimate figure in the plates section with its key and list of names.

Vol. 42A, 1946. (*Trans. Faraday Soc.*) Dielectrics Bristol, 24th–26th April.

Committee: *D. W. G. Style*; R. J. W. LeFevre; Willis Jackson; J. B. Moullin; H. Fröhlich.

Chair: W. E. Garner (President).

Total attending ca. 150. Participants from overseas 19; contributing 9: including P. Abadie (Bagneux); A. E. van Arkel (Leiden); E. Bauer (Paris); C. J. F. Böttcher (Leiden); A. von Hippel (M.I.T.); J. G. Kirkwood (Chicago); M. Magat (Paris); C. P. Smyth (Princeton).

Main headings: (I) Theoretical; (II) Relaxation Times; (III) Breakdown; (IV) Measurement Techniques and Terminology; (V) Applications Including Liquid Systems; (VI) Chemical Compounds; (VII) Solids.

Papers 32; discussion pages 36; total pages of text 244.

In terms of the Faraday Society, this was a successor to the Oxford Discussion of 1934. Although not large in numbers, the rank of the participants from the U.K. and overseas was at a high level. Of particular significance was the presentation of several frequency dispersion studies, that is, the emergence of dielectric spectroscopy.

During the war and in conjunction with all-important advances in the use of radar, the klystron had appeared as a reliable source of microwaves. Sources of adequate strength from 30 to 1.2 cm wavelength permitted absorption measurements not previously made in this region. It is interesting to see how university departments had quickly acquired what was usually war-surplus sources and used them to explore molecular spectra in a new region of the electromagnetic spectrum. At Oxford R. P. Penrose had made measurements at 1.2 cm; D. H. Whiffen and H. W. Thompson at 1.3 and 3.3 cm; C. H. Collie, R. M. Ritson and J. B. Hasted at 1.25 and 10 cm: at Cambridge F. J. Cripwell and G. B. B. M. Sutherland had results for 1.25 and 3.3 cm; but at Manchester Willis Jackson and J. G. Powles had made measurements at six microwave frequencies up to 25 GHz ($\lambda = 1.2$ cm). Their results served to define fully, for the first time, the contour of the Debye dipolar absorption for four simple polar molecules and the significantly small influence of the solvent viscosity on their dipole reorientation relaxation times.

This was the third of many Faraday Discussions to be located in the same Bristol University lecture theatre. In 1929 with Lowry as President there had been one on Molecular Spectra and Molecular Structure: in 1938, with Travers in the chair, the topic was Reactions Involving Solids.

Vol. 42B, 1946.   *Swelling and Shrinking*   London, 24th–26th September.
Committee: Not named, but members of the Colloid Committee.
Chair: W. E. Garner (President).
Total attending 300. Participants from overseas 32; contributing 18; including D. G. Dervichian (Paris); P. M. Doty (Notre Dame, Indiana); J. J. Hermans (Utrecht); M. Magat (Paris); J. Th. G. Overbeek (Eindhoven); P. Putzeys (Louvain).
Main headings: None given.
Papers 35; discussion pages 53; total pages of text 302.
Contributors not named above included J. D. Bernal, who presented a General Introduction; F. E. Eirich; J. W. McBain; and M. F. Perutz.

There were eight papers dealing with synthetic polymers: five on cellulose and related structures; others on classical colloids, on soaps, biological fibres and cells.

Vol. 1, 1947.   *Electrode Processes*   Manchester, 9th–10th April.
Committee: *W. F. K. Wynne Jones*; C. F. Goodeve; A. Hickling; J. B. Baxter.
Chair: W. E. Garner (President).
Total attending ca. 135. Participants from overseas ca. 20; contributing 15, including A. Frumkin (Moscow); M. Haissinsky (Paris); J. Heyrovsky (Prague); J. G. Hoogland (Netherlands); J. J. Lingane (Harvard).
Main headings: (I) General and Theoretical; (II) (a) Deposition of Hydrogen; (b) Deposition of Metals; (III) Anodic and other Electrodic Processes.
Papers 41; discussion pages 43; total pages of text 334.
J. N. Agar (4).

That pre-war conditions were not yet established is apparent from an attendance of only 135 participants. Three senior Russian authors sent papers and, by correspondence, contributed to the discussion exchanges. Inevitably much attention was focussed on conditions at and near the electrode surface. In addition to the formation of the diffusion layer, its depth and electrical capacitance, the major concern of the Discussion centred on the associated rate processes, the rates of ionic discharge and of post-discharge interactions. The over-voltages arising from hydrogen and oxygen evolution at electrodes in aqueous media were discussed. Quite different considerations arose in metal deposition: there the structure of the deposit and its relation to the substrate were important aspects which benefited from electron diffraction studies.

Progress in this field could be well assessed by comparing the contents of this Discussion volume with the report on Electrode Reactions and Equilibria, the title for a meeting held 24 years earlier.

Vol. 2, 1947. *The Labile Molecule* Oxford, 23rd–25th September.
Committee: H. W. Melville, M. G. Evans, C. N. Hinshelwood, C. A. Coulson, W. A. Waters.
Chair: W. E. Garner (President).
Total attending ca. 250. Participants from overseas 23+; contributing 20, including P. D. Bartlett (Harvard); L. F. Fieser (Harvard); M. Magat (Paris); F. R. Mayo (Passaic, NJ); A. Pullman (Paris); E. W. R. Steacie (Ottawa); A. V. Tobolsky (Princeton).
Main headings: (I) Theoretical; (II) Gas Phase; (III) Liquid Phase: (A) Electron Transfer Reactions; (B) Hydrocarbons in Solution; (C) Oxidation-Reduction Reactions; (D) Experimental Technique; (IV) Polymerisations: (A) Reactions of Radicals and Monomers: (a) Initiation; (b) Propagation; (B) Chain Transfer and Inhibition; (C) Degradation.
Papers 50; discussion pages 71; total pages of text 404.
C. A. Coulson (4); M. J. S. Dewar (4); F. R. Mayo (3).

This was a major, not to say a monster Faraday Discussion. The subject was well chosen as the importance of free radicals in reaction kinetics had been demonstrated in many areas. Long accepted in oxidation processes, free radicals were now seen to play a key role in very many polymer formation processes. The extensive interest in the

The Faraday Society General Discussions 289

latter, and the many other developments during the war years which were now available for presentation, led to the unusual breadth of the material considered: in more normal times it would have formed the subject matter for at least two Discussions. The meeting served to emphasise the initiative and status of the Faraday Society in the post-war years.

Vol. 3, 1948. *Interaction of Water and Porous Materials* Southampton, 31st March–2nd April.
Committee: *J. H. Schulman*, G. S. Hartley, R. D. Preston, R. K. Schofield.
Chair: E. K. Rideal, N. K. Adam.
Total attending ca. 125. Participants from overseas 17; contributing 6, including J. H. van den Honert (Leiden); H. Lundegårdh (Sweden).
Main Headings: (I) Fundamental Aspects; (II) Botanical Aspects; (III) Zoological Aspects; (IV) Permeability to Water and Water Vapour of Textile and other Fibrous Materials; (V) Discussion on Oleophobic Surfaces.
Papers 32; discussion pages 33; total pages of text 293.

The discussion was essentially in terms of the macroscopic physical properties of water and their role in its transport. The biological systems showed how recondite processes could be in them, especially for non-equilibrium states.

Vol. 4, 1948. *The Physical Chemistry of Process Metallurgy* Ashorne Hill, Leamington Spa, 23rd–25th September.
Committee: *C. F. Goodeve*, H. J. T. Ellingham, C. W. Dannatt, S. Robson.
Chair: A. J. Allmand (President); Sir Andrew McCance.
Total attending ca. 130. Participants from overseas 26; contributing 13, including S. Fornander (Surahammar, Sweden); A. Juliard (Brussels); E. A. Peretti (Notre Dame, IL); L. M. Pigeon (Toronto); M. J. N. Pourbaix (Brussels).
Main headings: (I) Metallic Solutions; (II) Roasting and Reduction Processes; (III) Slags and Refining Processes.
Papers 25; discussion pages 72; total pages of text 343.

This discussion is notable as representing a reprise of early interests of

Chair: Sir John Lennard-Jones (President).
Total attending ca. 250. Participants from overseas 33; contributing 18, including J. H. de Boer (Delft); O. Beeck (Emeryville); A. Eucken (Göttingen); H. Eyring (Utah); K. J. Laidler (Washington); J. A. Morrison (Ottawa); H. S. Taylor (Princeton); J. Turkevich (Princeton).
Main headings: (I) Theories of Adsorption and Properties of Surface Layers; (II) Adsorption and Catalysis on Metals; (III) Adsorption and Catalysis on Oxides; (IV) Techniques.
Papers 44; discussion pages 31; total pages of text 365.
H. S. Taylor gave the fifth Spiers Memorial Lecture.
Beeck (3); Eley (3); T. J. Gray (3).

This Discussion topic drew together a significant number of its most important researchers. Its importance in industrial chemistry ensured that much study was always in progress. The new techniques and insights from solid-state physics were clearly in evidence. What not many years earlier had been relatively simple questions of molecules adsorbed and interacting on surfaces were now seen to involve a multiplicity of features: the lattice structure on the surface: lattice defects, their mobility and that of electrons or ions, leading to questions of band structure, conduction, semi-conduction and insulation – all had to be considered, at least in appropriate cases. To the outsider there seemed to be a multiplicity of adsorption states and a considerable variety of mechanisms in heterogeneous catalysis. However, in systems of one type the facts seemed to be well established.

Vol. 9A, 1950. *Spectroscopy and Molecular Structure* Cambridge, 25th–27th September.
Committee: J. T. Randall, G. B. B. M. Sutherland, W. C. Price.
Chair: Sir John Lennard-Jones (President): H. W. Thompson.
Total attending 250+ (A+B). Participants from overseas 75 (A+B); contributing 21, including G. Herzberg (Ottawa); D. H. Hornig (Providence, RI); J. Lecomte (Paris); S. Leach (Paris); M. Magat (Paris); R. Mecke (Freiburg); H. H. Nielsen (Columbus, Ohio); A. and B. Pullman (Paris); E. K. Plyler (Washington); E. Bright Wilson (Harvard).

Main headings: (I) Electronic Spectra; (II) Vibrational and Rotational Spectra; (III) Vibrational Spectra of Complex Molecules.
Papers 33; discussion pages 44; total pages of text 350.

This was the first part of a 'double' General Discussion arising from the need to give the Colloid and Biophysical Committee their annual due: see Vol. 9B below. Of this Molecular Spectroscopy meeting all is said by writing down the names of the first five authors in section (II): H. H. Nielsen, G. Herzberg, E. K. Plyer, E. Bright Wilson, D. F. Hornig. If one were choosing a 'first eleven' in molecular spectroscopy in 1950 these five would be in the team and Herzberg would be the captain. The other contributors included many only marginally less distinguished. Notable too was the transition of Thompson from the ultraviolet to the infrared and Porter's significant presentation of flash photolysis.

Vol. 9B, 1950. *Optical Methods of Investigating Cell Structure* Cambridge, 27th–28th September.
Committee: *J. T. Randall*, G. B. B. M. Sutherland, W. C. Price.
Chair: J. T. Randall.
Total attending 250+ (A+B). Participants from overseas 75 (A+B), contributing 9, including J. R. G. Bradfield (Chicago); H. B. Catchpole (Illinois); B. Commoner (St Louis); A. Engström (Karolinska Institutet); B. Thorell (Karolinska Institutet).
Main headings: (I) Microscopical Apparatus – Design and Techniques; (II) Application of Micro- and Macro-Spectrography.
Papers 17; discussion pages 25; total pages of text 151.

This was the second part of the Cambridge meeting. What the record shows was a lengthy discussion of spherical mirror microscopes, and that, of the 126 pages of the text given to papers, 47 pages are taken by 7 papers from Randall's King's College, London, Laboratory. This was the meeting of which the organisation caused so much discussion in the Faraday Council because of the involvement of the Colloid and Biophysical Committee. Although Randall was appointed convenor by the Faraday Council, he had been three times absent when it was discussed for a fourth time, and the difficulties were only resolved by a further special meeting of the Council (15th December 1949).

Vol. 10, 1951. *Hydrocarbons* Oxford, 11th–13th April.
Committee: *C. E. H. Bawn*, R. P. Bell, M. G. Evans.
Chair: Sir Charles Goodeve (President).
Total attending ca. 350. Participants from overseas 31; contributing 24, including J. G. Aston (Pennsylvania); S. W. Benson (California); R. D. Brown (Australia); Bryce Crawford (Minneapolis); G. Glockler (Iowa); E. C. Kuoyman (Amsterdam); M. Magat (Paris); W. A. Noyes Jr (Rochester, NY); K. S. Pitzer (Berkeley); J. Turkevich (Princeton).
Main headings: (I) Hydrocarbon Structure and Bond Properties; (II) Hydrocarbon Reactions: (A) Thermal Reactions; (B) Oxidation Reactions.
Papers 30; discussion pages 54; total pages of text 338.

Papers by M. G. Evans, Hinshelwood (2), Kistiakowsky, Lennard-Jones (2), Melville, Norrish and Pitzer show the quality of this Discussion. This is one of the Discussions for which L.E.S.'s summary extends to over 2000 words, but it is doubtful whether the substantial impact of its papers could be conveyed in significantly fewer.

Vol. 11, 1951. *The Size and Shape Factor in Colloidal Systems* Ashorne Hill, Leamington Spa, 18th–20th July.
Committee: *J. H. Schulman* and the members of the Colloid and Biophysical Committee.
Chair: Sir Charles Goodeve (President); Sir Eric Rideal.
Total attending ca. 220. Participants from overseas 38; contributing 19, including D. O. Dervichian (Paris); R. M. Fuoss (Yale); O. Kratky (Graz), A. Szent-Gyorgyi (Woods Hole, MA).
Main headings: (I) Classical Colloids; (II) Colloidal and Polyelectrolytes; (III) Fibrous Macromolecular Systems; (IV) Viruses.
Papers 26; discussion pages 43; total pages of text 252.

Debye submitted a paper, but as it had been published elsewhere it was withdrawn. Read today, Rideal's introductory survey serves to illustrate how much progress there had been. The size and shape of particles was determined by direct observation, apart from some X-ray diffraction studies.

Vol. 12, 1952. *Radiation Chemistry* Leeds, 8th–10th April.
Committee: *H. J. Emeleus*, F. S. Dainton, C. B. Allsopp and two from

the Colloid and Biophysical Committee.

Chair: Sir Charles Goodeve (President).

Total attending ca. 200. Participants from overseas 47, contributing 16, including A. O. Allen (Brookhaven); P. Bonet-Maury (Paris); M. Burton (Notre Dame, IL); W. M. Garrison (Berkeley); M. Haissinsky (Paris).

Main headings: (I) The Primary Act; (II) Actinometry and Radiolysis of Pure Liquids; (III) Indirect Action – Aqueous Systems with Single Solute; (IV) Protection and Sensitization.

Papers 23; discussion pages 88; total pages of text 318.

Dainton gave an introductory paper in which he itemized the features to be considered if an adequate account of typical radiation processes were to be elucidated. These included: (i) Primary energy injection measurement; (ii) Primary radiation yield; (iii) Track density patterns; (iv) The space uniformity or non-uniformity of the reaction processes; (v) Specific processes due to interaction with water and therefore present in all biological systems.

These general features established how far more involved radiation chemistry can be than its forerunner, photochemistry. This volume carries a photograph of the participants of whom 147 are identified, including F. C. Tompkins (Secretary) and Beatrice Kornitzer (Assistant Secretary). It is surely to be regretted that this visual record of the participants was so frequently neglected.

Vol. 13, 1953. *The Physical Chemistry of Proteins* Cambridge, 6th–8th August.

Committee: *J. H. Schulman*, Paley Johnson and two from the Colloid and Biophysical Committee.

Chair: J. H. Schulman.

Total attending ca. 400. Participants from overseas 85; contributing 24, including J. T. Edsall (Harvard); E. Cohn (Harvard); P. Doty (Harvard); T. L. Hill (Bethesda, MD); L. Pauling (Pasadena); K. O. Pedersen (Uppsala); H. Tiselius (Uppsala).

Main headings: (I) Experimental Techniques; (II) Low Molecular Weight Proteins; (III) High Molecular Weight Systems; (IV) Protein Interactions; (V) Conjugated Proteins.

Papers 27; discussion pages 42; total pages of text 287.

At this meeting Edsall gave the sixth Spiers Memorial Lecture. The report states 'about 400 members and visitors were present', making it one of the largest attendances at a Faraday General Discussion. There were many senior physical chemists present who are not reported as taking part in the discussions. There was much discussion of biological interest, including muscle-contraction. Within nine months and a hundred yards from where this meeting was held, the solution of another problem was established by Crick and Watson, the structure of DNA.

Vol. 14, 1953. *The Reactivity of Free Radicals* Toronto, Canada 8th–9th September.
Committee: Melville coordinated the Discussion with E. W. R. Steacie and Canadian colleagues.
Chair: Not recorded.
Total attending 200+. Participants from the U.K. were 44 named members and officers, of whom 31 contributed: from the U.S.A. 25 named participants, including M. Burton (Notre Dame); A. Farkas; M. S. Kharasch (Chicago); G. B. Kistiakowsky (Harvard); V. K. La Mer (Columbia University); L. Pauling (Pasadena); O. K. Rice (N. Carolina); W. H. Rodebush (Illinois).
Main headings: (I) Reactions in the Gas Phase; (II) Reactions in the Liquid Phase.
Papers 24; discussion pages 39; total pages of text 256.

This was the first General Discussion held in North America. Those from the U.S.A. named above are only 8 out of 25 American members of the Faraday Society who attended. As the total attendance was over 200, more than half were Canadian nationals. The inception of the meeting was due largely to E. W. R. Steacie, then President of the National Research Council of Canada and, later President of the Faraday Society.

Vol. 15, 1953. *The Equilibrium Properties of Solutions of Non-electrolytes* 16th–18th April, London.
Committee: *M. G. Evans*: no others are named.
Chair: Sir Hugh Taylor.

Total attendance ca. 250. Participants from overseas 57; contributing 23, including J. H. Hildebrand (Berkeley); J. G. Kirkwood (Yale); J. Koefoed (Copenhagen); I. Prigogine (Brussels); R. L. Scott (Los Angeles); I. Timmermans (Brussels); B. Vodar (Bellevue, France).
Main headings: (I) General Theory; (II) Experimental Work.
Papers 28: discussion pages 59: total pages of text 292.

Hildebrand gave the seventh Spiers Memorial Lecture. This was the one hundredth General Discussion of the Faraday Society. It was appropriately located in Faraday's scientific home, The Royal Institution of Great Britain, and was the occasion for celebrating the 50th Anniversary of the founding of the Society. The quality of the Discussion matched these circumstances.

A principal interest was the calculation of entropy and energy factors in the liquid state, and hence other physical properties. In the pre-computer era simple models of a quasi-lattice form were used for entropy calculation and, often, the Lennard-Jones 12–6 interaction energy function was introduced as a workable approximation. It is not a matter of hindsight to be surprised that Guggenheim felt he was making a significant point in qualifying the adequacy of the 12–6 function. Since 1928 the Morse function indicated that when atoms came into close contact an exponential function was needed the better to express the repulsive energy.

Both these features – the quasi-lattice model and the 12–6 potential function – were soon to be abandoned in developments which were greatly accelerated by the arrival in most laboratories of early versions of the electronic computer. In the 1936 meeting on the same theme Bernal's use of the distribution function and Fritz London's exposition of dispersion forces had given new pathways in liquid structure studies; no equally fruitful concepts appear from the 1953 Discussion. Substantial advances were again recorded at the Exeter meeting on liquids in 1967.

Vol. 16, 1954. *The Physical Chemistry of Dyeing and Tanning* Leeds, 8th–10th September.
Committee: *D. D. Eley*, J. H. Schulman, J. B. Speakman, J. A. Kitchener, K. G. Pankhurst.

Chair: Sir Hugh Taylor (President).
Total attending 154. Participants from overseas 17; contributing 13, including K. H. Gustavson (Stockholm); B. Olofsson (Gothenburg); W. R. Remington (Delaware); H. J. White (Princeton).
Main headings: None given.
Papers 23; discussion pages 29; total pages of text 251.

This meeting could well be regarded as a missionary effort by the Faraday Society to promote an analytic scientific study of important traditional crafts. In that sense it was undoubtedly successful.

Vol. 17, 1954.   *The Study of Fast Reactions*   Birmingham, 7th–9th April.
Committee: *F. J. W. Roughton*, H. W. Melville, R. P. Bell.
Chair: R. G. W. Norrish (President).
Total attending ca.250. Participants from overseas 34, contributing 17, including S. H. Bauer (Cornell); M. Eigen (Göttingen); H. S. Johnston (Stanford); W. Jost (Darmstadt); G. B. Kistiakowsky (Harvard); I. M. Kolthoff (Minneapolis).
Main headings: (A) Gas Reactions; (B) Solution Reactions.
Papers 28; discussion pages 28; total pages of text 234.

The 'Study' of the title is taken as experimental study, and after an introductory paper by Melville, section A consists of eight accounts, all on different techniques developed by major schools at Stanford, Birmingham, Harvard, Cambridge, Pasadena and Cornell. A paper by Bull and Moon reported the first ever study of chemically reactive collisions in a molecular beam. Section B on solution studies is equally varied and impressive. It contains a brief paper by Eigen summarizing the methods which led him to sharing a Nobel Prize in Chemistry with Norrish and Porter, who are represented in both sections A and B.

Vol. 18, 1954.   *Coagulation and Flocculation*   Sheffield, 15th–17th September.
Committee: *D. D. Eley*, D. C. Henry, R. K. Schofield, J. M. Marrack.
Chair: R. G. W. Norrish (President).
Total attending ca.250. Participants from overseas 34; contributing 24, including D. V. Derjaguin (Moscow); D. G. Dervichian (Paris); P. J.

Flory (Stanford); P. S. Prokhorov (Moscow); J. Th. G. Overbeek (Utrecht): B. Tezak (Zagreb).
Main headings: (I) Classical Coagulation (II) Coacervation (III) Biological Systems.
Papers 32; discussion pages 70; total pages of text 364.
A. S. C. Lawrence (4): J. Th. G. Overbeek (4).

This Discussion saw the presence of senior scientists from Eastern Europe and from the U.S.S.R. There were introductory papers for each of the three divisions in the meeting. The introduction of quantitative measurements and the consideration of the factors in antigen–antibody precipitation deserve mention.

Vol. 19, 1955. *Microwave and Radio-Frequency Spectroscopy* Cambridge, 4th–6th April.
Committee: *H. C. Longuet-Higgins*, R. G. W. Norrish, R. E. Richards, T. M. Sugden.
Chair: R. G. W. Norrish (President).
Total attending ca.250. Participants from overseas 82; contributing 19, including J. Duchesne (Liège); W. Gordy (Durham, NC); W. D. Gwinn (Berkeley); D. F. Hornig (Providence, RI); G. B. Kistiakowsky (Harvard); D. M. Kozyrev (Kazan); W. N. Lipscomb (Minneapolis); R. S. Mulliken (Chicago); A. M. Prokhorov (Moscow); C. H. Townes (New York).
Main headings: (I) Microwave Spectroscopy; (II) Microwave Absorption; (III) Nuclear Magnetic Resonance.
Papers 32; discussion pages 42; total pages of text 281.

This is an instance of the Faraday Society staging a General Discussion on a topic just as its basic scope and impact were being established. It resulted in the largest attendance of overseas participants up to its date and was an outstanding survey enhanced in its value by the critical contributions in the reported discussion exchanges.

Vol. 20, 1955. *The Physical Chemistry of Enzymes* Oxford, 10th–12th August.
Committee: *F. J. W. Roughton*, D. D. Eley, C. N. Hinshelwood, M. Dixon, A. G. Ogston, H. Gutfreund.

Chair: R. G. W. Norrish (President).

Total attending 250+. Participants from overseas 85; contributing 26, including F. Bergmann (Jerusalem); M. J. Fraser (Halifax, Nova Scotia); J. G. Kirkwood (Yale); K. J. Laidler (Ottawa); H. Neurath (Seattle); J. Polonovski (Paris); H. Theorell (Stockholm).

Main headings: (I) Characterization and Physical Properties; (II) Kinetics and Mechanisms.

Papers 28; discussion pages 76; total pages of text 316.

An even larger number of overseas participants attended this Discussion, principally from: U.S.A. 33; Netherlands 14; Sweden 5; Germany 5; Israel 4; Denmark 4; India 2. From the report of the meeting one acquires the impression that it was of very high quality. In addition to those already named above, the contributors included R. L. Baldwin; Malcolm Dixon; D. D. Eley; Philip George; R. K. Morton; A. G. Ogston; F. J. W. Roughton; J. W. Williams.

Vol. 21, 1956.   *Membrane Phenomena*   Nottingham, 10th–12th April.
Committee: *D. D. Eley*, J. H. Schulman, R. M. Barrer.
Chair: R. P. Bell (President).

Total attending ca. 250. Participants from overseas 45; contributing 24, including K. F. Bonhöffer (Göttingen); F. Helfferich (Göttingen); G. Manecke (Berlin); M. Nagasawa (Nagoya); G. Scatchard (M.I.T.); H. J. Staverman (Delft).

Main headings: (A) Fundamental Studies; (B) Properties of Particular Membranes; (C) Properties of Biological Membranes.

Papers 23; discussion pages 63; total pages of text 280.

The eighth Spiers Memorial Lecture was given by E. T. Teorell (Uppsala).

Membranes had now come to include, at least for the purpose of this meeting, ion-exchange systems which might be purely inorganic in character. This had general significance in that it drew attention to the importance of initial interactions at the surface of the membrane.

Some of the ion-exchange systems could be treated in terms of well defined models allowing quantitative calculations which could be compared with experimental data. This provided a stimulus towards comparable treatments of natural membranes. It became clear that fuller structural detail was a prerequisite to the adequate understanding of these recondite biological entities.

Emeritus Professor F. G. Donnan, one of the founder members of the Society and a notable contributor to the study of ionic factors in membranes, was sent greetings from the meeting.

Vol. 22, 1956. *The Physical Chemistry of Processes at High Pressures* Glasgow, 20th–21st September.
Committee: *A. R. Ubbelohde*, F. P. Bowden, D. Traill.
Chair: R. P. Bell (President)
Total attending not given. Participants from overseas 49; contributing 15, including M. A. Cook (Salt Lake City); H. G. Drickamer (Illinois); A. H. Ewald (Sydney); S. D. Hamann (Sydney); B. Vodar (Bellevue, France).
Main headings: (A) Equations of State, Physical Properties and Thermodynamic Transformation; (B) Chemical Reactions and Transformations at High Pressure; (C) Detonation and Other High Temperature Phenomena at High Pressures.
Papers 27; discussion pages 38; total pages of text 220.
B. Vodar (3).

Section C was largely concerned with explosion processes. Amongst the extreme conditions explored in the papers was that of argon at 72 000 atm. This pressure was generated by explosions and the gas properties examined by using X-ray flash techniques.

Vol. 23, 1957. *Molecular Mechanism of Rate Processes in Solids* Amsterdam, 15th–18th April.
Committee: *R. G. W. Norrish*, F. P. Bowden, A. R. Ubbelohde, F. C. Tompkins; C. J. F. Böttcher acted as the liaison with the Dutch members.
Chair: R. P. Bell (President); Mr. Gerdling.
Details of participants are not given.
Main headings: (A) Relaxation Processes; introduced by C. J. F. Böttcher (Leiden); (B) Steady State Processes with no Lattice Rearrangement; introduced by J. S. Koehler and F. Seitz (Urbana, Illinois); (C) Steady State Processes Involving Lattice Rearrangement, introduced

by J. H. de Boer (Geleen, Netherlands).
Papers 22; discussion pages 50; total pages of text 234.

This General Discussion was held at the invitation of, and in conjunction with the Royal Dutch Chemical Society and was notable as an instance of the strongly developed, international character of the Faraday Society. Members and visitors, who included scientists from America, Australia, Belgium, France, Germany, Italy, Netherlands, Sweden, Switzerland and the United Kingdom, were welcomed at a reception given by the Burgomaster of Amsterdam at the Rijksmuseum and at the Koninklijke Institut by the President of the Faraday Scoiety, Mr R. P. Bell, F.R.S., and the President of the Royal Dutch Chemical Society, Mr. Gerdling.

An obvious comment is that by arranging such an overseas meeting the Faraday Society acquired a Discussion planned on a different basis from the U.K. meetings. This programme was, almost entirely, based on suggestions by the Dutch organisers. Their outlook would have been different and, in relation to developments throughout Europe, better informed than that of U.K. Faraday members. This advantage was gained thanks to the accepted international character of the Society.

Vol. 24, 1957. *Interactions in Ionic Solutions* Oxford, 17th–19th September.
Committee: R. P. Bell, C. W. Davies, D. H. Everett, G. J. Hills, J. E. Prue.
Chair: R. P. Bell (President).
Total attending 250. Participants from overseas 45; contributing 25, including M. Eigen (Göttingen); H. Falkenhagen (Rostock); H. S. Harned (Yale); L. Onsager (Yale); R. A. Robinson (Singapore).
Main headings: (I) General Theory; (II) Incomplete Dissociation; (III) Ion–Solvent Interaction.
Papers 23; discussion pages 59; total pages of text 232.
Harned gave the ninth Spiers Memorial Lecture.

This Discussion was on a topic much cultivated by physical chemists since the time of Arrhenius, but after the Debye–Hückel treatment and its refinement by Onsager, little qualitatively new had developed: the solvation of ions was as open a question as it had long been.

A major concern in the meeting was the wide departure from the Debye–Hückel relations. Much evidence indicated that ion association of various degrees of complexity occurred. C. W. Davies summarized data from a number of sources which led to reasonably consistent ion-pair equilibrium constants representing initial departures from complete dissociation. Bjerrum had envisaged this as early as 1926.

Manfred Eigen described his pulsed-field method, the development of which led to his sharing a Nobel Prize and which allowed estimates of relaxation times for the ionic interactions.

Vol. 25, 1958. *Configuations and Interactions of Macromolecules and Liquid Crystals* Leeds, 15th–17th April.
Committee: Not recorded
Chair: Sir Harry Melville (President).
Total attending 200+. Participants from overseas 38; contributing 15, including S. E. Bresler (Leningrad); D. G. Dervichian (Paris); K. V. A. Kropatshev (Leningrad); M. Kryszewski (Torun); W. Maier (Freiburg); S. A. Rice (Chicago); D. F. Waugh (Harvard).
Main headings: (I) Liquid Crystals; (II) Macromolecules.
Papers 22; discussion pages 57; total pages of text 229.

There were lengthy contributions to the discussion sessions: one of S. A. Rice's occupies eight pages of text, being longer than several of the papers. It must be realised that many items in the record of these discussions will have been communicated after the meetings, sometimes by authors not actually present during the discussion.

The meeting took place when many aspects of long-chain behaviour were being profitably studied and there were common features with anisotropic liquids and biomolecular systems. It seemed clear that generalised models such as that considered by Longuet-Higgins would need to be considered carefully in view of the great differences in interaction energies and flexibilities illustrated by polythene and cellulose derivatives.

In his introductory survey, Bernal showed his ability to coordinate a wide range of observations. He provided an overall framework for the discussion of the molecular systems in terms of their primary, secondary, tertiary and, finally, quaternary structures. This served to include major biological instances where he emphasized the role of the hydrogen bond. This was appropriate as a bridge to the presentation of macromolecules which Astbury made in terms of protein structures.

Vol. 26, 1958. *Ions of the Transition Elements* Dublin, 9th–11th September.
Committee: *J. W. Linnett*, M. Pryce, L. E. Orgel.
Chair: J. W. Linnett, J. H. Van Vleck (Harvard), L. Pauling (Pasadena).
Total attending ca.150. Participants from overseas: not recorded; contributing 7. The record merely states that 'about 150 members and visitors were present', without giving any figure for those from outside the British Isles.
Main headings: (I) Optical and Magnetic Properties; (II) Energetics of Complexes.
Papers 21; discussion pages 29; total pages of text 186.

The attendance was one of the smallest for a General Discussion. It was one of the few such meetings that Pauling attended and, remarkably, there is no record that he made a contribution. It was singularly appropriate for its date as it dealt with the new, theoretically sophisticated, electronic-state interpretations in inorganic chemistry. It was in this context that what had been empirical chemistry was being transformed by analytical valence theory.

Vol. 27, 1959. *Energy Transfer with Special Reference to Biological Systems* Nottingham, 14th–16th April.
Committee: *F. S. Dainton*, G. Porter, D. D. Eley, J. A. V. Butler and members of the Colloid and Biophysical Committee.
Chair: E. W. R. Steacie (President).
Total attending 200+. Participants from overseas 53; contributing 20, including S. Claesson (Uppsala); Z. R. Grabowski (Warsaw); F. H. Johnson (Princeton); J. Olin (Zürich); E. Rabinowitch (Urbana); A. Szent-Gyorgyi (Woods Hole).
Main headings: (I) Modes of Energy Transfer from Excited and Unstable Ionized States; (II) Energy Migration in Organized Biological Systems.
Papers 25; discussion pages 59; total pages of text 266.
Th. Förster (Stuttgart) gave the tenth Spiers Memorial Lecture.

This meeting attracted many biophysicists. Alternative modes of photo-excitation formed much of the subject matter. There was much speculation on radiationless transfer and photo-electronic pathways in the biological systems but not, it would seem, experimental confirmation for most of the proposals advanced in such contexts.

Vol. 28, 1959. *Crystal Imperfections and Chemical Reactivity of Solids* Kingston, Ontario, 2nd–4th September.
Committee: *E. W. R. Steacie*, J. A. Morrison, J. W. Mitchell, D. D. Eley, D. J. E. Ingram, W. E. Garner.
Chair: E. W. R. Steacie (President).
Total attending 250+. The record names 41 members and officers from the U.K., one from Dublin and one from Erlangen who attended.
Main headings: I; II; III; IV: but no further titles.
Papers 24; discussion pages 41; total pages of text 243.

This was the second General Discussion held in Canada, seven years after the Toronto meeting. It owed much to the President, who was also the senior scientist in the National Research Council of Canada. Of the twenty-four papers, nine came from the U.K., seven from the U.S., five from Canada, one each from the U.S.S.R. and West Germany, and one was partly U.S. and partly Canada.

Vol. 29, 1960. *Oxidation–Reduction Reactions in Ionizing Solvents* Durham, 11th–13th April.
Committee: *W. F. K. Wynne Jones*, J. H. Baxendale, F. S. Dainton.
Chair: Sir Harry Melville (President).
Total attending ca.200. Participants from overseas 30; contributing 14, including A. W. Adamson (California); E. de Boer (Amsterdam); V. Caglioti (Rome); L. J. Csanji (Szeged); R. W. Dodson (Brookhaven); G. J. Hoijtink (Amsterdam); R. A. Marcus (Brooklyn); L. J. Oosterhoff (Amsterdam); J. Sutton (Paris); A. A. Vleck (Prague).
Main headings: (I) Exchange Reactions and Electron Transfer Reactions Including Isotopic Exchange; (II) Oxidation–Reduction Reactions Involving Inorganic Substrates; (III) Oxidation–Reduction Reactions Involving Organic Substrates.
Papers 23; discussion pages 42; total pages of text 255.

Whilst many General Discussion topics might appear to be more physics than chemistry, this subject was of the reverse character, being almost pure chemistry.

Vol. 30, 1960.   *The Physical Chemistry of Aerosols*   Bristol, 13th–15th September.
Committee: R. Lessing, G. A. H. Elton, A. R. Ubbelohde, J. A. Kitchener.
Chair: Sir Harry Melville (President).
Total attending 100+. Participants from overseas 10+; contributing 8, including F. P. Buff (Rochester, NY); F. T. Gucker (Indiana); C. Kaziz (Paris); M. Kerker (Potsdam, NY); B. R. Stein (Frankfurt); V. B. Vouk (Zagreb).
Main headings: (I) Nucleation: Homogeneous and Heterogeneous; (II) Growth of Particles; (III) Physical and Chemical Properties.
Papers 23; discussion pages 32; total pages of text 222.

War Department scientists were clearly involved in studies of aerosols, but the attendance was modest, 'over 100 members and guests were present'. Three papers came from Moscow, two having Derjaguin amongst the authors, the other having Prokhorov. The record does not suggest they were present at the meeting. The volume of papers shows there was much in common with the discussion on Coagulation and Flocculation in 1954.

Vol. 31, 1961.   *Radiation Effects in Inorganic Solids*   Saclay (France), 11th–12th April.
Committee: F. S. Dainton, G. Porter, J. A. V. Butler [M. M. Caillat, M. Magat, M. Guinier, M. Haissinsky].
Chair: Sir Cyril Hinshelwood (President).
'Over 200 members (i.e. of the Faraday Society) and visitors were present: these included representatives from Austria, Belgium, Czechoslovakia, Denmark, France, Germany, Italy, Netherlands, Poland, Switzerland, the U.K., and the U.S.A.'
Main headings: (I) Metals and Alloys; (II) Non-Metallic Solids.
Papers 26; discussion pages 30; total pages of text 269.

A General Introduction was given by the American physicist, G. H. Vineyard, of the Brookhaven National Laboratory. There was a notable representation from national Radiation or Atomic Energy Laboratories. In addition to Brookhaven (U.S.A.), these were Oak Ridge (U.S.A.), Harwell (U.K.), Saclay (France), Grenoble (France), Radium Institute (Paris), Academy of Sciences (Moscow). From the last Voevodsky attended to present a paper.

The topic was essentially one of solid-state physics. The lattice and its defects, holes, excitons etc. and their mobilities, were the subjects of sophisticated physical studies.

Vol. 32, 1961. *The Structure and Properties of Ionic Melts* Liverpool, 5th–7th September.
Committee: *A. R. Ubbelohde*; no other names given.
Chair: Sir Cyril Hinshelwood (President).
Total attending ca.150. Participants from overseas 45; contributing 30, including A. Bloom (Tasmania); J. O'M. Bockris (Pennsylvania); A. S. Dworkin (Oak Ridge, TN); T. Førland (Trondheim); K. Furukawa (Sendai, Japan); A. Klemim (Mainz) M. Shimoji (Hokkaido); B. R. Sundheim (New York); J. Zarzycki (Paris).
Main headings: (I) Structure and Spectra; (II) Cryoscopy; (III) Thermodynamics; (IV) Transport Properties.
Papers 26; discussion pages 53; total pages of text 262.

As a General Introduction Bloom gave the eleventh Spiers Memorial Lecture.

The meeting was one incorporating significant theoretical and experimental presentations, often in the same paper. Studies of equilibrium and transport properties were reported. For the former, X-ray and neutron diffraction provided basic aspects; amongst other methods the electrical conductivity was prominent in the latter, frequently indicating that complexes of ions appeared in the melt. Thermodynamic functions could be calculated for melts of simple ions and the results compared with experimental data as a means of approaching adequate models for their structure. It scarcely needed experiment to show that mixed melts were less tractable. Had the frequency dependence of the conductivity been explored, it would have shown that a jump process was usually involved, the temperature dependence of its frequency providing evidence of the associated energy.

This small meeting was largely the result of Ubbelohde's prompting. The 45 overseas scientists formed almost half of the total attendance. Of the 24 papers, only 5 were contributed by U.K. authors.

Vol. 33, 1962.   *Inelastic Collisions of Atoms and Simple Molecules*   Cambridge, 10th–12th April.
Committee: *J W Linnett*, H. C. Longuet-Higgins, and co-opted members.
Chair: Sir Cyril Hinshelwood (President).
Total attending 100+. Participants from overseas 67; contributing 16, including G. Amat (Paris); D. Herschbach (California); K. F. Herzfeld (Washington); I. K. Larin (Moscow); S. Leach (Paris); E. E. Nikitin (Moscow); J. C. Polanyi (Toronto); B. Widom (Cornell).
Main headings: (I) Collision Processes not Involving Chemical Reactions; (II) Collision Processes Involving Chemical Reactions.
Papers 25; discussion pages 68; total pages of text 294.

It would seem that the total attendance – 'over 100 members and guests' – was not large, but more than one half came from overseas. Amongst those distinguished visitors, it is noticeable that R. H. Cole (Providence, RI); H. Eyring (Utah); D. F. Hornig (Princeton); W. Jost (Göttingen); do not appear to have contributed to the Discussion. This Discussion is clearly cognate to that on Fast Reactions of 1954. Advances are seen in the experimental data becoming available on energy exchange between different molecular modes – translation, vibration and rotation. Electronically excited states were included in these studies. Each different molecule had its own behaviour in these energy exchanges and this lack of pattern did not allow of general conclusions coming from the meeting.

Vol. 34, 1962.   *High Resolution Nuclear Magnetic Resonance*   Oxford, 17th–19th September.
Committee: *J. A. Pople*, H. C. Longuet-Higgins, R. E. Richards.
Chair: Sir Cyril Hinshelwood (President).
Total attending ca. 250. Participants from overseas 65; contributing 15, including B. P. Dailey (New York); H. S. Gutowsky (Urbana, IL); J.

D. Roberts (Pasadena); J. N. Shoolery (Palo Alto, CA); J. S. Waugh (M.I.T.).
Main headings: (I) N.M.R. in Diamagnetic Materials; (II) N.M.R. in Paramagnetic Materials; (III) Applications to the Determination of Molecular Structure; (IV) Applications to the Study of Kinetic Processes.
Papers 24; discussion pages 22; total pages of text 196.

Nuclear magnetic resonance was in a highly developed state before it became the subject of a General Discussion of the Faraday Society – three major texts had already been published in 1959. Usually such major developments in physical chemistry received earlier placing in the Discussions. Possibly this delayed interest arose from its chemical physics character. Theory and technique were presented at a high level, one development being Andrews's introduction of the rotating crystal and rotating field procedures.

Vol. 35, 1963. *The Structure of Electronically Excited Species in the Gas-Phase* Dundee, 2nd–3rd April.
Committee: *A. D. Walsh*, D. H. Everett, D. W. G. Style, T. M. Sugden.
Chair: A. R. Ubbelohde (President).
Total attending 135+. Participants from overseas 30; contributing 14, including A. E. Douglas (Ottawa); D. A. Dows (Los Angeles); H. Hering (Paris); G. Herzberg (Ottawa); A. Kuppermann (Pasadena); D. A. Ramsay (Ottawa); B. Rosen (Liège); J. H. van der Waals (Amsterdam).
Main headings: None given.
Papers 23; discussion pages 28; total pages of text 233.

This was a General Discussion of the highest quality. It had an introduction in the form of the twelfth Spiers Memorial Lecture given by the world leader in the study of electronically excited molecular states, Gerhard Herzberg. His school at Ottawa was well represented at the meeting. Unusual was the absence of any sub-division of the contributions. Although classifiable as spectroscopic and, as such, in a distinct minority amongst Faraday Discussions, the topic was fundamental to much in chemistry and, in particular, to that regular concern of the Society, reaction kinetics.

Vol. 36, 1963. *Fundamental Processes in Radiation Chemistry* University of Notre Dame, Indiana, 2nd–4th September.
Committee: F. S. Dainton, G. Porter, W. Wild (in U.K.).
Chair: A. R. Ubbelohde (President) and six other senior participants.
Total attending 200+. Essentially half the papers came from the U.K. (10), France (1) and Germany (1); the others came from Canada (1) and the U.S.A. (9).
Main headings: None given.
Papers 22; discussion pages 93; total pages of text 319.

The discussion must have been vigorous; it occupied 30% of the text. Dainton contributed Concluding Remarks.

This Discussion bears resemblance to that in 1959 on Energy Transfer in Biological Systems. The activating agents – photons and charged particles – rarely producing a simple effect: the sequence of events was usually very short-lived and whilst the end result may have been adequately defined, there appeared to be much room for speculation as to how it was reached.

Vol. 37, 1964. *Chemical Reactions in the Atmosphere* Edinburgh, 2nd–3rd April.
Committee: R. G. W. Norrish; G. Porter; T. L. Cottrell.
Chair: A. R. Ubbelohde (President).
Total attending 143. Participants from overseas 38; contributing 15, including L. V. Berkner (Dallas); W. L. Fite (San Diego); M. Nicolet (Brussels); H. I. Schiff (Montreal).
Main headings: (I) Reactions and Photochemistry of Atoms and Molecules; (II) Reaction of Charged Species.
Papers 22; discussion pages 17; total pages of text 218.

Of the papers eight were from the U.K., eleven from the U.S.A. and only one of those from an institute of higher education (Rensselaer Polytechnic, NY): others came from the Geophysics Corporation of America; Aerospace Corporation, Los Angeles; Air Force Research Laboratories, Massachusetts; etc.

Norrish had long pressed this topic in Faraday Council considerations of future General Discussions. It related to his long developed interests in photochemistry and anticipated, even foreshadowed, the later greatly expanded studies of the subject.

Vol. 38, 1964. *Dislocations in Solids* Göttingen, West Germany, 15th–17th April.
Committee: *A. R. Ubbelohde*; C. J. F. Böttcher; W. Jost; D. H. Everett; A. H. Cottrell.
Chair: A. R. Ubbelohde (President).
Total attending 164; participants: see below.
Main headings: (I) General Dislocation Theory; (II) Work Hardening – Surface Effects; (III) Dislocation Mobility and Generation; (IV) Structure.
Papers 28; discussion pages 73; total pages of text 314.

This was the first Faraday Society General Discussion held in West Germany. It was greatly facilitated by Professor W. Jost's presence on the Council. If the total contributions (i.e. papers and individual contributions) to the discussion sections are counted, their origins are approximately the following: West Germany 14; U.K. 11; U.S.A. 9; Australia, Belgium, Canada, Czechoslovakia, Netherlands, South Africa, Switzerland one each. The principal contributing centres were Stuttgart, Göttingen and Cambridge (England).

Dislocations were a major theme in an earlier Faraday Discussion – it was at Bristol in 1949, when F. C. Frank's new insights were much to the fore. In Vol. 38, of over 470 references, only five were noticed to Frank's work. The volume is notable in having 38 pages of photoprints, many of them providing striking illustrations of structural defects in solids.

Vol. 39, 1965. *The Kinetics of Proton Transfer Processes* Newcastle-upon-Tyne, 12th–14th April.
Committee: *V. Gold*; R. P. Bell; J. N. Agar; W. F. K. Wynne Jones; E. F. Caldin.
Chair: F. S. Dainton (President).
Total attending 200+. Participants from overseas 47; contributing 20,

including B. E. Conway (Ottawa); M. Eigen (Göttingen); E. U. Franck (Karlsruhe); M. M. Kreevoy (Minneapolis); A. H. Kresge (Chicago); F. A. Long (Cornell); D. B. Mathews (Philadelphia); H. W. Nürnberg (Jülich).
Main headings: None given.
Papers 21; discussion pages 78; total pages of text 272.

In a brief, masterly General Introduction, Eigen makes references to significant earlier Faraday Society Discussions bearing on the present topic: a Cambridge meeting on Homogeneous Catalysis when, *inter alia*, Brönsted's quantitative treatment of acidic and basic catalysis was presented: then, 'another famous meeting' (London, 1941), where the 'illuminating papers given by Ingold, Hughes and others certainly respresent a landmark in establishing the discipline of physical-organic chemistry!'. In studying the elementary steps of hydrogen-ion catalysis 'again it was the Faraday Society who opened the discussion of such possibilities at their Birmingham meeting (1954).' That was the meeting where Eigen himself expanded the principles and scope of his methods of studying relaxation times.

Four different major methods are illustrated in the present meeting. There was much (and at times apparently heated) discussion on whether proton tunnelling occurred in a number of the reaction steps studied. Certainty in this much anticipated feature only came later: most clearly from infrared spectra of individual molecular structures.

Vol. 40, 1965. *Intermolecular Forces* Bristol, 14th–16th September.
Committee: *D. H. Everett*; H. C. Longuet-Higgins; J. S. Rowlinson; R. H. Ottewill.
Chair: F. S. Dainton (President).
Total attending 276. Participants from overseas 76; contributing 30, including J. A. Barker (Melbourne); W. Byers Brown (Wisconsin); F. Danon (Buenos Aires); N. R. Kestner (Stanford, CA); D. P. Poshkus (Vilnius); O. K. Rice (North Carolina); H. A. Scheraga (Cornell).
Main headings: None given.
Papers 26; discussion pages 44; total pages of text 284.
H. C. Longuet-Higgins gave an introduction to the Discussion which formed the thirteenth Spiers Memorial Lecture.

In his Summarizing Remarks, C. A. Coulson emphasized how broad and deep was the theme of intermolecular forces – 'one of the few great threads which serve to bind us together... in the exploration of chemistry.' This claim is supported by the attendance which included, from the Netherlands 20; U.S.A. 14; West Germany 8; Belgium 6; France 5; Italy 5; Austria 3; Australia, Bulgaria, New Zealand, Spain, Switzerland 2 each; and others from Argentina, Denmark, Canada, East Germany, Israel and Poland.

Coulson's comment is illustrated by the major sequence provided by the Faraday Society's earlier General Discussions on liquids: at Edinburgh (1936) and London (1953). This Bristol meeting was the last before the computer took over. Significant new experimental data were reported from the collisional interactions observed in crossed molecular beams. This offered an approach to a more direct check on molecular potential functions than had previously been available.

Quantum models, including molecular orbital representations, had established the electronic structures of molecules and served to emphasize that even molecular pair interactions would need a multiple-term potential function. That this would not generally be deducible from experimental data was a consequence of the variety of site interactions in condensed phases. The gas phase offered the possibility of justifiable averaging.

The interaction of polymer, including protein molecules, presented complexity of a different order of magnitude. Here an essential feature was the appropriate choice of approximations. The computer was in evidence in several evaluations, but no fully computer based treatments were presented.

Vol. 41, 1966. *The Role of the Adsorbed State in Heterogeneous Catalysis* Liverpool, 4th–6th April.
Committee: *C. E. H. Bawn*; F. S. Dainton; D. D. Eley; C. Kemball.
Chair: F. S. Dainton (President).
Total attending 342. Participants from overseas 105; contributing 47, including G. K. Boreskov (Novosibirsk); A. Cimino (Perugia); G. Ehrlich (Schenectady); Z. Knor (Prague); V. Ponec (Prague); L. L. van

Reijen (Amsterdam); G. M. Schwab (Munich); J. Turkevich (Princeton).
Main headings: None given.
Papers 23; discussion pages 111; total pages of text 410.

This topic had a long history in Faraday Society Discussions and it produced one of the larger meetings then recorded: '342 members and others were present.' Of this number one-third were from overseas and contributed substantially to the meeting. In the first discussion session there are 27 items reported, 17 of them from overseas visitors. The overseas visitors who can be counted, came as follows: Netherlands 25; U.S.A. 21; France 14; West Germany 12; Belgium 5; Czechoslovakia 5; Switzerland 4; Denmark, East Germany, Italy 3 each; Canada, Spain, U.S.S.R. 2 each; Hungary, Poland, Thailand one each. This was a repeat and an updating of the 1950 Discussion: even the location, Liverpool, was the same. New techniques were in evidence: field emission microscopy: infrared spectroscopy of adsorbed layers: electron-spin resonance spectroscopy: Hall effect studies. Whilst techniques had advanced, there did not seem to have been a corresponding increase in certainty for interpretations.

Vol. 42, 1966. *Colloid Stability in Aqueous and Non-Aqueous Media* Nottingham, 26th–28th September.
Committee: *D. D. Eley*; G. D. Parfitt; D. H. Everett; J. A. Kitchener; R. H. Ottewill.
Chair: F. S. Dainton (President).
Total attending 247. Participants from overseas 64; contributing 30, including D. V. Derjaguin (Moscow); Yu. M. Glazman (Kiev); V. K. La Mer (New York); J. Lyklema (Wageningen); E. Matijevic (Potsdam, NY); K. J. Mysels (Los Angeles); J. Th. G. Overbeek (Utrecht); L. Romo (Quito); B. Tezak (Zagreb).
Main headings: None given.
Papers 24; discussion pages 95; total pages of text 316.

Interspersed between the papers, the record shows nine discussion sessions, taking on average ten pages each. This pattern of discussion for small groups of papers seems to have generated much valuable discussion. Overbeek's introductory paper offered a systematic survey of the many factors having a role in colloid stability: these were so many as to leave uncertainties which would dominate in a particular system. Wilhelm Ostwald would have been quite at home in the meeting.

Vol. 43, 1967. *The Structure and Properties of Liquids* Exeter, 11th–13th April.
Committee: *D. H. Everett*; J. S. Rowlinson; H. N. V. Temperley; M. L. McGlashan; G. Allen.
Chair: C. E. H. Bawn (President); D. H. Everett; M. L. McGlashan; J. S. Rowlinson.
Total attending 275+. Participants from overseas 91; contributing 24, including B. Dorner (Julich); M. Fixman (Yale); E. U. Franck (Karlsruhe); D. Henderson (Waterloo, Ontario); F. Kohler (Vienna); C. J. Pings (Pasadena); S. A. Rice (Chicago); J. Stecki (Warsaw).
Main headings: None given.
Papers 22; discussion pages 53; total pages of text 242.
G. S. Rushbrooke gave the fourteenth Spiers Memorial Lecture.

Of this Discussion it could well suffice to say that it was as significant a contribution to the study of liquids as the earlier one (1936) in Edinburgh. In the meantime Lennard-Jones and his associates had achieved notable success in calculating the properties of the inert gas solids. Now the computer was coming in to play a role which soon became a dominant one. This role, together with the various representations in terms of mathematical models which could be envisaged as helpful in understanding liquids, was outlined in a notable introductory survey by Rushbrooke, himself one of the Lennard-Jones school.

Distinct advances were demonstrated, thanks firstly to the use of new experimental techniques. X-Ray diffraction by liquids had been refined: N.M.R. studies were reported, and both dielectric and far-infrared spectroscopy provided insights.

The question of appropriate intermolecular energy functions necessarily had a special part in the discussions. Rowlinson has pointed out that one development which was of major significance appeared only in the actual discussion sessions: that is, the modifications of the hard-sphere molecular representation by the addition of an attractive force term. This approximation by Barker and Henderson of Melbourne proved to be a singularly valuable contribution to the theory of liquid studies.

Vol. 44, 1967. *Molecular Dynamics of the Chemical Reactions of Gases* Toronto, 5th–7th September.
Committee: G. Porter: the other members are not named.
Chair: C. E. H. Bawn (President); D. J. Le Roy (Toronto); J. C. Robb (Birmingham).
Participants from the U.K. numbered 46; from France (1), from Germany (2) and from India (1): in a total of over 200 attending.
Main headings: None given.
Papers 23; discussion pages 71; total pages of text 301.
J. C. Polanyi (Toronto) was himself an author of two of the papers presented, and there were others from his laboratory. In addition he presented a summary of the situation for these studies. Three previous Faraday Society Discussions in the post-quantum theory period had made relevant contributions. In 1937 at Manchester (where Michael Polanyi and his Department were fully involved), in 1938 at Oxford, and in 1962 at Cambridge, specific aspects had been considered. A principal concern, the energy surface representing the reaction process, was being better approximated by more sophisticated theoretical treatments and by a greater variety of experimental methods.

In view of the complexity of energy states and energy exchanges of interacting molecules, specific observations on such exchanges and states were not only essential but had now become possible by a variety of methods well illustrated at the meeting. The methods included: crossed and direct molecular beam interactions; mass spectroscopic analysis; infrared emission analyses of molecular states. There was a corresponding development in the theoretical understanding of these processes.

Vol. 45, 1968. *Electrode Reactions of Organic Compounds* Newcastle-upon-Tyne, 2nd–4th April.
Committee: *A. R. Ubbelohde*; A. North; J. N. Murrell; M. Willis.
Chair: Not given.

Total attending 200+. Participants from overseas 51; contributing 24, including A. J. Bard (Austin, Texas); B. E. Conway (Ottawa); Z. R. Grabowski (Warsaw); X. de Hemptinne (Leuven); J. Jordan (Pennsylvania); R. A. Marcus (Urbana, IL); S. Tsumtsumi (Osaka).
Main headings: None given.
Papers 22; discussion pages 62; total pages of text 272.

A considerable variety of reactions were revealed and discussed. In quality too, there was a wide range in the standard of science presented. To an outsider there did not seem to be a substantial potential towards organic synthesis projected by these accounts.

Vol. 46, 1968. *Homogeneous Catalysis with Special Reference to Hydrogenation and Oxidation* Liverpool, 17th–19th April.
Committee: *R. H. Nyholm*: no other names given.
Chair: C. E. H. Bawn (President).
Total attending 208. Participants from overseas 83; contributing 24, including W. Brackman (Amsterdam); J. Halpern (Chicago); B. R. James (Vancouver); V. D. Komissarov (Moscow); F. Marta (Szeged); L. Ogata (Tokyo); R. Ugo (Milan).
Main headings: None given.
Papers 19; discussion pages 44; total pages of text 222.

With a total attendance of '208 members and others' this Discussion attracted a high proportion of overseas visitors. The majority of papers reported on the use of transition metal or rare metal complexes as catalysts and there was clearly much interest shown by industrial research laboratories. There is a marked simplification and corresponding clarity of the mechanisms when comparison is made with the Discussions on Heterogeneous Catalysis in 1950 and 1966.

Vol. 47, 1969. *Bonding in Metallo-Organic Compounds* Cambridge 25th–27th March.
Committee: *N. N. Greenwood*, D. W. J. Cruickshank, J. Lewis, R. E. Mason, L. E. Sutton.
Chair: G. Gee (President).
Total attending 110. Participants from overseas 24; contributing 10, including T. L. Brown (Urbana, IL); F. A. Cotton (M.I.T.); L. F. Dahl

(Madison, WI); J. A. Ibers (Evanston, IL); W. Zeil (Ulm).
Main headings: None given.
Papers 21; discussion pages 28; total pages of text 199.

Some simple figures relating to this General Discussion are significant. The most relevant is that despite the Faraday Society's lack of enthusiasm for discussions on valency theory (as central a topic as any in physical chemistry) the opportunity offered by this meeting only resulted in four papers in which theoretical treatments of bonding were a principal concern.

The papers in their entirety show a wide range of experimental methods being deployed. In addition to X-ray and electron diffraction the forms of spectroscopy included infrared, ultraviolet, photoelectron, Mössbauer, electron impact and N.M.R.

The attendance was notably modest, '110 members and others.'

For the Faraday Society it was exceptional in that of the 24 papers presented only six came from overseas. In the discussion contributions the figures are even more remarkable. There were four discussion sessions, and in these, the number of contributions from the U.K. in relation to the totals were: 17/17; 7/11; 15/15; 14/17; or overall 53/60.

In the annals of the Faraday Society the meeting was notable in being centred on chemistry in its classical form, that is, on the formation of new compounds and their properties. If the meeting was not a pioneering one on its topic, it did offer Faraday Society members an overview of one aspect of the new chemistry of metal compounds.

Vol. 48, 1969. *Motions in Molecular Crystals* Oxford, 16th–18th September.
Committee: *D. H. Whiffen*, P. A. Egelstaff, M. M. Davies, D. W. J. Cruickshank, J. W. White.
Chair: G. Gee (President).
Total attending 152. Participants from overseas 58; contributing 29, including C. Brot (Paris); J. S. Dryden (Sydney); J. Janik (Krakow); W. Jost (Göttingen); D. W. McCall (Murray Hill, NJ); J. A. Morrison (Ottawa); O. Schnepp (Los Angeles).
Main headings: None given.
Papers 21; discussion pages 39; total pages of text 215.

Neutron scattering and neutron diffraction studies were amongst major contributions to the experimental data discussed, which also included spectroscopic (optical, infrared, Raman, but not far-infrared) studies, electrical and thermodynamic properties. Guest molecules in clathrates, crystalline hydrogen halides, and ice were some of the systems considered.

Vol. 49, 1970. *Polymer Solutions* Manchester, 15th–17th April.
Committee: *G. Gee*; G. Allen; C. Booth; S. F. Edwards; J. Lamb; J. S. Rowlinson.
Chair: G. Gee (President)
Total attending 210. Participants from overseas 54; contributing 22, including K. Bak (Denmark); H. Eisenberg (Israel); M. L. Huggins (Stanford, CA), F. E. Karasz (U.S.A.); J. Koefoed (Denmark); R. L. Scott (Los Angeles); T. Shimanouchi (Tokyo).
Main headings: None given.
Papers 21; discussion pages 41; total pages of text 280.

Studies of polymer solutions had moved beyond experimental evaluations of solubilities, viscosities and other simple physical properties. The meeting offered advanced theoretical analyses: thermodynamic, molecular statistical, and mathematical model. A major concern was the adequate statistical representation of a partially flexible long molecule in all its randomly and not-so-randomly coiled forms. Amongst the experimental data discussed were measurements of diffusion, ultrasonic scattering, N.M.R. spectra, fluorescence and light scattering. The contributions included ones from C. Domb (London), S. F. Edwards (Manchester), O. B. Ptitsyn (Moscow) and D. Patterson (Montreal).

In the fifteenth Spiers Memorial Lecture Flory delineated many factors (molecular sizes, configurations, flexibilities, interaction energies, entropies of melting) necessarily involved in a quantitative analysis of polymer solutions. It was, perhaps, a typical substantial Faraday General Discussion: without revealing a new area for physico-chemical advances, it presented authoritative accounts of the prospects in an established major interest. In retrospect, it was much to be regretted that Volkenstein of Leningrad was not a participant.

Vol. 50, 1970. *The Vitreous State* Bristol, 22nd–24th Sepember.
Committee: D. H. Everett, A. J. Leadbetter, H. W. Douglas, P. N. Gaskell.
Chair: G. Gee (President).
Total attending 160. Participants from overseas 59; contributing 17, including N. S. Andreev (Leningrad); H. Krebs (Stuttgart); O. V. Mazurin (Leningrad); G. F. Nelson (Ohio); M. Pollak (Israel); S. A. Rice (Chicago); M. Shimaji (Hokkaido); D. Turnbull (Harvard); E. Whalley (Ottawa); J. Zarzycki (Montpellier).
Main headings: None given.
Papers 23; discussion pages 33; total pages of text 234.

Disorder is the molecular characteristic of the vitreous state so perhaps it is not surprising that there was no obvious defined pattern to the contributions. Of course the studies pursued two aspects, the locations and separations of the constituent particles, and their motion, vibrational or diffusional. To these ends a notable variety of physical methods of diagnosis were brought to bear: spectroscopy of many kinds–infrared, Raman, radiofrequency or dielectric; electric parameters–conductivity, Hall effect, thermoelectric power, electron excitation; X-ray, neutron and visible light scattering. There were other methods. An aspect which received repeated consideration was that of phase separation in glasses, one feature associated with the instability of the vitreous state.

Vol. 51, 1971. *Electrical Conductivity in Organic Solids* Nottingham, 14th–16th April.
Committee: D. D. Eley, R. J. P. Williams, J. M. Thomas, E. P. Goodings, R. B. Leslie, M. R. Willis.
Chair: J. W. Linnett (President).
Total attending 120. Participants from overseas 38; contributing 16, including E. L. Frankevich (Moscow); L. Glasser (S. Africa); H. Inokuchi (Tokyo); G. R. Johnson (Australia); M. M. Labes (U.S.A.); H. A. Pohl (U.S.A.); S. Walker (Canada); W. Siebrand (Ottawa).
Main headings: None given.
Papers 20; discussion pages 43; total pages of text 222.

Electrical conductivity had been one of the earliest and strongest interests of the Faraday Society, but that was principally in the context of aqueous salt solutions. Since 1945 solid-state physics had grown enormously, particularly in relation to metals, oxides and semiconductors. It was Eley who proposed and organized the discussion of organic solids. It must, in fact, be recorded that scarcely anyone played a more active role than Eley in promoting and taking responsibility for Faraday Society General Discussions.

The subject involved some quasi-chemical but more particularly solid-state physical concepts. The electronic structures of molecules, and the delocalization of electrons, were key factors. Photoconductivity had been a major stimulus. The mobile charges could be represented by electrons, protons, holes or excitons: the last had been recognized as pockets of electronic excitation energy at localized sites in solid aromatic structures. The equivalent of doping in inorganic systems could be attained by incorporating alkali metal ions as the salt form of an organic compound. The possibility of organic semiconductors was coming into view.

Professor T. J. Lewis can be quoted for the opinion that this meeting 'was a landmark on a path which, via the 88th Faraday Discussion on Charge Transfer in Polymeric Systems almost 20 years later, has led to the present day study of Molecular Electronics: this interdisciplinary activity, with its rich diversity of applications and sophisticated technology has the potential to supercede all that has been achieved in silicon electronics'.

Vol. 52, 1971. *Surface Chemistry of Oxides* Brunel University, 13th–15th September.
Committee: G. W. Parfitt: no other names given.
Chair: J. W. Linnett (President).
Total attending 250+. Participants from overseas 109; contributing 27, including J. H. de Boer (Delft); R. O. James (Melbourne); A. V. Kiselev (Moscow); J. Lyklema (Wageningen); J. P. Quirk (West Australia); Z. G. Szabo (Budapest); J. B. Uytterhoeven (Leuven); A. Zecchina (Turin).
Main headings: None given.
Papers 24; discussion pages 115; total pages of text 322.

This, the last General Discussion held under the aegis of the Faraday Society as such, strongly maintained an international character. Those from overseas included, from The Netherlands 15; France 14; Belgium and Italy 10; U.S.A. 9; Canada and West Germany 7; Australia 3; Egypt, Poland, Switzerland 2; Bulgaria, Eire, Hungary, Israel, New Zealand, Portugal, Spain, U.S.S.R. 1 each.

As a consequence, the origins of the papers contributed also showed an unusual pattern. Seventeen were from overseas: from Belgium, The Netherlands, U.S.A. three each; Australia two; Canada, Egypt, France, Italy, West Germany and the U.S.S.R. one each. Only four papers came from U.K. universities and two from U.K. industrial and one from a Government Institute. This was distinctly unusual. At an early stage of the planning the organizing committee had been told by the Faraday Council to reduce the number of review-type papers they seemed prepared to accept. Usually when a committee found that far fewer than half the papers (i.e. fewer than say eight) would come from the U.K. they either dropped the subject entirely or recast it for the Discussion being planned.

Not surprisingly, the oxide surfaces exhibited great differences even in structure and composition: the depth of the surface layer, the concentration of OH groups and of various cations, were shown to be important variables. Many different measurements were discussed: surface area; adsorption; electric conductivity; infrared and E.S.R. spectra; X-ray diffraction studies; catalysis; microporosity; *et al.* No general principles were likely to be illustrated by so heterogeneous a group of systems. In scientific terms it could perhaps not be considered an outstanding General Discussion of the Faraday Society. However, the Secretary, a solid-state specialist, reported to Council that the meeting 'was excellent in every way'.

THE SYMPOSIA

Throughout the 1945–71 period, informal meetings are noted in Council minutes. They quickly became Informal Meetings, usually on specific, limited topics, organized by an active protagonist often at his own university or institute. Some such meetings might last, at most, six hours. The Faraday Society's name would be attached only if Council were happy with the status of the convenor and the worthiness of the topic.

Details of these Informal Meetings are not listed here. They increased in number and in extent. One, on the Mechanisms of Adsorption from Solution, had expanded to two days, 19th–20th September 1963, at Imperial College, London. It attracted a total of 144 participants, 81 from industrial organizations, 63 from academic institutes, and, amongst the total, 16 from overseas. This was equivalent to a medium-sized General Discussion. Such developments led to Council forming the Standing Committee on Conferences and the emergence of Symposia as a separate category.

The details of the Symposia are listed below.

Vol. 1, 1967. *Mössbauer Effect* London, 12th–13th December.
Total attending 98+. Participants from overseas 22; including J. Danon (Rio de Janeiro); J. F. Duncan (New Zealand); P. K. Gallaher (Murray Hill, NJ); K. P. Mitrofanov (Moscow); M. Pasternak (Israel); V. S. Skpinel (Moscow); R. Wäppling (Uppsala).
10 papers, each with separate discussion; total pages of text 134.

Vol. 2, 1968. *Molecular Wave Functions* London, 12th–13th December.
Total attending 144. Participants from overseas 35; including D. M. Bishop (Canada); Lady Brigitte Dufour (Belgium); J. A. Pople (Pittsburgh); M. Randic (Yugoslavia); A. Sgamellotti (Italy).
12 papers; discussion pages 10; total pages of text 100.

Vol. 3, 1969. *Magneto-Optical Effects* London, 11th–12th December.
Total attending 112. Participants from overseas 33; including J. Badoz (Paris); R. C. Fay (Ithaca, NY); R. M. Hochstrasser (Philadelphia); L. J. Oosterhoff (Leiden); P. S. Pershan (Harvard); P. J. Stephens (Los Angeles); J. H. van der Waals (Leiden).
14 papers; discussion pages 16; total pages of text 154.

Vol. 4, 1970. *Optical Studies of Adsorbed Layers at Interfaces* London, 14th–15th December.

Total attending 140+. Participants from overseas 39; including R. Adzic (Yugoslavia); A. Marseille (Netherlands); G. Urbain (France); K. Vetter (Germany).
13 papers with 5 discussion periods; total pages of text 206.

Vol. 5, 1971.   *Liquid Crystals*   London, 13th–14th December.
Total attending 142. Participants from overseas 43; including S. Chandresekhar (India); A. Djeransk (Bulgaria); P. G. de Gennes (Paris); J. Janik (Krakow); G. Meier (Freiburg); V. S. V. Rajan (Canada); K. Raksani (Hungary).
13 papers; discussion pages 49; total pages of text 180.

These Symposia were clearly successful mini-General Discussions, as evidenced by the number of overseas participants. Their topics were significant and the contributed papers were consonant with the best standards of the Faraday Society. The criticism might be that at least two of the subjects could have been chosen for General Discussions. In their best phase prior to 1960, Faraday Discussions had offered early presentations of new themes which became of major interest in chemical science. On that criterion the Mössbauer effect, discovered in 1960, could have been a worthy General Discussion: and it has already been pointed out that molecular orbital theory, established as the dominant presentation by valence theory from the mid-fifties, was never discussed by the Faraday Society. The Symposium in 1968 dealt with details of the treatment and it was essentially for specialists in the field. Finally, magneto-optical effects was a topic which J. H. van der Waals, as a member of Council, had advocated long before the Symposia were instituted.

THE ASSESSMENT OF THE GENERAL DISCUSSIONS

Some assessment of the General Discussions must be offered. This is necessitated by the fact that for most physical chemists, and perhaps particularly for those outside the U.K., the Faraday Society was known and judged, as much as anything, by its General Discussions.

Amongst the many who wrote their appreciation of the Faraday

Society and its Discussions few did so at greater length than Lord Dainton. It is appropriate to record a typical voice from the host of partisans: it can only be a brief sample from much that came to hand. As an undergraduate at Oxford he was early given literature references for an essay: 'From that day forward I regarded the Faraday Society as the owner of two treasure houses in the *Transactions* and *Discussions* which together provide a record of the making of physical chemistry in much of this century. The Discussions also give insights into the personalities of the contributors (who) . . . could be, on occasions, as temperamental and irrational as any prima-donna or thespian. It was a relief to discover that all scientists, however great, make mistakes and therefore for progress everyone, however experienced or distinguished, must submit to criticism by others, however young and inexperienced. The photographs of those who were present at Discussions added to this reader's interest. Physiognomy allied to the style and manner of the contributions as written in the record gave a more complete idea of the individual participant. I recall the olympian Donnan, the mandarinesque Sidgwick, the combative Travers, the impressive Debye, the courteous Bonhöffer and many others. . . .

'Of course the actual Discussions themselves were not always unalloyed sweetness and light. Cleverness and modesty are not invariable bedfellows and there is competition in science as well as cooperation. There were occasions when the style of the criticism went beyond what was considered acceptable, indeed one such behavioural lapse (by E. A. Guggenheim) was the subject of a hastily summoned meeting of Council members during a Discussion: but such happenings were the rare exception. . . . Much depended on the firmness and tact of the chairman of the individual sessions and in this connection I must record how the chairmen were supported by the "traffic lights": Green–Amber (one minute more)–Red, which were remorselessly but fairly operated by the Secretary'.

In assessing the Discussions it is firstly necessary to consider what were the aims of those who determined the activities of the Faraday Society. Certainly their priority was not to develop the study and scope of physical chemistry as such. They needed a high-class journal to present their research to the scientific world: and they provided the further excellent medium, the General Discussions, to have their work considered and stimulated by their professional colleagues. Undoubtedly these meetings would have continued had no overseas scientists attended, but it was a notable bonus, always aimed for and much appreciated, that many came from outside the U.K. Not only did they attend, but overseas visitors regularly added to the status of the Discussions by their substantial contributions.

Over the 1945–71 period the topics of the General Discussions were numerically distributed as: Reaction Kinetics and Molecular Reaction Dynamics 14; Colloid Science 12; Crystalline State 7; Spectroscopy and Structure 7; Equilibrium Systems 7; Electrochemistry 4; Radiation Chemistry 3.

It is a reasonable suggestion that these figures provide an outline of major research activities as assessed by the Council of the Faraday Society. It scarcely needs emphasizing that the scientific standard of the Discussions was generally very high. Space need not be taken to detail the claim that 'they were the best of their kind'. However, the writer would not claim that the subject weighting represented a fair reflection of the developments in Physical Chemistry over this period. That would be unlikely in view of the fact that the Faraday Society Council was so predominantly composed of members, not necessarily immediately from, but with strong associations – one might say derived from – Oxford, Cambridge and London universities. Any such comment is rightly open to qualification, but it is worth making. The only outstanding school of Physical Chemistry not dependent upon the south of England was that of Polanyi at Manchester: the Scottish universities were also substantially self-sustaining.

It is from this background that the dominant places of Reaction Kinetics and Colloid Science arise. Two of the most influential and supportive members of Council, each over many years, were E. K. Rideal and R. G. W. Norrish. They each acquired very notable protégés who took away and developed their interests.

That being so, the question arises: what does this distribution of topics represent? It is necessary to emphasize that it does not equably represent the developments in the science of Physical Chemistry.

In the science of molecules, the basic aspects are their structure and an understanding of how that arises. There was, however, minimum attention given to molecular structure in Faraday Discussions: it did appear, incidentally, in the few considerations allotted to molecular spectra. No paper in all the Discussions discusses the determination of molecular structure by X-ray, electron or neutron diffraction. The first has, to an overwhelming degree, established the geometry and precise scale of molecules and their bonds. The Discussions on crystals could not have taken place without the use of results from X-ray diffraction studies. But perhaps these had no significant place in Chemistry schools in the U.K.? Not so; Glasgow, The Royal Institution and Birmingham immediately come to mind. Neutron diffraction also had notable practitioners in the U.K.

For many years microwave spectroscopy was providing the most precise details on gaseous molecules. This development was never followed in the General Discussions, nor its later expansion into that unique extension of molecular science, the recognition of molecules and radicals of medium complexity in outer space.

With the molecular structure fixed, what comes next in its study? On the factual side, its stability, and the energy needed for its distortion and disruption. These data are derived very largely from aspects of molecular spectroscopy never adequately considered in Faraday Discussions. In fact, the gap now suggested is encapsulated in the title of Kronig's volume *The Optical Basis of the Theory of Valency*. This was published in a series edited by Eric Rideal.

Now, the theory of valency has been mentioned. It does not appear in our volumes. Perhaps this is the most surprising, if not also the most important omission in the coverage of Physical Chemistry offered by the Discussion volumes. Surprising because one of the Society's Presidents was a pioneer in the field. J. E. Lennard-Jones, as he then was, played a major part in the development of molecular-orbital theory and his students, most especially Coulson, but also Pople and others, later had international status as its leading exponents. Some special reasons(s) must account for the omission of valency theory from Faraday Discussions. It is central to molecular science. It could be that Lennard-Jones was that *rara avis* amongst professors, one who did not see his own subject as central to the science. At the 1931 British

Association Meeting he gave, to the Chemistry Division, an account of valency of masterly clarity, after insisting he knew less of chemistry than anyone else present. But then he and his protégés were natives of the vast continent of mathematics who, nevertheless, did not wish to flaunt Dirac's dictum of 1930 in Faraday Society circles.

An informed comment on the question of valence theory presentation by the Faraday Society comes from Christopher Longuet-Higgins. Firstly, he quotes L. E. Sutton's 'excellent account of the transition from classical to modern valence theory in *The First Fifty Years* published by the Faraday Society in 1953': 'No single development in the field was more important than that of *Molecular Spectra in relation to Molecular Structure* [the Faraday Discussion at Bristol, 1929]. . . . "The discussion was held on 24th and 25th September, and was an outstanding success". The participants included Herzberg, Hund and Mulliken. The meeting must have been the inspiration for Lennard-Jones's LCAO approximation, published in the *Transactions* just two years later, and might therefore be regarded as the "diaspora" of British theoretical chemistry.'

'Why, you ask, did the Society never arrange a General Discussion on Valence Theory? Could the answer be that we felt uneasy with the more baroque theories of Linus Pauling and others, and preferred our ideas about molecular structure and reactivity to be linked as closely as possible with physical and chemical phenomena?'

Are there any physical chemists who accept that colloid interests occupy more than one fifth of their domain? Again, in suggesting an over-play of Reaction Kinetics one recalls the comment that it has long been the field in which facts are readily collected but satisfying insights come only sparingly. One indisputable fact is that the Faraday Society's concern with Reaction Kinetics, as judged from the General Discussions, only marginally included reactions in solution. Why was that? It was not for the want of interesting developments. Much analysis of (what could have been regarded as too readily available) solution data such as the Polanyi–Evans satisfying account of the Walden inversion, similar insights into *cis–trans* conversions, others into catalysis, especially with its extension of the acid–base function (already initiated by Lowry) which demystified many molecular changes – none of this was presented in Faraday Discussions. On the only occasion when the Ingold school's use and interpretation of reaction rates in solution received some consideration, it was seen how unhappy many physical chemists were with the acceptance of *rate coefficients* as discriminating criteria for reaction mechanisms. A rate coefficient, it was emphasised, resulted from the combination of an energy factor with a quasi-entropy (probability) term: analytical significance should not be attached to it. This instance leads to the suggestion that Faraday Society *afficionados* might well have associated solution kinetics with the simple forms of physical chemistry pursued by members of the Chemical Society. That Society at one time was so averse to theory that no paper without new experimental data (presumably a new melting point would suffice) was acceptable for publication.

More speculative comment on the Discussions would be unwise. What is established is that the record of the Faraday Society is not the history of Physical Chemistry, even in the U.K. One can hope that historians of chemistry will take note.

8

The Council of the Society

NECESSITY DICTATES THAT a record be presented of the Faraday Society Council over the years 1945–71. To make it brief, this cannot be other than a catalogue, and the following pattern has been adopted.

For the years adjacent to 1945, the Society's year started in October. The Officers and Members of Council are given for October 1945, with the dates for their retirement. Then the replacements are given each year, as they occur: also with the dates for retirement. The reader will become aware that for the early years of this period, there was little regularity in the sequence of changes. Normally, Presidents were in office for two years: Vice Presidents and Members of Council for three years, but Past Presidents (PP) became Vice Presidents for at least six years, and sometimes almost indefinitely longer. Elected Vice-Presidents, on their retirement, could be immediately returned to Council as Ordinary Members.

Of the nominated Officers the most important were: the (essentially executive) Secretary; the Treasurer (also Chairman of the Finance Committee); the Chairman of the Publications Committee (which oversaw all aspects of the publications); the Editor, for all publications. In addition, the Faraday Society appointed representatives on up to 20 other committees or panels. The only representatives listed here are those on panels whose function impinged most directly on the Faraday Society. Where no retirement date appears, it signifies the appointment was for an undetermined period, terminated by the appointment of a successor.

A full listing of the Council is given for 1945, 1955, 1965 and 1971. It became established that the Treasurer and Chairman of the Publications Committee were, *ex officio*, full members of the Council.

From October 1945 the Officers of the Faraday Society were:

1945 President: W. E. Garner (1947: replacing E. K. Rideal)
Vice Presidents: A. J. Allmand (1947); A. Findlay (1948); C. F. Goodeve (1948); R. O. Griffith (1948); M. Polanyi (1948); E.

K. Rideal (PP); S. Sugden (1946); M. W. Travers (PP).
Members of Council: T. Beacall (1946); G. M. Bennett (1948); A. Caress; C. W. Davies (1948); C. H. Desch (1946); M. G. Evans (1946); R. W. Lunt (1946); H. W. Melville (1946); F. I. G. Rawlins (1948); R. E. Slade.
Hon. Secretary and Editor: G. S. W. Marlow; Assistant Secretary: Miss B. E. Kornitzer
Treasurer: R. E. Slade; Chairman Publications Committee; A. J. Allmand.
Editorial Board *J. Physical Chemistry*: C. N. Hinshelwood; J. R. Partington.

1946 Vice Presidents (to make up to maximum number): J. E. Lennard-Jones (1949); H. W. Melville (1949).
Members of Council: R. P. Bell (1949); G. Gee (1949); G. S. Hartley (1949); Sir Robert Robertson (1949).
Assistant Editor: F. C. Tompkins.

1947 President: A. J. Allmand (for one year, by his choice).
Vice President: C. N. Hinshelwood (1950).
Members of Council: A. E. Alexander (1950); J. Ferguson (1950); W. G. Penney (1950).
Chairman Publications Committee: E. A. Guggenheim.
Editor: F. C. Tompkins.

1948 President: Sir John Lennard-Jones (1950).
Vice Presidents: R. P. Bell (1950); G. M. Bennett (1949); M. G. Evans (1951); H. J. Emeleus (1951); G. S. Hartley (1951).
Members of Council: C. E. H. Bawn (1951); C. F. Goodeve (1951); R. Lessing (1951); J. T. Randall (1951); G. B. B. M. Sutherland (1951).
Secretary and Editor (from 3rd June 1948): F. C. Tompkins.

1949 Vice Presidents: C. E. H. Bawn (1952); G. Gee (1952).
Members of Council: F. P. Bowden (1952); C. A. Coulson (1952); J. H. Schulman (1952); R. E. Slade (1952); H. W. Thompson (1952).
Treasurer: O. H. Wansborough-Jones.

1950   President: Sir Charles Goodeve (1952).
Vice Presidents: H. W. Melville (1953); Sir Alfred Egerton (1953).
Members of Council: J. S. Anderson (1953); F. S. Dainton (1953); E. A. Guggenheim (1953); C. K. Ingold (1953).
Representatives on the Chemical Council: R. Lessing (1950); G. M. Bennett (1951); G. Gee.

1951   Vice Presidents: R. P. Bell (1954); R. Lessing (1954); J. T. Randall (1954).
Members of Council: D. D. Eley (1954); A. G. Evans (1954).
Chairman Publications Committee: R. P. Bell.

1952   President: H. S. Taylor (for one year, during which he was knighted).
Vice Presidents: M. G. Evans (1955); J. H. Schulman (1955).
Members of Council: R. M. Barrer (1955); E. J. Bowen (1955); H. C. Longuet-Higgins (1955); A. R. J. P. Ubbelohde (1955).
Editorial Board *J. Physical Chemistry*: C. N. Hinshelwood; E. J. Bowen; R. P. Bell; E. A. Moelwyn-Hughes.

1953   President: R. G. W. Norrish (1955).
Vice Presidents: F. P. Bowden (1956); E. J. Bowen (1956); G. Gee (1956); C. K. Ingold (1956).
Members of Council: C. E. H. Bawn (1956); J. A. V. Butler (1956); D. H. Everett (1956); G. H. Twigg (1956); W. F. K. Wynne Jones (1956).

1954   Vice Presidents: C. A. Coulson (1957); H. W. Melville (1957).
Members of Council: R. P. Bell (1957); R. Lessing (1957).

1955   President: R. P. Bell (1957).
Vice Presidents: Sir John Lennard-Jones (PP); Sir Charles Goodeve (PP); Sir Hugh Taylor (PP); R. G. W. Norrish (PP); Sir Eric Rideal (PP); F. P. Bowden (1956); E. J. Bowen (1956); G. Gee (1956); C. K. Ingold (1956); C. A. Coulson (1957); H. W. Melville (1957).

The Council of the Society 333

Members of Council: C. E. H. Bawn (1956); J. A. V. Butler (1956); D. H. Everett (1956); G. H. Twigg (1956); W. F. K. Wynne Jones (1956); R. Lessing (1957).
Secretary and Editor: F. C. Tompkins; Assistant Secretary: B. E. Kornitzer.
Treasurer: O. H. Wansborough-Jones; Chairman Publications Committee: R. P. Bell.

1956 Vice Presidents: C. E. H. Bawn (1958); J. A. V. Butler (1958); E. W. R. Steacie (1959); W. F. K. Wynne Jones (1958)*.
*Having just served on Council in another capacity, invited to serve two years only.

1957 President: Sir Harry Melville (1958).
Vice Presidents: R. Lessing (1960); E. A. Moelwyn-Hughes (1960).
Member of Council: L. E. Sutton (1960).

1958 President: E. W. R. Steacie (1959).
Vice Presidents: D. H. Everett (1961); G. Gee (1961); J. W. Linnett (1961).
Members of Council: M. M. Davies (1961); H. C. Longuet-Higgins (1961); R. Spence (1961); D. W. G. Style (1961); A. R. J. P. Ubbelohde (1961).

1959 President: Sir Harry Melville (1960).
Vice Presidents: C. E. H. Bawn (1962); B. Topley (1962).
Members of Council: J. A. V. Butler (1962); D. D. Eley (1962); W. Jost (1962); J. A. Pople (1962).

1960 President: Sir Cyril Hinshelwood (1962).
Vice Presidents: T. M. Sugden (1963); L. E. Sutton (1963).

1961 Vice Presidents: C. J. F. Böttcher (1964); H. C. Longuet-Higgins (1964); A. R. J. P. Ubbelohde (1964).
Members of Council: C. H. Bamford (1964); T. L. Cottrell (1964); C. Kemball (1964); G. Porter (1964); R. E. Richards (1964).

1962   President: A. R. J. P. Ubbelohde (1964).
Vice Presidents: D. D. Eley (1965); D. H. Everett (1965); W. Jost* (1965).
Members of Council: J. N. Agar (1965); C. E. H. Bawn (1965); V. Gold (1965); J. S. Rowlinson (1965).
Chairman Publications Committee: C. E. H. Bawn.
*In the Council Minutes this appears as A. N. Other.

1963   Vice Presidents: C. H. Bamford (1966); F. S. Dainton (1966).
Members of Council: P. G. Ashmore (1966); D. H. Whiffen (1966).

1964   President: F. S. Dainton (1966).
Vice Presidents: J. W. Linnett (1967); G. Porter (1967); R. E. Richards (1967).
Members of Council: A. D. Buckingham (1967); M. L. McGlashan (1967); J. C. Robb (1967); M. C. R. Symons (1967).

1965   President: F. S. Dainton (1966).
Vice Presidents: R. G. W. Norrish (PP); R. P. Bell (PP); H. W. Melville (PP); C. N. Hinshelwood (PP); F. F. Musgrave (Treasurer); A. R. J. P. Ubbelohde (PP); C. H. Bamford (1966); J. W. Linnett (1967); G. Porter (1967); R. E. Richards (1967); C. E. H. Bawn (1968); A. D. Walsh (1968); Lord Wynne-Jones (1968).
Members of Council: P. G. Ashmore (1966); D. H. Whiffen (1966); A. D. Buckingham (1967); M. L. McGlashan (1967); J. C. Robb (1967); M. C. R. Symons (1967); J. N. Agar (1968); V. Gold (1968); J. S. Rowlinson (1968).
Secretary and Editor: F. C. Tompkins; Assistant Secretary: B. E. Kornitzer.
Treasurer: F. F. Musgrave; Chairman Publications Committee: C. E. H. Bawn.

1966   President: C. E. H. Bawn (1968).
Vice Presidents: The only change arose from an imposed restriction to have no more than five Past Presidents, which led to R. G. W. Norrish retiring.

The Council of the Society

Members of Council: P. Gray (1969); C. Kemball (1969); J. E. Prue (1969); N. Sheppard (1969).
Chairman Publications Committee: C. Kemball.

1967 Vice Presidents: D. H. Everett (1970); F. F. Musgrave (1970); J. S. Rowlinson (Treasurer).
Members of Council: H. M. Frey (1970); J. Lamb (1970); G. Porter (1970); A. Carrington (1970); F. F. Musgrave retired from the Treasureship.

1968 President: G. Gee (1970).
Vice Presidents: R. P. Bell (PP); H. W. Melville (PP); A. R. J. P. Ubbelohde (PP); J. W. Linnett (1971); T. M. Sugden (1971); J. H. van der Waals (1971).
Members of Council: M. M. Davies (1971); G. J. Hoytink (1971); K. J. Ivin (1971); B. A. Pethica (1971).

1969 Vice Presidents: no change.
Members of Council: P. Gray (1972); J. N. Murrell (1972).

1970 President: J. W. Linnett (1972).
Vice Presidents: C. A. Coulson (1973); P. Gray (1973); G. Porter (1973).
Members of Council: D. H. Everett (1973); E. U. Franck (1973); G. J. Hills (1973); F. F. Musgrave (1973); J. H. Purnell (1973).

1971 These are the officers elected (or confirmed) in 1971 and so to serve in the first year of the Faraday Division of the Chemical Society.
President: J. W. Linnett (1972).
Vice Presidents: C. E. H. Bawn (PP); Sir Frederick Dainton (PP); Sir Harry Melville (PP); A. R. J. P. Ubbelohde (PP); G. Gee (PP); M. M. Davies (1974); B. A. Pethica (1974); F. C. Tompkins (1974).
Members of Council: G. Allen (1974); E. F. Caldin (1974); H. A. Skinner (1974); J. S. Rowlinson (1974).
Secretary and Editor: F. C. Tompkins; Treasurer: J. S. Rowlinson:
Chairman Publications Committee: D. H. Everett.

The election of F. C. Tompkins as Vice President was to ensure his place on the Council. Since the incorporation of the Faraday Society, as Secretary, he was obliged to attend Council meetings but had no vote. That would not apply after amalgamation. Likewise, the nature of the office of Treasurer of the Faraday Society would change. Its holder was, *ex officio*, a member of the Council. J. S. Rowlinson's election ensured his status in the soon-to-be-formed Faraday Division; P. Gray succeeded him as Treasurer.

At its penultimate meeting on 1st June 1971 the Faraday Society Council nominated its representatives for the offices in the new Chemical Society: these were confirmed at the last A.G.M. held at Brunel University, Uxbridge, 14 September 1971. For the new Chemical Society the principal offices were membership of:

> Council: J. W. Linnett, J. S. Rowlinson, D. H. Everett.
> Executive Committee: J. W. Linnett.
> Interdivisional Committees: J. W. Linnett, J. S. Rowlinson, D. H. Everett.
> National Conference Committee: P. Gray, F. C. Tompkins.
> Finance Committee: J. S. Rowlinson.
> Industrial Division: B. A. Pethica.
> External Relations Committee: M. M. Davies.
> Publications Service Board: F. C. Tompkins.
> Publications Committee for *J. Chem. Soc. Faraday Transactions* and *Discussions*: Chairman: D. H. Everett.
> To the Library Committee D. A. Young was appointed and to other Publications Committees: G. J. Hills, D. H. Everett, N. Sheppard, F. C. Tompkins, J. N. Murrell, J. E. Prue, J. H. Purnell, G. Allen.

In the previous chapters L.E.S. has included biographical accounts of the members of Council of the Faraday Society over the years 1903–45. For 1945–71 they receive a more restricted notice. For those Council members not already dealt with, the aim has been to give merely skeletal information, allowing the individuals to be identified. Briefest of all is the mention of their research interests: this must not be taken as adequate, but as indicating some significant part of their concerns. Usually the school of initial graduation is given. This follows from a

Welsh proverb which translates: the man clearly shows the character of his roots. The names now come only within the period of years when they first appeared on Council.

BIOGRAPHICAL NOTES ON MEMBERS OF THE FARADAY SOCIETY COUNCIL 1945–1954

Alexander, A. E.   Reading and Cambridge graduate; Fellow of King's College, member of the Colloid Science Laboratory: Professor at Sydney, Australia: colloid scientist.
Anderson, J. S.   Imperial College, London graduate and lecturer. Senior lecturer, Melbourne. Professor of Inorganic Chemistry, Oxford: solid-state chemistry.
Barrer, R. M.   New Zealand graduate: research at the Colloid Science Laboratory, Cambridge: later Professor at Imperial College, London: solid-state catalysis and zeolites.
Bawn, C. E. H.   Pre-1939 Bawn was at Bristol and Manchester: shortly after the war he became Professor at Liverpool: a kineticist with interests in the polymer field and in the use of spectroscopy in kinetic problems.
Bell, R. P.   Oxford graduate and Fellow of Balliol College: later, Professor, Stirling University: solution and quantum chemistry, both structural and kinetic aspects.
Bennett, G. M.   Professor of Chemistry at Sheffield and King's College, London. Later became the Chief Government Chemist.
Bowden, F. P.   Tasmanian graduate: Fellow, Caius College Cambridge, and Reader in the Physical Chemistry Department: electrochemistry and tribophysics.
Bowen, E. J.   Oxford photochemist: Fellow, University College and well known as a resourceful experimentalist.
Butler, J. A. V.   From Reading he became the resident physical chemist at the Chester Beatty Cancer Research Institute: solution and colloid electrochemist.
Coulson, C. A.   Cambridge graduate: Rouse Ball Professor of Mathematics at Oxford. Best known as author of *Valence* and as a Founder of Oxfam: molecular orbital theory.
Dainton, F. S.   Oxford and Cambridge graduate: Professor at Leeds

and Oxford: Vice-Chancellor of Nottingham University: Chancellor of Sheffield University: first knighted, then Life Peer: photochemical and polymerisation kinetics and radiation chemistry.

Davies, C. W.   Aberystwyth graduate and later Professor there: authored research texts on electrochemistry: ion-exchange mechanisms.

Eley, D. D.   Manchester and Cambridge graduate: member of the Colloid Science Laboratory: Professor at Nottingham University with special interests in surface chemistry and biophysical chemistry.

Emeleus, H. J.   London graduate: Reader in Inorganic Chemistry at Imperial College, London and then Professor at Cambridge: wide ranging inorganic chemist.

Evans, A. G.   Manchester graduate: researched with M. Polanyi: Professor, University of Wales at Cardiff: kinetics and reaction mechanisms.

Evans, M. G.   Manchester graduate of Polanyi school: Professor at Leeds and later Manchester: elder brother of A. G. Evans: quantum chemistry and transition state developments.

Everett, D. H.   Reading graduate: tutor at Oxford then Professor at Dundee and Bristol: special interests in solution, interfacial and colloid thermodynamics but also in the thermodynamics of hysteresis.

Gee, G.   Manchester and Cambridge graduate: member Colloid Science Laboratory: Research Director British Rubber Producers Association before succeeding M. G. Evans as Professor, Physical Chemistry at Manchester.

Griffith, R. O.   Liverpool graduate and lecturer: associate of W. McC. Lewis and photochemist.

Guggenheim, E. A.   Cambridge graduate. Imperial College: later Professor at Reading: thermodynamics and statistical mechanics.

Hartley, G. S.   An associate of Donnan's at University College, London, with a special interest in colloidal electrolytes and the application of Debye–Hückel theory to double layers.

Ingold, C. K.   External London graduate and Professor at Leeds and University College London where he inaugurated one of the largest and liveliest schools of physico-organic and inorganic chemistry. Knighted.

Lennard-Jones, J. E.   Cambridge mathematician: Professor of Theoretical Physics at Bristol and then of Theoretical Chemistry at Cambridge. Vice-Chancellor of Keele University. Knighted.

Longuet-Higgins, H. C.   Oxford graduate, tutored by R. P. Bell: Professor of Theoretical Physics at King's College, London and then of Theoretical Chemistry at Cambridge: theoretical chemistry.

Moelwyn-Hughes, E. A.   Liverpool and Oxford graduate: joined the Colloid Science Laboratory, Cambridge before becoming a lecturer in the Physical Chemistry Department: solution reaction kineticist.

Penney, W. G.   London–Cambridge mathematician who became Professor of Theoretical Chemistry in London and later Director of the U.K. nuclear bomb programme. Awarded knighthood and life peerage.

Polanyi, M.   Graduated in medicine at Budapest: after years in Berlin, came to Manchester as Professor of Physical Chemistry. A founding father of molecular reaction dynamics. Polymath who later held Chair of Political Economy at Manchester.

Randall, J. T.   London graduate: industrial research director before becoming Professor of Physics at King's College, London: founded the unit in which Wilkins and Franklin contributed to the determination of the DNA structure. Knighted.

Schulman, J. H.   Brazilian graduate who became Assistant Research Director at the Colloid Science Laboratory, Cambridge: biophysical chemistry.

Sutherland, G. B. B. M.   St. Andrews and Cambridge graduate: via Physical Chemistry Department and the Colloid Science Laboratory, Cambridge, became Director, N.P.L.: knighted: special interests in infrared spectroscopy.

Taylor, H. S.   Liverpool graduate: later Head, Frick Chemical Laboratory and Graduate College, Princeton: heterogeneous catalysis. Knighted.

Twigg, G. H.   St. Andrews graduate: Ph.D. at the Colloid Science Laboratory, Cambridge: became a Director of Distillers Co. Ltd.: reaction catalysis at surfaces.

Ubbelohde, A. R. J. P.   Oxford graduate: later Professor at Queen's University, Belfast and Imperial College: thermodynamics and solid state.

Wansborough-Jones, O. H. Cambridge graduate: worked with Haber in Berlin: Assistant Director at the Colloid Science Laboratory Cambridge: eventually Chairman of Directors of Messrs. Albright and Wilson: gas reaction catalysis and isotope effects. Knighted for his War work.

Wynne Jones, W. F. K. Aberystwyth and Oxford graduate: associate of Eyring's at Princeton: Professor at Reading and Newcastle-upon-Tyne where he built up a school of fuel chemistry. Life Peer.

FARADAY SOCIETY COUNCIL 1955–1964

Agar, J. N. Cambridge graduate: whole academic life spent in Cambridge: electrochemist of wide knowledge: an invaluable critic.

Ashmore, P. G. Cambridge graduate who did research with Norrish and became Professor at University of Manchester Institute of Science and Technology: reaction kineticist.

Bamford, C. H. Cambridge graduate: research with Norrish: Director of Courtauld's Research Laboratory before becomming Professor at Liverpool: photochemist and kineticist.

Böttcher, C. J. F. Professor at Leiden whose main research interests were in the molecular theory and experimental study of dielectrics.

Buckingham, A. D. Sydney University and Cambridge graduate: Oxford University Lecturer, Bristol University Professor before becoming Professor of Theoretical Chemistry at Cambridge combined with experimental studies.

Cottrell, T. L. Edinburgh and Oxford graduate: Professor at Edinburgh and Vice-Chancellor of Stirling University: inelastic gas–molecule collisions.

Davies, M. M. Aberystwyth and Cambridge graduate: member of the Colloid Science Laboratory and Department of Physical Chemistry, Cambridge: Professor at Aberystwyth: infrared and dielectric spectroscopist.

Gold, V. London graduate: Professor at King's College, London: physical organic chemist and kineticist.

Jost, W. Held positions in several German universities, finally at Göttingen: noted for his synoptic view of physical chemistry and clarity of thought.

Kemball, C.   Cambridge graduate: initial research with Rideal: Professor at Belfast and then at Edinburgh: kinetics of surface reactions.

Linnett, J. W.   Oxford graduate and lecturer, then Professor and later Vice-Chancellor at Cambridge: spectroscopy and forcefields, later kinetics.

McGlashan, M. L.   Graduate of New Zealand and Reading: research with Guggenheim, being his only successful Ph.D. student: became Professor at Exeter and later at University College, London.

Pople, J. A.   Cambridge graduate and research student with Lennard-Jones: Head of Molecular Science unit at N.P.L. before becoming Professor at Carnegie–Mellon University, Pittsburgh: molecular orbital and structural theory.

Porter, G.   Leeds and Cambridge graduate: Professor at Sheffield and Director of the Royal Institution: President of the Royal Society: shared a Nobel Prize with Norrish and Eigen: knighted and Life Peer: O.M., flash photolysis and photosynthesis.

Richards, R. E.   Oxford graduate and eventually Professor of Physical Chemistry there: molecular spectroscopist and NMR specialist: became Warden of Merton College Vice-Chancellor of Oxford and Director of the Leverhulme Trust. Knighted.

Robb, J. C.   Aberdeen graduate: Professor of Physical Chemistry at Birmingham: polymer studies.

Rowlinson, J. S.   Oxford graduate: Lecturer at Manchester before becoming Professor at Imperial College London, then Oxford: Fellow of Exeter College: statistical mechanics and the theory of liquids.

Spence, R.   Leeds graduate, attached to Chalk River (Canada): Head of the Chemistry Division at Harwell and eventually succeeded Cockcroft as Director there.

Steacie, E. W. R.   Royal Military College, Kingston, Ontario, then McGill, and King's College, London, before a career in N.R.C. Canada, of which he became president: gas kinetics.

Style, D. G. W.   King's College, London, graduate who became Professor of Physical Chemistry there: protégé of Allmand: photochemist.

Sugden, T. M.   Cambridge graduate and research student of Price:

became Research Director for Shell at Thornton Research Centre and later Master of Trinity Hall, Cambridge, Knighted: atomic and ionic mechanisms in flames.

Sutton, L. E. Oxford graduate: became Fellow and Tutor at Magdalen College: main research interests in dielectrics and electron diffraction.

Symons, M. C. R. From Battersea Polytechnic and Southampton University: became Professor at Leicester: solution spectroscopy, especially the E.S.R. spectra of free radicals.

Topley, B. University College, London: associate of Goodeve: solid-state kineticist before joining Messrs. Albright and Wilson.

Whiffen, D. H. Oxford graduate and research associate of H. W. Thompson: after a period at Birmingham University and N.P.L., became Professor at Newcastle-upon-Tyne: research activities in infrared, microwave and E.S.R. spectroscopy.

FARADAY SOCIETY COUNCIL 1965–1971

Allen, G. Leeds chemistry graduate, Manchester Faculty member, and Professor: Divisional Research Director Messrs. I.C.I. Ltd.: knighted: Head of S.E.R.C.: Personnel Director of Unilever: polymer physics and physical chemistry.

Caldin, E. F. Oxford graduate: reader at Leeds and Professor at University of Kent: thermodynamicist and solution chemist.

Carrington, A. Southampton and Cambridge graduate: Royal Society Professor at Southampton, Oxford and again Southampton: E.S.R. Spectroscopy and spectroscopy of molecular ions.

Franck, E. U. Professor at Karlsruhe: experimental and theoretical study of gases and liquids, including water at very high pressures.

Frey, H. M. Student of Hinshelwood's at Oxford and of Kistiakowsky's at Princeton, then at Southampton and Professor and Deputy Vice Chancellor at Reading: gas kinetics and surface chemistry.

Gray, P. Cambridge graduate, Fellow and Faculty member: Professor at Leeds before becoming Master of Gonville and Caius College, Cambridge: reaction kinetics, explosions and oscillatory systems.

Hills, G. J. Birkbeck and Imperial Colleges: Professor at Southampton, then Principal and Vice-Chancellor Strathclyde University, Glasgow: electrochemist: knighted.

Hoytink, G. J.   Netherlander who succeeded Porter as Professor at Sheffield: free-radical spectroscopy in the solid state.

Ivin, K. J.   Cambridge graduate and associate of Dainton's: Professor at Belfast: polymer chemistry.

Lamb, J.   Imperial College, London and Willis Jackson associate: Professor of Electrical Engineering at Glasgow: experimental and interpretative studies of dielectric and viscoelastic relaxation in oils and polymers.

Murrell, J. N.   Cambridge graduate who became Professor of Theoretical Chemistry at Sussex University: molecular electronic structure, spectroscopy and reactivity.

Musgrave, F. F.   Born Sydney, Nova Scotia: graduate of Dalhousie University: Matriculated from Christ Church: studied under Hinshelwood: gas kineticist.

Pethica, B. A.   Cambridge graduate and member of the Colloid Science Laboratory: research manager Messrs. Unilever Ltd. at Port Sunlight before leaving for posts in U.S.A.

Prue, J. E.   Oxford graduate, student of R.P. Bell's: lecturer at Reading: died young: electrolyte solutions.

Purnell, J. H.   Cardiff and Cambridge graduate: later Professor at Swansea: President of the Royal Society of Chemistry: special interests in kinetics and chromatography.

Sheppard, N.   Cambridge graduate, member of Colloid Science Laboratory, and Fellow of Trinity College: Professor at University of East Anglia at Norwich: infrared spectroscopist.

Skinner, H. A.   Oxford University graduate and Senior Faculty member Manchester University: research in thermochemistry.

van der Waals, J. H.   Leiden University Professor of Physics: research interests in electric, magnetic and optical aspects of molecular behaviour, related to *the* van der Waals.

Walsh, A. D.   Cambridge graduate: lecturer at Leeds, then Professor at Dundee: spectroscopist.

Dorothy Jordan Lloyd

It comes as a shock to realize that in the total record of the Faraday Society, 1903–72, only one woman appears in any significant elected role. There is a minute of the Finance and Publications Committee meeting of 9th August 1942: in attendance were 'the Chairman (i.e. the Treasurer), the Secretary (Marlow), Dr. Dorothy Jordan Lloyd and Dr. Lessing on the phone.' It was an interesting way of achieving a quorum in wartime.

Dorothy Jordan Lloyd (1889–1946), the daughter of a prominent Birmingham medical family, studied zoology at Newnham College, Cambridge. She had a most distinguished research career—see *Nature*, 1947, **159**, 190; *J. Chem. Soc.*, 1948, 1727. Stimulated by Gowland Hopkins and W. B. Hardy, her work was centred on colloid aspects of fibrous proteins (including gelatine, collagen, hair and silk) on which she made pioneering contributions. From 1921 she led research for the British Leather Manufacturers. Her record marks her as probably the premier British woman in traditional twentieth century physical chemistry. The establishment of a Memorial Fund in her name is minuted for the Faraday Council of 13th July 1949.

COMMENTS ON THE PRESIDENTS AND MEMBERSHIPS OF THE FARADAY SOCIETY COUNCIL

In commenting on those elected to the Faraday Society Council, it is appropriate to start with the Presidents. Their status as scientists is reflected in the fact that of the fifteen occupying that office over 1945–71, including two from overseas, all were Fellows of the Royal Society and, in their capacity as scientists, seven were knighted, two created Life Peers and two were Nobel Prizewinners: only Goodeve and Steacie were not primarily in university chemistry.

Recurring to the seventeen Presidents from 1903–45, there was one only (Rintoul), not elected Fellow of the Royal Society; seven were knighted, and Lord Kelvin received the Order of Merit. This close duplication in the two periods is not maintained in one other relevant aspect. In the earlier group, nine (Swan, Kelvin, Swinburne, Glazebrook, Hadfield, Robertson, Desch, Mond and Rintoul) had attachments and experience other than academic: for 1945–71, only Goodeve was not a professor of chemistry. For the Society, this marks the lack of interest in applied science after 1945.

In the first half of the Society's history of the eight who were essentially pure academics (Perkin, Lodge, Porter, Donnan, Lowry, Sigwick, Travers, Rideal) four can be labelled Oxbridge. Of the fifteen Presidents in the 1945–71 period, seven can be counted as pure Oxbridge, whilst two others (Melville and Gee) incorporated distinct Cambridge elements.

Similar observations can be made for the members of Council. Fifty-five are counted for 1945–54, of whom nine can be rated as at least partially non-academic (Beacall, Caress, Desch, Goodeve, Lessing, Lunt, Rawlins, Slade and Wansborough-Jones). This figure (16.4%) is further reduced for 1955–71 when, at most, eight (Cottrell, Goodeve, Lessing, Musgrave, Pethica, Topley, Twigg and Wansborough-Jones) out of a total of 64 had distinct non-academic attachments (12.5%).

If in 'Oxbridge' are counted those whose initial or research graduation (often leading to a postdoctoral stay) was at Oxford or Cambridge, then a dominant representation is found for the 1945–71 members of Council. For 1945–54, there are 25 Oxbridge entries in the 55 total: 45.5%. For 1955–71 it is at least 40 in a total of 64: 62.5%. The next largest representation is formed by the London University members who appear to be, at maximum, 12 out of 55 or 21.8% for 1945–54: 10 out of 64, or 15.6% for 1955–71.

What must appear as a disbalance is more sharply defined if the Cambridge representatives are taken separately. They form 20 out of 55, or 36.4%, for 1945–54; 28 out of 64 for 1955–71, or 43.8%. In these counts three members were themselves combined 'Oxbridge': Moelwyn-Hughes, Dainton and Linnett: on the basis of the years spent, the approximate allocation is Oxford 1, Cambridge 2. There can be little justification for this bias in the Faraday Council membership, but it is not difficult to see how it came about. Rideal and Norrish each played unusually important roles in the Society: their protégés often became notable physical chemists whose names came to mind when new nominations were being considered by the few senior officers of the Society so involved: and that ensured their appearance on Council. It goes without saying that many significant U.K. physical chemists did not become members of the Faraday Council. Perhaps the most important consequence of this restriction of choice was the non-appearance of certain topics in the list of General Discussions.

9
Honorary Life Membership

THE COUNCIL MEETING at which Marlow's death was announced (18th March 1948) was one of importance also in other details. The Assistant Secretary was asked to put considerations of instituting an Honorary Life Membership on the agenda for the next meeting: this is item 14 of the minutes, indicating how many items had forestalled its mention. Substantive consideration did not occur until a new Council met in October 1948. Then the same sub-committee as was considering the memorial to Marlow was asked to report on how this other intention should be implemented. An earlier consideration of this intention came to nothing (see Chapter 3.16(f)).

The institution of Life Memberships was supported, and, with minimal changes, the conditions suggested for the award were accepted by Council (16-3-49). The award should be made to any of its members who had rendered outstanding services to the Society and to Physical Chemistry, the total number at any time not to exceed ten. This addition to the Rules of the Society would need to be approved at a Special General Meeting.

It is seen that Honorary Life Membership, the principal honour in the Society's bequest, was restricted to members of the Society. One can ask how far-sighted was this? Unaware of the rule, it so happened that the author wrote in 1965 asking Council to consider awarding this distinction to Professor Peter J. Debye. There was no regret expressed in the Secretary's reply that the award was precluded as Debye was not a member of the Faraday Society. Who was the loser here?

The first election to Honorary Life Membership is recorded in the Council minutes for 24th November 1949: at a Special General Meeting associated with the A.G.M. at Reading University on 23rd September 1949, Council was empowered to make elections on the terms it had accepted. The following members were unanimously elected:

Honorary Life Membership

Professor M. W. Travers
Professor C. H. Desch
Professor F. G. Donnan
Professor A. G. Ferguson

No other names are mentioned. This first election to Honorary Life Membership was to be reported in *Nature* and other journals.

In June 1951 Council unanimously recommended for Honorary Life Membership

Professor Nevil Vincent Sidgwick

The names of Professors Tizard and McBain and Dr. A. McCance were to be considered 'by a sub-committee' (sic). The reader, if not the writer, is free to speculate on the eminence as physical chemists or the outstanding services to the Faraday Society which might apply in these cases. In June 1952 Professor J. W. McBain was so elected.

A Committee was nominated in November 1952 to guide Council on the election of further Honorary Life Members to be associated with the 50th birthday of the Society. The Committee consisted of the officers of the Society and Professor Goodeve, the immediate Past President. At the next meeting in March 1953 on the proposal of Dr. Schulman, seconded by Professor Dainton, the list drawn up by the officers of the Society was accepted unanimously as it stood:

Professor N. Bjerrum
Professor W. E. Garner F.R.S.
Sir Harold Hartley F.R.S.
Professor J. H. Hildebrand
Professor E. K. Rideal F.R.S.
Sir Henry Tizard F.R.S.

Further elections were now restricted by the maximum of ten members. Death would provide vacancies. It is reasonable to complete here the list of all those elected to Honorary Life Membership before the Society as such disappeared.

Sir Hugh Stott Taylor in 1955
Dr. R. Lessing in 1957

In 1959 the names of Egerton, Kendall, Hinshelwood and Ingold were mentioned for consideration. Only one of these was subsequently elected:

> Sir Owen Wansborough-Jones in 1959
> Sir Alfred Egerton in 1959
> Professor N. K. Adam in 1960
> Sir Charles Goodeve in 1961
> Professor H. S. Harned in 1965
> Professor R. G. W. Norrish in 1966
> Professor E. A. Guggenheim in 1967

A total of 21 were elected Honorary Life Members. Of these 16 had served as U.K. residents who were at least members of Council of the Faraday Society. Some were both distinguished physical chemists and outstanding servants of the Society. It can be accepted that any who were not obviously distinguished scientists had been chosen for their great service to the Society, of which the Council would be the best informed and most critical of assessors. There remain five overseas members: McBain, Bjerrum, Hildebrand, H. S. Taylor and Harned. Taylor, having served as President, was especially qualified for his distinction. The others had done, in Faraday Society terms, no more than contribute to the General Discussions. They may therefore be taken as the Faraday Council's choice of the most eminent physical chemists from amongst its overseas members. The distinction was instituted only in 1949, and from that date to 1972 the list can be taken as much as a comment on the Faraday Society's members and its Council as on the elected physical chemists. Amongst those not listed are Michael Polanyi; C. N. Hinshelwood; G. N. Lewis; Linus Pauling; Manfred Eigen; George Porter.

Having given some time to studying the Society's affairs over the relevant years, one is prepared to question – and the reader also can judge – whether some of those chosen, presumably for their contribution to the Society, are not, at least, fully matched in that regard by a number of others. Of those chosen for their distinction as physical chemists, one can perhaps offer a fair comparison. From its inception in 1841, the Chemical Society elected their choice of distinguished chemists as Foreign Fellows, clearly also for life. By 1848 they had elected Liebig, Bunsen, Redtenkacher, Will, Fresenius, Hofmann, Schönbein, Kolbe and Dumas. Of these nine only three (one third) are not immediately identified (a century and a half later) by an informed chemist. The same will not be said of the Faraday Society's choice.

10

The Bourke Lectureship, the Marlow Medal and the Spiers Memorial Lectureship

THE READER WILL recall the generous bequest to the Society known as the Bourke Fund (see Chapter 3.8). The income from this was partly accumulated as a capital reserve and partly used to cover the Society's general activities. At the first Council meeting with H. S. Taylor in the chair (4-6-53) Dainton 'asked whether Council would consider using the income from the Bourke Fund for establishing some form of visiting lectureship for the benefit of provincial universities. He was asked to prepare a memorandum on the subject, bearing in mind an annual expenditure of, say, £150. The chairman said that the offer of such lectureships was particularly valuable to those whose reputations were not already firmly established.'

Coming as it did 20 years after the bequest was received, this was an inspired suggestion. Not only did it ensure members being perpetually reminded of the Society's main benefactor but also, as the President indicated, the Society would promote deserving reputations. Perhaps even more significantly, it would provide a positive contribution from the Society to provincial universities, where members were otherwise aware of it only as publishing the *Transactions* and organising General Discussions. Furthermore, it offered the means of bringing newly emerging themes and techniques in physical chemistry to the attention of U.K. audiences.

Draft rules presented to Council by a committee (Bell, Dainton, Eley and Longuet-Higgins) led to the Bourke Lectureships being established in June 1954. The essential aspects were:

> One or two lecturers per year to be given 75 guineas (£78.75) for expenses.
> Each lecturer to deliver not fewer than two lectures. Preference should be given to younger scientists.
> The Council reserved the right to publish the lectures (which implied a script could be asked for).
> The chairmen for the lectures to be appointed by Council.

There was immediate selection of a lecturer for the year, 1954. Dr. F. Mayo, Professor J. G. Overbeek, Professor D. F. Hornig, and Dr. van Krevelen were suggested, and Professor Hornig was selected. He had two principal research interests: the interpretation of the infrared spectra of crystal lattices and shock-wave propagation in gases. He later became Chief Scientific Adviser to the President of the U.S.A.; he had sat overnight in a small tent around the first atom-bomb at Los Alamos. He lectured at Cambridge, Manchester and King's College, London for the Faraday Society: but he also lectured at Aberystwyth.

The list of Bourke Lecturers and the places at which they lectured is given below. It may be noted that the considerable eminence of many in this list only became widely recognised in the years after their appointment to the Bourke Lectureship.

THE BOURKE LECTURERS

Year	Lecturer	Venues
1955	Professor D. H. Hornig (U.S.A.)	Manchester, Cambridge, King's College, London
1956	Professor J. J. Hermans (The Netherlands)	Liverpool, Birmingham, Imperial College, London
1957	Professor A. J. Rutgers (Belgium)	Bristol, Newcastle upon Tyne, Reading
	Professor W. J. Jost, (West Germany)	Oxford, Cambridge, Imperial College, London
1958	Professor W. H. Stockmayer (U.S.A.)	Manchester, Leeds, Aberdeen
	Professor R. H. Stokes (Australia)	Newcastle upon Tyne, Aberystwyth, Glasgow
1959	Dr. R. Gomer	Imperial College, London, Belfast, Bristol
	Dr. V. V. Voevodsky (U.S.S.R.)	Edinburgh, Manchester
1960	Dr. H. J. Bernstein (Canada)	University College, London, Oxford, Dundee
	Professor A. Perterlin (West Germany)	Birmingham, Nottingham, Liverpool
1961	Professor H. S. Johnston (U.S.A.)	Oxford, Aberystwyth, Edinburgh

1962	Dr. M. Eigen (West Germany)	Bristol, Cambridge, Reading
	Professor J. H. van der Waals (The Netherlands)	Oxford, Nottingham, Sheffield
1963	Professor A. M. Liquori (Italy)	King's College, London, Liverpool, Keele
1964	Professor S. A. Rice (U.S.A.)	University College, London, Sheffield, Glasgow
1965	Professor H. Gerischer (West Germany)	Southampton, Belfast, Newcastle upon Tyne
1966	Professor C. Sadron (France)	Leeds, Strathclyde, Liverpool
	Professor H. M. McConell (U.S.A.)	University College, London, Cambridge Keele
1967	Professor D. R. Herschbach (U.S.A.)	Manchester, Bristol, Edinburgh
	Professor G. Wilse Robinson (U.S.A.)	Dundee, East Anglia, The Royal Institution
1968/9	Professor W. Klemperer (U.S.A.)	Southampton, Newcastle upon Tyne, Leeds
1970	Professor E. U. Franck (West Germany)	Leicester, Southampton, Imperial College, London
1971	Professor H. Wolf (West Germany)	Leicester, Southampton, Imperial College, London

It is appropriate to emphasize here how much care the Council took in the elections to the Bourke Lectureship and in the award of the Marlow Medal. Members were prepared to offer their own choice for the former and, on occasions, five or six names were considered and even voted on before a decision was made. The Lectureship acquired a notable international prestige.

THE MARLOW MEDAL

It was at the Council meeting which ratified the conditions for the Bourke Lectureship in June 1954, and at which Coulson raised the question of overseas members of Council, that the treasurer (Lessing)

The Bourke Lectureship

made a significant suggestion: 'that it was the practice of some other societies to offer annual prizes for papers of outstanding merit' and he wished Council to consider doing so to encourage the younger physical chemists.

The group of four who had advised on the Bourke Lectureship reported on 4th March 1955. The 'premium papers' award would include the presentation of a medal to be called the Marlow Medal. With the cooption of the Treasurer, the sub-committee was asked to present specific details.

These details were presented and, with slight modification, accepted on 2nd June 1955. The essentials then agreed can be summarised:

1. The Marlow Medal 'with an accompanying grant, not exceeding £50' to be made annually to a candidate of sufficient merit. A second award could be recommended.
2. Applicants must be members of the Faraday Society for at least three years and they must apply before their thirty-third birthday.
3. Papers on topics appropriate for the *Transactions*, but not necessarily published in the *Transactions*, must have been 'received' before the candidate's thirty-second birthday.
4. Joint publications could be considered with appropriate consultation with any senior authors.
5,6. Members of the Faraday Society could apply themselves or on behalf of others.
7,8. Council would appoint special assessors and the arrangements would be reviewed after five years.

For the Marlow Medal a minimum of three and sometimes five members of Council or senior colleagues were appointed to assess the applications. Again, the standard established was of the highest. This is partly reflected by the withholding of the award when sufficient merit did not appear, but more obviously it is established by the later careers of the medallists. Some years the applicants provided up to three candidates 'each of them worthy of the award' but usually only one was chosen. Why this was so is not clear, as 'a second award can be recommended' (2nd June 1955). When this opportunity was not taken the 'fully worthy' candidates who were missed out usually lost their recognition subsequently by the age restriction. Occasionally there was concern that too few were applying: contenders could themselves apply or they could be suggested by members of the Society: in particular, members of Council were encouraged to put forward applicants.

RECIPIENTS OF THE MARLOW MEDAL

1957	J. S. Rowlinson
1958	J. A. Pople
1959	P. Gray
1960	No award
1961	J. S. Griffith
1962	J. C. Polanyi
1963	S. A. Rice
1964	No award
1965	A. M. North
1966	A. Carrington
1967	C. N. Ramachandra Rao
1968	N. M. Atherton
1969	J. W. White
1970	M. A. A. Clyne
1971	G. R. Luckhurst

THE SPIERS MEMORIAL LECTURE

This, the senior Faraday Society lectureship, was instituted in 1927 [see Chapter 3.16(a)]. It is appropriate to list the lecturers here. Of the fifteen elected during the life of the Faraday Society, three can be recognized as Nobel prizewinners. No more need be said than that the standard was of that character.

SPIERS MEMORIAL LECTURERS

1. 1928 Sir Oliver Lodge
    (Given at the Royal Institution)
2. 1931 Sir Robert Mond
3. 1934 Sir William Bragg (Royal Institution)
    (Given at the Royal Institution)

The Bourke Lectureship

4. 1937 Professor W. T. Astbury (Leeds University) (Given at the Royal Institution)
5. 1950 Sir Hugh Taylor (Princeton University, U.S.A.)
6. 1952 Dr. J. T. Edsall (Harvard University, U.S.A.)
7. 1953 Professor J. H. Hildebrand (University of California, Berkeley, U.S.A.)
8. 1956 Professor E. T. Teorell (Uppsala University, Sweden)
9. 1957 Professor H. S. Harned (Yale University, U.S.A.)
10. 1960 Professor Th. Forster (Technische Hochschule, Stuttgart, West Germany)
11. 1961 Professor H. Bloom (University of Tasmania, Australia)
12. 1963 Professor G. Herzberg (National Research Council, Canada)
13. 1965 Professor H. C. Longuet-Higgins (University of Cambridge)
14. 1967 Professor G. S. Rushbrooke (University of Newcastle upon Tyne)
15. 1970 Professor P. J. Flory (Stanford University, U.S.A.)

Nothing more estimable can be written of the Faraday Society than that three non-academics to whom the Society owed so much are worthily commemorated in the honours bestowed by the Spiers Memorial Lecture, the Bourke Lectureship and the Marlow Medal.

11
The Chemical Council

THE RELATIONS BETWEEN the Faraday Society and the Chemical Council mostly concerned minor advantages offered by joint membership of a committee coordinating the U.K. chemical societies. The Chemical Society and the Society for Chemical Industry had formed the Council to promote favourable inter-relations. In March 1945 the Faraday Society was asked to cooperate in their system of promoting joint membership: this was based on a points system whereby members of one Society acquired a credit towards reduction of the subscription to another. To this end the Chemical Council was prepared to coopt two members from the Faraday Society into its Council.

The immediate reason for this approach was the threat presented by the American Chemical Society's offer of very favourable terms for Canadian chemists, provided all Canadian members were to be recruited. However, other details such as the common indexing of U.K. chemical publications and the precise definition of terms in chemical science (this was a Royal Institute of Chemistry venture) were helped by a coordinated approach. Arrangements were also established whereby the Chemical Council collected the subscriptions from members of two or more of the Societies.

Probably the most important consequence following on its membership of the Chemical Council was that the Faraday Society benefited from its Publications Fund. This was subscribed to by those institutions, including industry, anxious to support publication of British chemical research. From this source the Faraday Society received significant support. In this context should be mentioned the Faraday Society's contributions to The Chemical Society's Library, an important facility of which its members acquired useage. The Faraday contribution for 1948 was £117. At the same time, the Treasurer approached the Chemical Council for a grant of £700 towards publication costs which, at that time, were running the Society into deficiencies. A grant of £500 was forthcoming.

Changes in the Chemical Council led to the termination of the collection of joint subscriptions at the end of 1954 and the reduction in the Faraday Society's representation from two to one.

Nevertheless, in 1961 the Chemical Council asked the Faraday Society to forecast its grant applications for the following four years, despite the fact that Council minutes do not record any such sums in the immediately recent years. In 1957 a request for £1750 was made, but the result is not recorded. A request in 1961 from the Faraday Society for £6000 over four years was changed in 1962 to £7000 over two to three years. This, of course, was 'in aid of publication'. The grant allocated was £2200 for 1962, accompanied by the Chemical Council's hope that the Faraday Society's finances would soon be self-balancing.

The next reference to the Chemical Council comes on 4th November 1965. It merely records that in place of F. F. Musgrave as the Faraday representative, 'G. H. Twigg is appointed for the remainder of its (the Chemical Council's) existence'. A premonition of this termination can be read in a a message to the Faraday Society. On 1st June 1961 it is reported that the Chemical Council wished to give thought to its future. It averred that changing circumstances necessitated a reconsideration of the support being given to the publications of chemical societies. It is in this context that the link with the Faraday Society disappeared.

12

The Annual General Meetings

UP TO 1954 the Faraday Council's year ran from October 1st to September 30th; that is, changes in the elected officers and members of Council took place on October 1st. However, at the June Council 1954 it was decided that the Council's year should run from January 1st to December 31st, 'to conform with the Society's financial year'. Although it is not recorded, it could well be that this change was suggested on behalf of the Finance Committee by the Treasurer or, again, it could have been the Secretary. It would require a change in the Society's rules, and that required a Special General Meeting.

It should be explained that Annual General Meetings, which for many societies and most companies are events of special importance, were usually arranged for the Faraday Society to take place during (often on the first morning of) a General Discussion meeting coming in July, August or September. The Annual General Meeting was thus a floating event, both in date and location. The reason was obvious. This placing of the A.G.M. would bring it where, usually, a minimum of 50, and frequently more than 100 members would be attending the Discussion: and that in a (university?) neighbourhood which would itself include the homes of a further significant number. Even so, such was the minimal concern of the general membership, that these business meetings – involving the election of officers, consideration of the Society's finance, charges for membership, questions of publication in the *Transactions*, 'and any other business', – usually attracted fewer than 20 from the general membership.

More significant is the *modus operandi* by which Council arrived at the recommendations to be put at the A.G.M. Of most interest was the election of new officers. Every other year, a new President; each year, when, as usual vacancies occured, new Vice-Presidents, and also replacements for retiring members of Council. A small group of members, or any one member of Council, could send in their nominations for these offices, but there is no *record* that this ever occurred.

Accordingly, at the June meeting, Council were presented with the names of those suggested as nominees. These came from the President in the Chair following prior discussion between the President, some Past Presidents and the Secretary. Members of Council scarcely ever did anything other than accept the names announced. These then went forward to the A.G.M. for formal election. Even so, to the writer's knowledge, no dissatisfaction with the way the Society was run ever surfaced. If it had, it should have been recorded in the minutes of the Annual General Meetings for 1945–71: no such minutes have been found.

The absence until 1960 of any legal status for the Society afforded it notable pliability. The assumption of an indifferent, not to say docile, general membership is illustrated by the notice for one (of several) overseas Annual General Meeting:–'to take place at the Botanische Anstellen, Universität Göttingen at 9 a.m. on 16th September 1964.'

All that was found in the Burlington House Faraday Society archives for A.G.M.s in post-1945 years were printed notices of the A.G.M.s to be held with listings of current officers and their replacements nominated by Council, together with summary financial statements. (L.E.S. also comments on his futile search and the inadequacy of the record for Faraday Society A.G.M.s: Chapter 3.5.) From this evidence it can only be assumed that the Faraday Society was an institution as much accepted by its members as royalty was by the U.K. public. And, perhaps, for the same reasons: the members saw no need to interfere with a system which seemed to work smoothly, generally satisfied their wishes, and gave them no cause for concern.

13
Miscellaneous Items of Faraday Society Council Business

IN REPORTING THE concerns of the Faraday Council the reader will be aware that most of its major themes are dealt with separately: Finance, Publications, General Discussions, Legal Status, Amalgamation, etc. What remains is a series of often disconnected items which arose at particular times. And there is the general *modus operandi*.

To base an account of the Council's procedures on the Minutes could be inadequate, but they do serve to confirm practical experience. Business was always conducted very efficiently: agenda were meaningfully ordered, and worked through by individuals who had little time to waste. Sometimes items gave rise to exchanges, and when there were multiple alternatives, as in subjects for General Discussions, or choice of Bourke Lecturers, a majority opinion would be established and accepted. It could have been good practice that votes were rarely involved. The Minutes evidence the very few instances when this occurred: a number can be mentioned.

When Wansborough-Jones expressed the intention of resigning the Treasureship, Council were asked to consider four names: at this time they also chose a new member of the Finance Committee (3rd March 1960). Such elections usually took place by the acceptance of names presented by the Chair. Occasionally, when several names were put forward, e.g. for Honorary Life Membership, they were voted on, sometimes leading to no choice being made: on one occasion, when there were two vacancies, only Sir Charles Goodeve was elected (2nd March 1961). Single names were presented to Council and accepted without demur. In only one instance was the election of President voted on (29th August 1945). Two other cases can be noted.

In 1951 Council were reminded that the fiftieth anniversary of the Society would occur in 1953, giving added significance to the choice of President. Members were asked and did submit five names which were not recorded: rather, the President in office (Goodeve) proposed that they nominate H. S. Taylor of Princeton (6th March 1952). This had clearly been discussed with Taylor, as it was reported he would have leave of absence to undertake his duties for one year in the U.K. This arrangement was so acceptable as to be received unanimously.

A similar situation arose in 1957, when the President (Bell) explained to Council that as they planned a General Discussion in Toronto in 1959 (very much with the support of Dr. E. W. R. Steacie, President of the National Research Council of Canada) 'it would be very nice to have Dr. Steacie as their President'. This again was so obviously appropriate as to be instantly acceptable. But it meant that the normal two-year Presidency would be split, and names were requested 'to be considered' for the President in 1958. No alternatives appeared at the next meeting (6th June 1957) when Dr. H. W. Melville was listed for nomination.

It cannot be suggested that the Faraday Council operated other than in the way of most committees: but it does mean the Society had many key decisions taken by a self-maintained enlightened oligarchy. It would perhaps be generous to suggest that the results were as obviously celebrated as those of the Council of Ten at Venice.

An interesting point was raised by Coulson in November 1953 in a letter to the President (Norrish), who immediately declared his agreement. The question was whether Council should include overseas members. In the ensuing discussion it was emphasized that the members of the Society had always had the power to nominate anyone of their number for election to Council. It could be that explicit concern had arisen from the Society having recently had an overseas President (H. S. Taylor, 1952–53). One consequence was that over the rest of its life, the Faraday Council frequently had one and sometimes two non-U.K. members. The writer recalls mentioning this fact as 'one or two appointed members from the continent' in discussing the Society's activities with Jan Ketelaar at Amsterdam. 'Oh', he said, 'that's interesting: where are the others from?' It was agreed such membership should not be thought of on a territorial basis; it did not involve representation but should be determined by the suitability of the candidate as an individual member of Council. Later the Société de Chimie Physique at Paris regularly appointed members from Germany and the U.K. to its Conseil. The French Société was initially more generous than the Faraday, it refunded such foreign members' total travelling expenses. The Faraday Society provided only travelling expenses within the U.K. This must have limited the choice to those who could afford or who could acquire coverage of expenses, *e.g.* for return flights to London up to three times a year.

One transient item on which Council spent time was the proposal by the Borough of Southwark in London to promote a national appeal for the formation of a Faraday Trust. It transpired that the aim (please accept this) was to construct two islands on the Elephant and Castle site: one to accommodate an underground substation, the whole to be known as Faraday Square. After meetings at which the Royal Institution was also represented, the proposal petered out, ending as an intention to erect a plaque in Faraday's memory to cost £3000 (5th June 1958).

A similar negative conclusion was arrived at on the question of recognizing the centenary of Faraday's death (1967). A Committee was convened with representatives of the Royal Society, the Royal Institution, the Faraday Society and the Institute of Electrical Engineers. Its decision was that it would be inappropriate to commemorate the death (now frequently done for artists) 'and they could find no precedent for doing so' (17th March 1966).

Later (3rd June 1969) it was reported that a bronze plaque of Faraday had been presented to the Society. This had come from Sir Robert Hadfield, a founder member and President of the Society 1913–20, *via* the widow of Mr. Leonard Hill, to whom he had given it.

Here too can be mentioned the shoulder-length bronze bust of Faraday which stands on a plinth in the Council Chamber of the (now) Royal Society of Chemistry. It was bought soon after amalgamation by the Faraday Division Council. Thanks to the effective offices of the Executive Secretary of the Chemical Society, Mr. J. Ruck-Keene, an appropriate plaster cast was borrowed from the Royal Institution. The bronze casting was made, essentially as a student exercise, at one of the London art schools, so that it cost only £120. One member of Council took advantage of this to acquire two copies on the same exceedingly

Miscellaneous Items of Faraday Society Council Business 363

favourable terms. As a consequence, one of these has been on loan in the small dining room at 10 Downing Street since 1989.

In 1963 a comment critical of the Secretary (Tompkins) appears. In its context, reference must be made to an important Council minute pre-dating his appointment. It relates to appointments to be made by Council, more specifically to the offices of President, Vice President and to its own membership. The Secretary is asked (23rd October 1946) to prepare *a memo for each member of Council* of the offices becoming vacant and 'asking members to suggest in writing those they recommend for nominations.' There is no evidence in the Council minutes that this practice was established when Marlow died eighteen months later. Nor does it appear later. Certainly it was not in operation at any time after 1958. It was a notable omission in practice. The comment now referred to is: 'Whilst supporting the recommendations made, Council considered the fact (sic) that the Secretary had assumed responsibility for appointing [note: not recommending] sundry representatives of the Society without prior consultation with the President or Council.' This does suggest that the benign oligarchy could, on occasions, become an autocracy. Even so, the Council merely 'considered the fact'. The reader can be reminded that Marlow had been responsible for inserting a notice in the *Transactions* which had not been sanctioned by Council (Chapter 3.9.)

Mention has already been made of the existence of Faraday Society local representatives in the U.K. When first mentioned in the context of recruitment (7th June 1956) a characteristic 'no action' resulted. However, they came into existence at universities. In 1967 it was decided to publicize their names in the *Transactions* as they were asked to distribute notices of Faraday Society activities in their neighbourhood, to include industrial laboratories. In March 1968 on Council, Ashmore suggested a check should be made of the effectiveness of the local representatives when it was revealed that they existed in all the U.K. universities except Durham and Sussex.

In the last years of the Faraday Society, as such, there was a return to its earliest roots, that is to physical chemistry in its applied aspects. In May 1950 the minutes of the Publications Committee record an approach by Bawn, M. G. Evans, Melville and Ubbelohde advocating the publication of a journal of applied physical chemistry. They emphasized that there was a real need for such a journal. The Committee 'could not recommend that the Society undertake it at present'. It is not mentioned in the Council minutes.

It was in 1969 after preliminary discussions largely due to the initiative of B. A. Pethica of Messrs. Unilever, who was a member of Council, that the Faraday Society set up an Industrial Committee. Its particular concern was defined 'to make recommendations to the Standing Committee for Conferences on themes of industrial interest judged to be timely and of high scientific merit in fields of interest to the Society'. The Industrial Committee was to have no fewer than nine members appointed by the Faraday Council (three being *ex-officio*), and it would liaise with the Society for Chemical Industry. All this appears at the same time as a major discusssion on the conditions for amalgamation with the Chemical Society (November 1969) was taking place. Dr. Pethica was appointed the first chairman of the Industrial Committee, which was already organizing a meeting on *Thin Liquid Films and Boundary Layers* at Cambridge in September 1970.

14

Amalgamation

THE FARADAY SOCIETY was founded in 1903, 62 years after the Chemical Society, whose first President, Thomas Graham, is best remembered as a physical chemist. Such a designation or differentiation did not become established within chemistry until later. It is reasonable to accept the traditional date of 1887 as that for the recognition of an independent physical chemistry, with the first appearance of the Ostwald–Van't Hoff *Zeitschrift für Physikalische Chemie*.

It is noteworthy that our greatest physical chemist had little association with our Chemical Society.

'One misses the name of Faraday in the list of original members, and his election took place only in January, 1842. On different occasions subsequently he was made a Vice-President or a member of Council, but it does not appear from the minutes that he attended meetings in either capacity. Later, in 1851, Faraday was proposed as President, but he declined the honour for reasons of health. The still higher distinction of the Presidency of the Royal Society he likewise declined in 1857.'
(T. S. Moore and J. C. Philip, *The Chemical Society 1841–1941*, 1947, p.16.)

THE BACKGROUND SITUATION

By 1903, the Chemical Society had a preponderant commitment to organic chemistry. The contrast shown by the early Faraday Society could scarcely have been greater. Of the Faraday Society's first nine Presidents, only one was a chemist, the most eminent of the U.K.'s organic chemists, Sir William Perkin. The other eight consisted of four physicists, three others perhaps best described as electrical engineers, and one metallurgist. F. G. Donnan was the first recognizable physical chemist to be President (1924–26).

These observations are relevant as reminders of the substantial separation between the Chemical and the Faraday Societies during the early years of the latter. Chapter 1 provides the detailed picture. With the passage of the years the gap narrowed markedly. The Faraday Society lost its interest in the more applied electrical features, such as electroplating, and also in metallurgy. Likewise, in the first inter-war period, 1919–39, organic chemists came to depend upon and themselves to use physical methods of increasing sophistication: reaction rates, thermochemistry, dipole moments, spectroscopy, dispersion of optical rotation, were amongst topics taken up by organic chemists.

Thus, it became an established practice, certainly by the 1930s, for the Chemical Society to arrange (usually evening) meetings in which an appropriately related group of physico-chemical papers would be presented and discussed. This aspect developed further. Note has already been made of Garner calling attention in the Faraday Council to a full-blown discussion on Catalysis being organized by the Chemical Society (1st March 1957).

Perhaps even more significant was the development of many Subject Groups under the aegis of the Chemical Society. These consisted predominantly of enthusiastic young research chemists who themselves organized informal discussions on topics in their interests.

From 1946 to 1972 there were 13 Presidents of the Chemical Society. These included six who had been members of the Faraday Council: Hinshelwood, Rideal, Ingold, Emeleus, Melville and Porter. Only two of these (Emeleus and Ingold) did not also become Presidents of the Faraday Society, and it is noteworthy that Rideal, a great pillar of the Faraday Society, was President of the Chemical Society before Ingold.

With the advent of the computer into laboratories during the 1960s, all physical science research was greatly accelerated. To mention only spectroscopy – the variety and range of studies in molecular science were transformed. There was no obvious corresponding expansion in Faraday Society activities. Chemical Physics was, at best, weakly represented in the *Transactions*. Suggestions that the Society should extend its activities, that subject groups might be sponsored, met with no response either in 1961 or later in 1965. It was Eley, in a Publications Committee, who referred to the Chemical Society's intention to divide its *Journal* into several sections, including one specifically

Amalgamation 367

devoted to Physical Chemistry. The Chairman (Bawn) agreed that this was a serious development. The Secretary (Tompkins) opined there was little common ground with the Chemical Society, and that the Faraday Society could publish little more because of limited finances (Publications Committee minutes 30-5-65). The general inclination was not to get involved in new activities, perhaps especially if they were suggested from outside.

New activities of any description would mean expansion of the Faraday secretariat. In the writer's experience this was consistently opposed. One reason advanced was the limited accommodation available: another was Tompkin's negative attitude. His routine tasks as Secretary would not have been possible in one day per week.

One case can be quoted. In June 1969 the Assistant Secretary (Miss Kornitzer) reported a meeting at which the Physical Society, the Institute of Physics and the Institute of Electrical Engineers had also been represented. It concerned the formation of a Dielectrics Study Group 'which had already organized two successful meetings'. The Faraday Council decided that no action was needed. The particular interest of this example is that the writer was the Secretary of the Group, and L. E. Sutton became the Chairman of what developed into an independent Dielectrics Society. In terms of its origins and field of interest, it would not have been inappropriate for the Faraday Society to have been associated with this venture. But, as was seen earlier, even a modest increase in Faraday activities, such as the addition of an annual Symposium meeting, needed an expansion of the secretariat and the relocation of Council meetings from Gray's Inn Square, where the accommodation was so limited (3-11-66).

THE INITIAL STAGES AND THE PROPOSALS ACCEPTED

With these preliminary indications of the situation in the Faraday Society, the steps which took it towards and into amalgamation can be recorded. This account will be entirely of the responses within the Faraday Society. The extensive discussions which took place between the Chemical Society, The Royal Institute of Chemistry, the Society of Chemical Industry and the Society of Analytical Chemistry, in many phases of which representatives of the Faraday Society were engaged, have been amply well delineated in D. H. Whiffen's *The Royal Society of Chemistry, The First 150 Years* (1991).

For a full documentation of the Letters of Intent, the Heads of Agreement and The Proposals finally accepted for amalgamation, the reader can consult the 33-page account by Sir James Taylor: *The Amalgamation Proposals*.

Those seeking indications of future events might claim to find one in a Council minute of 9th March 1967. 'The President (Bawn) said that the Treasurer (Musgrave) had left with him a paper in which he suggested that Council should consider the general future of the Society in all its aspects... It would need frank and open discussion amongst themselves, but this matter was to be treated as strictly confidential.' Some of this concern arose from the possibility of the Faraday Society relocating its activities in the extended premises being prepared at the Royal Institution. Here the Society was in a favourable position, as one of its senior members, Professor Porter, was Director of the Royal Institution.

However, the whole situation was changed by the initiative of a Past President of the Faraday Society. Sir Harry Melville had been Secretary of the Department of Scientific and Industrial Research and knew the positions and interests established throughout the U.K. chemistry world. As President of the Chemical Society he had used his Presidential Address in April 1967 to call for 'a closer association of the various societies covering the different aspects of chemistry'. That this call was made against a background favourable to such action is shown by the fact that the Chemical Society's Council voted unanimously, only two months later, in favour of an amalgamation of the three chartered bodies, the Chemical Society, the Royal Institution of Chemistry and the Society of Chemical Industry. A lawyer, Sir Eric Bingen, was given the task of making an independent assessment of this proposal and, when his health failed, it was passed to the hands of Sir James Taylor in April 1968. In the terms of reference for this consideration there appeared the clause 'to examine the fusion ... without excluding the possibility of kindred societies such as the Faraday Society and the Society of Public Analysts, due regard to be paid to the separate interests of the members'. Melville had himself inserted this reference to the other Societies. The Faraday Society owed to him its recognition in this all-important development.

When the situation was considered by the Faraday Council on 2nd November 1967), there was an essentially positive reaction. However, the President (Bawn) reported that no approach had yet been made to the Faraday Society. Fortunately, the reason for this was given by Whiffen, himself a member of Council. No formal invitation had been sent because, if it were turned down, it would have made it hugely more difficult for the Faraday Society to join at a later stage. A Select Committee was appointed to define the Faraday Society's position: the President, the Treasurer, Ubbelohde, Bell, Whiffen and the Secretary: with four to form a quorum.

The statement giving the Faraday Society's position was submitted to Sir Eric Bingen via Sir Harry Melville, President of the Chemical Society. The Select Committee had met representatives of the Chemical Society. Little progress could be reported (14-3-68). The Chemical Society had not thought out its position. The Faraday representatives had in mind some form of federation, in which the Society's identity would not be lost. As ever, the costs and viability of the *Transactions*, or of a replacement journal of Physical and Theoretical Chemistry in the new structure, were considered: as members of the joint organization would not be required to take any one of its several journals, the take-up of the *Transactions* 'would drop enormously' (Porter). The President said he would refer to the position at a luncheon to be given to Local Representatives at Liverpool in September. His statement there was, in essence, that an amalgamation was very unlikely.

The next occasion when amalgamation was discussed in the Faraday Council there was no progress to report from the Chemical Society (3-6-69). However, a number of other interesting details emerged.

The Institute of Physics had made an approach on incorporating the *Transactions* as one of the several parts of their *Journal of Physics*. To those who were unable to draw a line between chemistry and physics, this was a pleasing proposition, and Linnett made the obvious point that the U.K. lacked a Journal of Chemical Physics.

Pethica presented the case for the Faraday Society catering for its members in industry before they were recruited by other societies. This led to the formation of the Industrial Committee as already noted in the presentation on Discussions. The writer raised the question of relations between the Faraday Society and the related European Societies. Both the Société de Chimie Physique and the Bunsengesellschaft were anxious to envisage biennial joint meetings with the Faraday Society. The development of such liaisons would strengthen the Faraday Society's base. Local representatives could be appointed in European countries.

The situation developed rapidly in the five months from June to November 1969. The Select Committee had met various panels formed by the Chemical Society and the Royal Institute of Chemistry. Whilst they had been technically only observers, they now advocated becoming negotiators. There was general agreement that the Society should participate in what was spoken of as 'the new Chemical Association'. Some members of Council felt that whilst this was necessary, the Faraday Society might aim to retain its individuality, but Tompkins pointed out that the new Association would have up to 40 000 members and include its own Physical Chemistry Division. The Faraday Society would eventually not be able to compete if independent.

The decision was taken to enter into negotiations from what was regarded as a strong position, but without committment to amalgamate. Faraday representatives were appointed for the discussions on Publications; Group and Divisional Activities; Finance; General Negotiations. For the latter, which dealt with the major questions of the character of the new Society and the details of the amalgamation, the choice was the President (Gee), the Treasurer (Rowlinson) the two immediate Past Presidents (Bawn and Dainton) and the Secretary (Tompkins). The members at large were to be informed of this development and assured they would be consulted before any committment was undertaken. One noteworthy dissenting response was reported, from Dr. David McCall of the Bell Telephone Laboratories, Murray Hill, New Jersey. He had written 'expressing concern that the Society should not be submerged in a large amorphous body, and calling attention to the unhappy position of physical chemists in the American Chemical Society'.

Amalgamation

This sentiment was shared with some intensity by members of the Faraday Society, but the subsequent referendum revealed this view to represent a small minority only. The fact that it was an extremely busy American colleague who alone wrote it, suggests that the membership at large were insufficiently concerned to question the developments. It was certainly an aspect much considered by members of the Council. However, the overall acceptance of the proposal to amalgamate on terms yet to be finalized was well established by the results of a Members' Referendum (1970):

	Votes for amalgamation (%)	Votes against amalgamation (%)	Members recording a vote (%)
U.K. membership	86.2	13.8	71.3
Overseas membership	76.9	23.1	41.5
Total membership	82.4	17.6	55.9

It is relevant to note some of the detailed objections which were aired in Council when the proposals embodied in the new draft Charter were discussed in June 1970. Ubbelohde established that the Royal Institute of Chemistry was going to be much over-represented on the new Council and (surprisingly?) expressed doubt whether it would be appropriate for Chemical Science to be 'split into Physical, Organic, and Inorganic ... in, say, 15 years ahead'.

There was extended consideration of the retention of the title of the Faraday Society. It was known that many members wished for this, although the Referendum had been taken on the basis of it becoming the Faraday Division in the new Chemical Society. Specifically, it was accepted that the journal would be named *Transactions of the Faraday Division of the Chemical Society*. But on one detail, the Council were not prepared to accept any change. The refereeing system of the Faraday Society should be retained: thus, the Editor and Publications Committee appointed by the Faraday Council should operate as heretofore: the refereeing and acceptance of papers should remain with them. The acceptance achieved for this was undoubtedly important. It retained the status of the *Faraday Transactions* in the hands of professional physical chemists. Even the format of the *Transactions* was to be retained. The Chemical Society favoured a two-column A4 page. Any

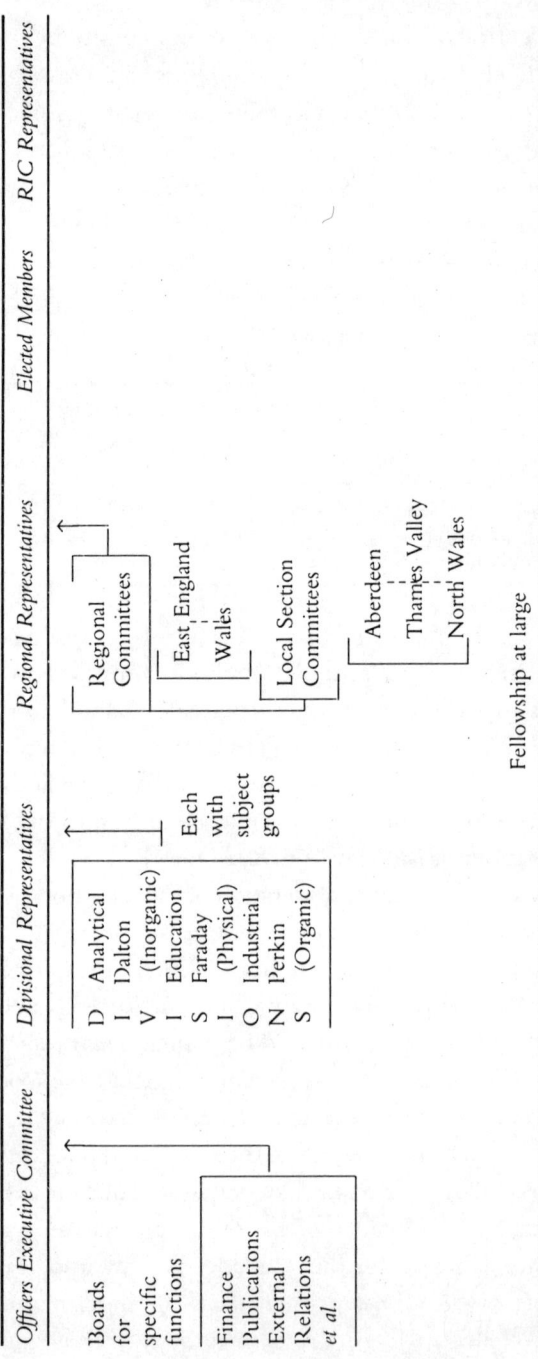

Fig. A. The Chemical Society Council

new journal(s) principally concerned with the chemistry–physics nexus were to come under the aegis of the Faraday Division. All these desiderata were later accepted.

As could be expected, the Faraday Society was particularly anxious to retain control of its publications. In the all-important matter of editorial control, this seemed adequately safe in Tompkins's hands, but the new system would lead to the loss of D. A. Young. He had played a significant role as Assistant Editor to Tompkins: especially in the busy transition year of 1971, his was a large part of the responsibility. His departure thereafter could have signalled a diminution in specifically Faraday control. However, he was eventually re-appointed as Assistant Editor, serving in this role until 1977, when he became Editor.* The significance of this lies in the continuance in office of personnel fully apprized and anxious to maintain Faraday Society practice and standards. In more changeable categories, such as the Faraday Council itself, a less certain insistence on Faraday control was likely to be (and apparently has been) the result. Only those well versed in a tradition can hope to maintain it.

One of the differences for the Faraday Society arose from the fact that the Chemical Society, the Royal Institute of Chemistry and the Society for Analytical Chemistry had long had very active local sections generating their own programmes of meetings. These provided a new benefit for Faraday members, who would become members of their local unified branches: of course, those many members who had also been members of the Chemical Society had already been involved at this level. Another positive development was the establishment of an External Relations Board for the new Society. To this the Divisional Councils elected their own representatives. Through this Board all formal relations with organizations external to the new Society were arranged. This assured the Faraday Division a permanent channel for communication with overseas societies building upon the superbly

---

* Tompkins remained Editor for the Faraday Division until 1977 when he made his final departure. On November 1 1977 there was a subdued presentation in the Hinshelwood Room of the Royal Society of Chemistry at which Magat was present. Tompkins would not have wished for anything more elaborate. Commenting on Laboratories being named after recent retired professors, he told the writer: 'I woudn't wish a gents' lavatory to be named after me.'

good Anglo–Dutch, Anglo–French and Anglo–German relations, and informal but lively Canadian and U.S. relations present before amalgamation. A further Sub-Committee of the new Council was formed to handle the activities of the Divisions with their overseas members or other overseas non-institutionalized professional groups. The writer represented the Faraday Society in the discussions which led to the local, regional and external affairs arrangements in June and October 1970.

THE STRUCTURE OF THE NEW SOCIETY AND THE CHANGES INVOLVED

The structure pattern in the new Society is shown in Fig. A, which is simplified from Taylor's *The Amalgamation Proposals*, p. 5. In it the RIC is shown as sending some of its own members to the Council. This arose from its retaining in 1972 independent control of professional (industrial) qualifications. This meant that in 1972 the RIC maintained a separate status, although it was already envisaged that unification, going beyond the amalgamation, was ultimately desirable. Many of the difficulties and the extended discussions, both before and after 1972, arose from the fact that the Chemical Society and the Royal Institute of Chemistry each had Charters. The section on the status of the Faraday Society indicates the legal complications which could arise from a Charter constitution. For the Faraday Society, as an Incorporated Company, the whole operation of amalgamation was simplified.

Brief mention must be given to the immediate changes for Faraday Society members which would follow amalgamation. From the first, it had been established that no disadvantage should accrue to members of the participating Societies for the first two years of the amalgamation. This meant that the basic fee for membership of the Chemical Society would be £2.50 for the first two years. Members of the Faraday Division would receive both *Faraday Division Transactions I (Physical Chemistry)* and *Faraday Division Transactions II (Chemical Physics)*. This will establish why it was anticipated that, after two years, the membership fee for the new Society would rise to about £16.00. For this Faraday members would acquire many benefits not previously theirs, e.g., receipt of the monthly house journal of the Chemical

Society, *Chemistry in Britain*, and participation in the much heavier programme of activities organized by the Chemical Society. The Faraday Treasurer (Rowlinson) had emphasized that, if remaining independent, the Faraday Society subscription would itself have to be increased to about £6.00.

In this context the transfer of all Faraday Society funds must be mentioned. There was the possibility of holding them independently as a Faraday Divisional Trust Fund, but all the specific allocations for maintenance of the Spiers, Bourke and Marlow commemorations were guaranteed. The Faraday Society agreed for its assets to be transferred to the new Society, to be kept in a separate account, and to be 'applied for objects as nearly as practicable or similar to those for which the Society was founded'. The key phrase, 'as nearly as practicable' replaced the Faraday Society's initial 'identical' in its paper to the Chemical Society. This change was described as 'an improvement' in a Memorandum on the Agreement by the Faraday Society's solicitors.

It is particularly relevant to establish what was the attitude of the Faraday Society membership to the amalgamation. Evidence has already been given of the pure formality of A.G.M.s at which members might have raised questions, on differences from the proposals of the Council. Naturally, the proposal to amalgamate had to be put to an Extraordinary General Meeting of the membership. Whiffen, who himself had been a member of the Faraday Council, puts it succinctly: 'In accordance with its traditions the Faraday Society could hardly be bothered with such business matters and the extraordinary meeting was held on April 15 (1971) before the day's session at a Discussion meeting at Nottingham; it attracted a vote of 26 to 3 in favour'. (*The Royal Society of Chemistry: The First 150 Years*, p. 39). These figures should be compared with the Society's U.K. membership of ca. 1150, i.e. 2.5% voted on the question.

THE TERMINAL STATE OF THE SOCIETY

The notice (not the minutes) for the last A.G.M. of the Faraday Society – on 14th September 1971 at 9.00 a.m. in the Lecture Centre of Brunel University, Uxbridge, Middlesex – was a more substantial item than usual. It records, as the Report of Council, on ten small pages

(13.5 by 21.0 cm), the details of the Society's programme and its officers for 1970. There is listed the membership of Council, the officers, publishers, auditors and bankers: the General Discussions and Symposia; membership numbers in general, the names of those in all committees as well as representatives on other bodies; a statement on finance, with income, expenditure and balance sheets, including those for the Spiers, Bourke, and Industrial Sub-committee accounts, plus a list of investments.

It is appropriate to give summaries of some aspects.

Publications: Papers/pages in the *Transactions*, 362/3288; in *Discussion No. 48*, 21/220, *No. 49*, 22/286; *Symposium No. 3*, 14/160: total pages, 3654.

Membership (corrected for the small changes before January 1st 1972): U.K.: Full members 1013; Junior members 123: total 1136.
Overseas: Full members 1189; Junior members 49: total 1238.
Total individual members 2374; non-membership subscriptions 2693. [An analysis of the Chemical Society membership for those seeking attachment to the Faraday Division, but not necessarily taking the *Transactions* or *Discussion* volumes, suggested, in March 1972, a total of 3800. By March 1974 the recorded membership of the Faraday Division was 5430.]

Deaths: Dr. G. W. Ferguson, Prof. P. Goldfinger, Prof. A. Michels, Mr. F. I. G. Rawlins, Prof. E. A. Guggenheim, Rev. L. M. Yeddanapalli, Prof. R. Brdicka, Prof. I. Amdur, Prof. K. H. Hansen, Dr. K. O. Kever, Dr. G. H. Twigg and Prof. C. K. Ingold.

Amalgamation: An Extraordinary General Meeting was held on the 15th April, 1971 at which a Special Resolution was passed to put the Society into voluntary liquidation, prior to amalgamation with the new Chemical Society, which will take effect as from 1st of January 1972. [No mention of the numbers voting.]

Finance: . . . the Society had a deficit in 1966 and the subscription rates for non-members were raised in January 1967. In each of the succeeding years there was an excess of income over expenditure, namely £10 591 in 1967, £5217 in 1968, £5515 in 1969 and £2289 in 1970 . . . This year, 1971, it is clear that we shall not be able to stem the tide any longer and a deficit of about £7000 is expected. Nevertheless, in view both of the recent surpluses and of the forthcoming amalgamation, Council decided not to change either members' subscriptions... or the non-members' subscriptions. It is clear, however, that had we not been about to join the new Society an increase must have come in 1972... We take into the new Society some substantial assets... These assets will be placed in a separate account and used for the purposes for which the Faraday Society was founded.

This statement can be followed by one from the minutes of the Council of the Faraday Division (6th June 1972).

The last year of the Faraday Society shows substantial excess of expenditure over income of £20 000. A deficit of about half this size was expected at the beginning of the year, the remainder being largely due to the unexpected size of the *Transactions*, *Discussions* and *Symposia*, and to the employment of an additional member of staff. The deficit can be set against the accumulated surpluses of the last five years of £22 000. The investments have increased in value by £41 000 during the past year and the Faraday Society brings to the amalgamated Society investments of £127 000. However, the Treasurer expected that the final amount handed over would be around £100 000 after outstanding debts had been paid and various trust funds established.

One further indication of the strength of the Society: local representatives had not only been established throughout the U.K. but also at a wide range of overseas centres. These can be listed: Australia: Adelaide, Canberra, Clayton, Armidale, Kensington, Sydney, Hobart, Nedlands. Belgium: Louvain. Canada: Edmonton, Vancouver, Ottawa, Kingston, Saskatoon, Toronto, London (Ontario). Uganda: Kampala. Eire: Dublin, Cork, Galway. Germany: Göttingen. India: Chandigarh. Netherlands: Leiden. South Africa: Cape Town, Durban. South America: Quito. U.S.A.: Boston, Los Angeles, Pasadena, Washington, Chicago, Boulder, Urbana, Bloomington, Ames, Ann Arbor, Minneapolis, Lincoln (Nebraska), Evanston, Notre Dame (Indiana), Stanford, College Station (Texas).

COMMENTS ON THE AMALGAMATION

In the absence of observations from Tompkins, the Faraday Society Secretary, who had been involved at all stages, comments on the nature of the exchanges which preceded the amalgamation come from John

Ruck-Keene who had been the Executive Secretary of the Chemical Society for many years. As usual, he has been very helpful and effective in providing the following:

'I relate my comments on C.S.-F.S. relationships to three phases which cover my period of service with C.S.

1. 1947–ca. 1967 Pre-Amalgamation.
2. 1968 – Amalgamation. Formal Discussions and Agreements.
3. Post-Amalgamation.

'In Phase 1 contacts were, in the main, between Presidents. These were then discussed at meetings of C.S. officers and usually reported in brief detail to C.S. Council. When both Presidents had similar views and aspirations it looked as if progress could be made. An example of this, I recall, were discussions between Eric Rideal and Sir John Lennard-Jones [Rideal was C.S. President 1950–52, Lennard-Jones was F.S. President 1948–50, so these discussion were as early as 1950. [The two Presidents were close Cambridge colleagues whose offices were a minute's walk apart. M.M.D.] These were well documented and reported fully to C.S. Council. However, there was no continuity and new Presidents might have different priorities. I was not directly involved and my information was therefore secondhand. I gained the impression – unfairly as I later discovered – that Tompkins would be "difficult".

'In Phase 2 I was involved, for the first time, in most of the committees and discussions with F.S. representatives. Initially C.S. and S.C.I. were involved. F.S. and S.A.C. joined in later. From my C.S.-oriented viewpoint I could see that all the other parties had legitimate interests to protect. However, it was with some surprise that I found that F.S. members presented theirs with an openness and candour which was not universal. In particular Tompkins, who I suspect had many personal reservations, did nothing to obstruct the progress his colleagues wished to make. In particular I believe it was Tompkins who, when the outline scheme was agreed, proposed that all funds and investments should be pooled and applied for the general benefit of the new Society (S.A.C. took a very different view!).

[Tompkins did not have the authority to make such a proposal on his own initiative: see below. M.M.D.]

'In Phase 3 I and other members of C.S. staff were involved with Miss Kornitzer and, at a different level, with Tompkins. I was agreeably surprised to find how easily any minor difficulties were resolved. Although this was true in both editorial and administrative departments it was ordinary routine matters which could have proved troublesome. Beatrice Kornitzer must have found it very strange to transfer to a much larger office with many staff around, and much credit is due to her, and to my colleague, John Gibson, for the friendly atmosphere that was established. Tompkins once told me that everything had worked out much better than he had expected. I accepted this as a compliment!'

Surviving memories suffice to establish that Tompkins, personally, was against amalgamation. It is to his credit that he 'did nothing to obstruct the progress'.

A closing comment on the Faraday Society again, with advantage, comes from without. Seeking the opinions of colleagues outside the U.K. brought the following from Professor H. Gg. Wagner at Göttingen: he writes firstly of his late colleague Professor Jost (9-2-94). 'Professor Jost served as Vice-President of the Faraday Society in the time from 1962–65 and before that he was ordinary member of the Council (1959–62).

'Professor Jost very much appreciated the way in which the Faraday Society acted, and I do remember that he very much liked the way in which the Faraday Society organized its meetings. He, in fact, considered that way to be probably the most effective one, and he tried to install it elsewhere. He often told me about the help he and Physical Chemistry in Germany in general got from British colleagues and the Faraday Society after World War II.

'When he came out of hospital in 1949, he tried to re-install the links between the Faraday Society and the Bunsengesellschaft für Physikalische Chemie, and it was a good deal of his initiative that these two societies organized Joint Meetings which later on became a good European habit and include the Physical Chemistry Societies of many European countries. I still remember that Joint Meeting with the

Faraday Society on the Interaction between Dislocation and Disordered Sites in Crystals, a meeting which I think was really a milestone for the understanding of those important processes.

'Jost also recruited members for the Faraday Society among his German colleagues, and that was the way how I and several others became members of the Faraday Society. Jost was very much interested in the activities which finally led to the foundation of The Royal Society of Chemistry in the U.K. and he considered that development to be an advantage for chemistry in general. He would have liked the German chemical societies to follow that development as closely as possible.'

If imitating is the significant form of appreciation, then our French colleagues offered that compliment. The writer was the U.K. representative on the Conseil de la Société de Chimie Physique when that proceeded to amalgamate with the Société de la Chimie de France.

Consideration of the further merging of the new C.S. and the R.I.C. was initiated in 1975, when it was already agreed that it should take the form of a new Charter for the Chemical Society. Its form was agreed in 1978, and it was granted by the Privy Council in 1980, accompanied by the award of royal patronage, so that the title became the Royal Society of Chemistry. In this, the structural pattern of the Royal Society of Chemistry became notably more involved by the need to deal with aspects of professional employment, including the establishment and maintenance of professional status at various levels (see *The First 150 Years*, p. 66). The only formal change for the Faraday Division was that it became a relatively smaller unit in the totality of the Society.

THE CONSEQUENCES OF THE AMALGAMATION

Observations under this heading, some of which are necessarily personal evaluations, can now be offered.

Firstly, it is apposite to quote again from Whiffen's *The First 150 Years*, p. 39, with *emphases* and insertions by the writer.

'One can summarise the position since then (i.e. 1972), a summary not significantly modified by the formation of the Royal Society of Chemistry. There is a Faraday Division which, through its Council, organises Faraday Discussions, is responsible for Faraday Groups, handles its finance including the subvention from the *residual* F.S. funds [e.g. the Bourke and Marlow Trust Funds. M.M.D.], still appoints the Spiers and Bourke lecturers and awards the Marlow medal and prize. The *Faraday Transactions* and *Discussions* continue with a scientific editor and advisory editorial board responsible to the Division. The Division can *comment* on any matter to the R.S.C. Council and is asked to do so on matters of special importance or special relevance to Faraday objects. [Colleagues now active in the Faraday Division stress that there are *five* other Divisions represented on the R.S.C. Council where all major decisions are taken. M.M.D.] Indeed most of those who were active in F.S. days still refer occasionally to the Faraday Society when they mean the Faraday Division.

'There were at the time of Amalgamation many members who regretted the passing of the Faraday Society, but voted for its liquidation believing this to be right. With hindsight they must agree that their altruistic vote for Amalgamation was wiser than they dared hope. Since at the end of 1988 the Faraday Division had 4061 registered members, including the President Elect for the [Sesquit] Centenary year, and thirteen affiliated Groups having collectively over 3700 members, the position of Physical Chemistry within the Society is a strong one.'

It is a common, if not universal finding that control of the purse strings controls overall activities: no U.K. voter is unaware of the fact that the Treasury can exert a veto on any Governmental policy. In this context it can be queried whether completely relinquishing control of its funds was a wise step for the Faraday Society to have taken (even though they were to be earmarked within the Chemical Society Funds); from the statement quoted of John Ruck-Keene's it might seem that Tompkins was responsible for this step. Certainly he did not object to it.

The Faraday Division has become one of six units within the R.S.C. In the course of a generation during which those familiar with and attached to Faraday Society procedures have gradually departed from the Council, it is inevitable that old practices and much status have disappeared. Such loss can only be felt by those accustomed to the earlier complete independence and, perhaps, a fast evaporating nostalgia is the painful only residue of the Faraday Society.

It must be accepted that the Faraday Society could not have survived bearing the restrictions within which it operated. Whilst there had never been a borderline between physical chemistry and the rest of science, the development in chemical physics and theoretical chemistry have completely flooded the plain between chemistry and physics. The area relevant to the Faraday Society's interests has far more than doubled since the 1950s. Furthermore, the rate of the developments within this much expanded field has accelerated. The Faraday Society showed no ability to cope with the challenge thus presented.

In the narrative of events repeated mention has been made of the negative attitude encountered by any suggested extension of the Society's activities. Unquestionably much of this arose from a situation in which the executive secretary solely responsible for supervising all the Society's activities was doing so on one day each week. It was an extension of the pattern in which, to control all significant happenings, Marlow called in the office for tea. One Treasurer was told by Tompkins: 'The Society cannot afford to have new members'. At the same time, during the rest of the week, these executive secretaries were the editors of the *Transactions* and of the reports of the *Discussions*. The extraordinary fact is that they succeeded so well.

From 1960 it was clear that a new and much expanded operation of Faraday Society interests was becoming necessary. The more this was resisted the more obvious the need became. Accordingly, the amalgamation was the appropriate and welcome opportunity for a remoulding process. That it took place so smoothly was due, in the first place, to Sir Harry Melville's key role, and to the good relations of senior Faraday members, including several Past Presidents, with relevant Chemical Society personnel.

There are, perhaps, two general regrets. By its disappearance into a chemical environment, the Faraday Society's role as a bridge between chemistry and physics has been weakened. It could be said that the Society was never very successful nor even active in that context. Certainly in Britain a weakness has existed through most of the century in the sponsorship of chemical physics studies. Exceptions appear in the

Amalgamation 383

form of the *Molecular Physics* journal which should have been, but never was, the responsibility of the Faraday Society:* and in more recent developments of solid-state chemistry.

Some reasons for this weakness may be suggested. Many students, including those of the major and influential Oxford school, were taught little physics. And, for more than 20 years after his death the dominant influence of Rutherford in British physics could be seen. The outstanding success of the Cavendish school and its protégés created an ambience in which for a physicist to become seriously interested in molecules meant he was going down market. The situation has been significantly different in the Netherlands, in Germany, and even in France, where, at least mathematics is adequately taught in the Lycées.

---

* It is relevant to quote what Professor David Buckingham wrote to the writer in 1994.
    It was a blow when in about 1957 Tompkins and particularly Ingold declined to support the idea of Christopher Longuet-Higgins, John van der Waals, and others that the Faraday Society should publish a new journal of chemical physics. Alas, that negative decision led to the birth in 1958 of *Molecular Physics* (published by Taylor and Francis). There is no doubt that *Molecular Physics* has done very well and continued to prosper, but it is hardly the European equivalent of the *Journal of Chemical Physics* that many of us wanted.

Envoi

THERE IS EVERY reason to claim that at its level – and there is no reason to elevate it above reality – the Faraday Society was a remarkably successful venture. The reader will be adequately aware of its achievements. To the present writer it appears as an expression of the English genius in deploying the empirical method. As in our national affairs, there was no constitution. This gave a flexibility which allowed the Society to deal with its interests and its problems on an *ad hoc* basis. With an equal claim to achievement, the same empirical approach pervades British science.

As these last paragraphs were being written a letter arrived from a senior colleague and friend: Sir Charles Frank had been invited to the Guest Night dinner of the 98th Faraday Discussion (on Polymers at Surfaces and Interfaces): and was asked to say a few words. 'I said my thanks for being invited to dine with the participants in a Discussion of what I can't help thinking of as the Faraday Society. It seemed absurd to me to think of Faraday as the representative of a *division* of Chemistry. I mentioned some of his fundamental contributions to various disparate branches of science, and finished by saying what a great pleasure it had been to dine with the practitioners of the chemical division of the science of Michael Faraday.'

Appendices

Appendix A

PRESIDENTS OF THE FARADAY SOCIETY, 1903-1971

1903–1904	Sir Joseph Swan, F.R.S.
1905–1907	Lord Kelvin, O.M., G.C.V.O., F.R.S.
1907	Sir William Perkin, LL.D., F.R.S.
1908–1909	Sir Oliver Lodge, D.SC., F.R.S.
1909–1911	J. Swinburne, M.INST.C.E., F.R.S.[a]
1911–1913	Dr. R. T. Glazebrook, C.B., F.R.S.[b]
1913–1920	Sir Robert A. Hadfield, Bart., D.MET., F.R.S.
1920–1922	Prof. Alfred W. Porter, D.SC., F.R.S.
1922–1924	Sir Robert Robertson, K.B.E., D.SC., F.R.S.
1924–1926	Prof. F. G. Donnan, C.B.E., PH.D., F.R.S.
1926–1928	Prof. C. H. Desch, D.SC., PH.D., F.R.S.
1928–1930	Prof. T. M. Lowry, D.SC., F.R.S.
1930–1932	Sir Robert L. Mond, LL.D.[c]
1932–1934	Prof. N. V. Sidgwick, SC.D., D.SC., F.R.S.
1934–1936	W. Rintoul, O.B.E.
1936–1938	Prof. M. W. Travers, D.SC., F.R.S.
1938–1945	Prof. E. K. Rideal, M.B.E., D.SC., F.R.S.[d]
1945–1947	Prof. W. E. Garner, C.B.E., D.SC., F.R.S.
1947–1948	Prof. A. J. Allmand, D.SC., F.R.S.
1948–1950	Prof. Sir John Lennard-Jones, K.B.E., D.SC., F.R.S.
1950–1952	Sir Charles Goodeve, O.B.E., D.SC., F.R.S.
1952–1953	Prof. Sir Hugh Taylor, K.B.E., D.SC., LL.D., F.R.S.
1953–1955	Prof. R. G. W. Norrish, SC.D., F.R.I.C., F.R.S.
1956–1957	R. P. Bell, M.A., F.R.S.
1958	Sir Harry Melville, K.C.B., D.SC., F.R.S.
1959	Dr. E. W. R. Steacie, O.B.E., D.SC., LL.D., F.R.S.

1960	Sir Harry Melville, K.C.B., D.SC., F.R.S.
1961, 1962	Sir Cyril Hinshelwood, O.M., D.SC., F.R.S.
1963, 1964	Prof. A. R. J. P. Ubbelohde, C.B.E., M.A., D.SC., F.R.S.
1965, 1966	Prof. F. S. Dainton, M.A., SC.D., F.R.S.[e]
1967, 1968	Prof. C. E. H. Bawn, C.B.E., PH.D., F.R.S.
1969, 1970	Prof. Geoffrey Gee, C.B.E., M.SC., SC.D., F.R.S.
1971	Prof. J. W. Linnett, M.A., D.PHIL., F.R.S.

[a]Succeeded as 9th Baronet in 1934. [b]Knighted in 1917, K.C.B., 1920, K.C.V.O., 1934. [c]Elected F.R.S. in 1938. [d]Knighted in 1951. [e]Knighted in 1971, created a life peer in 1986.

HONORARY LIFE MEMBERS

1950	Prof. F. G. Donnan, C.B.E., PH.D., F.R.S.
1950	Prof. C. H. Desch, D.SC., PH.D., F.R.S.
1950	Prof. A. Ferguson, M.A., D.SC.
1950	Prof. M. W. Travers, D.SC., F.R.S.
1951	Prof. N. V. Sidgwick, SC.D., D.SC., F.R.S.
1952	Prof. J. W. McBain, M.A., D.SC., F.R.S.
1953	Prof. Sir Eric Rideal, M.B.E., D.SC., F.R.S.
1953	Prof. W. E. Garner, C.B.E., D.SC., F.R.S.
1953	Prof. J. H. Hildebrand, PH.D., SC.D.
1953	Prof. N. J. Bjerrum.
1953	Sir Henry Tizard, G.C.B., K.C.B., C.B., A.F.C., M.A., F.R.S.
1953	Sir Harold Hartley, K.C.V.O., C.B.E., M.C., F.R.S.
1954	Sir Hugh Taylow, K.B.E., D.SC., LL.D., F.R.S.
1957	Dr. R. Lessing, C.B.E., PH.D.
1959	Sir Owen Wansborough-Jones, K.B.E., C.B., M.A., PH.D.
1959	Sir Alfred Egerton, Bart., M.A., F.R.S.
1960	Prof. N. K. Adam, M.A., S.C.D., F.R.S.
1961	Sir Charles Goodeve, O.B.E., D.SC., F.R.S.
1965	Prof. H. S. Hamed
1966	Prof. R. G. W. Norrish, SC.D., F.R.I.C., F.R.S.
1967	Prof. E. A. Guggenheim, M.A., SC.D., F.R.S.

Appendix B

THE TRANSACTIONS OF THE FARADAY SOCIETY

Actual year span	Vol. number	Number of pages of text[a]	Discussions Number (New Series)	Pages	Total
1905–1906	1	366			
1906–1907	2	241			
1907–1908	3	225			
1908–1909	4	214			
1909–1910	5	321			
1910–1911	6	234			
1911–1912	7	278			
1912	8	258			
1913–1914	9	338			
1914–1915	10	300			
1915–1916	11	284			
1917	12	328			
1917–1918	13	430			
1919	14	284			
1919–1920	15	580			
1920–1921	16	613 + 190 = 803[b]			
1921–1922	17	756			
1922–1923	18	403			
1923–1924	19	949			
1924–1925	20	626			
Index to volumes 1–20					
1925–1926	21	656			
1926	22	500			
1927	23	681			
1928	24	737			
1929	25	954			
1930	26	865			
1931	27	828			
1932	28	928			
1933	29	1336			

			Discussions		
Actual year span	Vol. number	Number of pages of text[a]	Number (New Series)	Pages	Total
1934	30	1192 + 6 = 1198[c]			
1935	31	1750			
1936	32	1752			
1937	33	1600			
1938	34	1548			
1939	35	1516			
1940	36	1252			
1941	37	804			
1942	38	520			
1943	39	450			
1944	40	580			
1945	41	788			
1946	42	800	A + B	560	1360
1947	43	822	1 + 2 + SC[d]	1081	1903
1948	44	1052	3 + 4	638	1690
1949	45	1174	5 + 6 + 7	870	2044
1950	46	1124	8 + 9	870	1914
1951	47	1370	10 + 11	591	1961
1952	48	1196	12 + 13 + 14	863	2059
1953	49	1506	15 + 16	548	2054
1954	50	1396	17 + 18	608	2004
1955	51	1758	19 + 20	599	2357
1956	52	1678	21 + 22	514	2192
1957	53	1676	23 + 24	480	2156
1958	54	1924	25 + 26	427	2351
1959	55	2230	27 + 28	525	2755
1960	56	1864	29 + 30	490	2354
1961	57	2322	31 + 32	543	2865
1962	58	2524	33 + 34	502	3026
1963	59	2902	35 + 36	565	3516
1964	60	2282	37 + 38	546	2828
1965	61	2833	39 + 40	569	3402
1966	62	3609	41 + 42	738	4347
1967	63	3103	43 + 44	556	3659
1968	64	3426	45 + 46	510	3936
1969	65	3376	47 + 48	426	3802
1970	66	3174	49 + 50	528	3702
1971	67	3649	51 + 52	610	4259

[a]Excluding incdexes: these are somtimes paginated with the main text, but sometimes not.
[b]The Report on Colloids published by the D.S.I.R. was 1900 pp.
[c]The obituary notice for Lt. Col. J. J. Bourke was 6 pp.
[d]Report on 'Surface Chemistry', published by Butterworths Scientific Publications, 1949.

Appendix C

GENERAL DISCUSSIONS HELD BY THE FARADAY SOCIETY

Date	No.	Subject	Place	Vol.	no. and pp.
29th January 1907	—	Osmotic Pressure	London	3	1–37
25th June 1907	—	Hydrates in Solution	London	3	123–163
26th April 1910	—	The Constitution of Water	London	6	71–123
23rd May 1911	—	High Temperature Work	London	7	136–173
23rd April 1912	—	Magnetic Properties of Alloys	London	8	94–219
12th March 1913	—	Colloids and their Viscosity	London	9	34–107
4th April 1913	—	The Corrosion of Iron and Steel	Manchester	9	108–139
12th November 1913	—	The Passivity of Metals	London	9	203–290
27th March 1914	—	Optical Rotary Power	London	10	44–138
23rd November 1914	—	The Hardening of Metals	London	10	207–293
19th October 1915	—	The Transformation of Pure Iron	London	11	125–182
8th December 1915	—	The Corrosion of Metals: Ferrous and Non-Ferrous	London	11	183–281
15th March 1916	—	Methods and Appliances for the Attainment of High Temperatures in a laboratory	London	12	1–63
8th November 1916	—	Refractory Materials	London	12	86–273
6th March 1917	—	Training and Work of the Chemical Engineer	London	13	61–118
1st May 1917	—	Osmotic Pressure	London	13	119–189
7th November 1917	—	pyrometers and Pyrometry	London	13	205–372
14th January 1918	—	The Setting of Cements and Plasters	London	14	1–69
14th February 1918	—	Electrical Furnaces	Manchester	14	70–104
7th May 1918	—	Co-ordination of Scientific Publication	London	14	105–131
12th November 1918	—	The Occlusion of Gases by Metals	London	14	173–263
21st January 1919	—	The Present Position of the Theory of Ionization	London	15	3–178
20th April 1919	—	The Examination of Materials by X-Rays	London	15 +App.	1–88 1–64
14th January 1920	—	The Microscope: Its Design, Construction and Applications	London	16	1–245
23rd March 1920	—	Basic Slags: Their Production and Utilization in Agriculture	London	16	261–335
25th October 1920	—	Physics and Chemistry of Colloids	London	16	Appendix 1–190
19th November 1920	—	Electrodeposition and Electroplating	Sheffield	16	471–553
11th February 1921	—	Capillarity	Manchester	17	369–399
6th April 1921	—	The Failure of Metals under Internal and Prolonged Stress	London	17	1–215
31st May 1921	—	Physico-Chemical Problems Relating to the Soil	London	17	217–368
28th September 1921	—	Catalysis with special reference to Newer Theories of Chemical Action	London	17	545–675
9th and 23rd March 1922	—	Some Properties of Powders with special reference to Grading by Elutriation	London	18	22–77
16th October 1922	—	The Generation and Utilization of Cold	London	18	137–273
13th April 1923	—	Alloys Resistant to Corrosion	Sheffield	19	156–230
28th May 1923	—	The Physical Chemistry of the Photographic Process	London	19	241–406

Date	No.	Subject	Place	Vol.	no. and pp.
13th and 14th July 1923	—	The Electronic Theory of Valency	Cambridge	19	450–543
26th November 1923	—	Electrode Reactions and Equilibria	London	19	667–838
17th December 1923	—	Atmospheric Corrosion. First (Experimental) Report to the Atmospheric Corrosion Research Committee (of the British Non-Ferrous Metals Research Association)	London	19	839–934
14th April 1924	—	Investigation on Oppau Ammonium Sulphate-Nitrate	London	20	45–83
28th April 1924	—	Fluxes and Slags in Metal Melting and Working	London	20	113–208
11th June 1924	—	Physical and Physico-Chemical Problems relating to Textile Fibres	Wembley	20	223–324
22nd October 1924	—	The Physical Chemistry of Igneous Rock Formation	London	20	413–501
9th December 1924	—	Base Exchange in Soils	London	20	550–617
8th June 1925	—	The Physical Chemistry of Steel-Making Processes	London	21	169–292
1st and 2nd October 1925	—	Photochemical Reactions in Liquids and Gases	Oxford	21	437–656
14th June 1926	—	Explosive Reactions in Gaseous Media	London	22	251–375
1st October 1926	—	Physical Phenomena at Interfaces, with special reference to Molecular Orientation	London	22	433–500
30th March 1927	—	Atmospheric Corrosion, Second Experimental Report to the Atmospheric Corrosion Research Committee and British Non-Ferrous Metals Research Association	London	23	113–204
22nd and 23rd April 1927	—	The Theory of Strong Electrolytes	Oxford	23	333–544
23rd November 1927	—	Cohesion and Related Problems	London	24	53–180
28th January 1928	—	Atmospheric Corrosion of Metals: Third Experimental Report of the British Non-Ferrous Metals Research Association by J. C. Hudson	London	25	177–252
28th and 29th September 1928	—	Homogeneous Catalysis	Cambridge	24	545–737
14th March 1929	—	Crystal Structure and Chemical Constitution	London	25	253–421
24th and 25th September 1929	—	Molecular Spectra and Molecular Structure	Bristol	25	611–949
25th and 26th April 1930	—	Optical Rotatory Power	London	26	265–461
29th September to 1st October 1930	—	Colloid Science Applied to Biology	Cambridge	26	633–865
17th and 18th April 1931	55	Photochemical Processes	Liverpool	27	357–572
12th and 13th January 1932	56	Adsorption of Gases by Solids	Oxford	28	129–447
21st to 23rd September 1932	57	The Colloidal Aspects of Textile Materials and Related Problems	Manchester	29	1–368
24th and 25th April 1933	58	Liquid Crystals and Anisotropic Melts	London	29	881–1085
28th to 30th September 1933	59	Free Radicals and Ions as Factors in Chemical Change	Cambridge	30	1–248
12th to 14th April 1934	60	Dipole Moments	Oxford	30	677–904
27th to 29th September 1934	61	Colloidal Electrolytes	London	31	1–422
29th and 30th March 1935	62	The Structure of Metallic Coatings, Films and Surfaces	London	31	1043–1290
26th to 28th September 1935	63	Phenomena of Polymerisation and Condensation	Cambridge	32	1–412
20th to 22nd April 1936	64	Disperse Systems in Gases; Dust, Smoke and Fog	Leeds	32	1041–1300
24th to 26th September 1936	65	Structure and Molecular Forces in (a) Pure Liquids and (b) Solution	Edinburgh	33	1–282
22nd and 23rd April 1937	66	The properties and Functions of Membranes, Natural and Artificial	London	33	911–1151
13th and 14th September 1937	67	Reaction Kinetics	Manchester	34	1–268
11th to 13th April 1938	68	Chemical Reactions involving Solids	Bristol	34	821–1087

Appendix C

Date	No.	Subject	Place	Vol.	no. and pp.
15th to 17th September 1938	69	Liminescence	Oxford	35	1–140
17th to 19th April, 1939	70	Hydrocarbon Chemistry	London	35	805–1092
(September 1939)	—	The Electrical Double Layer	(Cambridge)	36	1–322 +711–732
17th May 1940	—	The Hydrogen Bond	London	36	871–929
9th January 1941	—	The Oil–Water Interface	London	37	117–185
26th September 1941	—	Mechanism and Chemical Kinetics of Organic Reactions in Liquid Systems	London	37	601–806
29th May 1942	—	The Structure and Reactions of Rubber	London	38	269–388
24th September 1943	—	Modes of Drug Action	London	39	319–446
12th January 1944	—	Molecular Weight and Molecular Weight Distribution in High Polymers	London	40	217–280
2nd January 1945	—	The Application of Infra-Red Spectra to Chemical Problems	London	41	171–297
27th and 28th September 1945	—	Oxidation	London	42	99–398
24th to 26th April 1946	—	Dielectrics	Bristol	42A	1–256
24th to 26th September 1946	—	Swelling and Shrinking	London	42B	1–304
9th and 10th April 1947	1 (New series)	Electrode Processes	Manchester	— (New series)	1–338
23rd to 25th September 1947	2	The Labile Molecule	Oxford	—	1–409
5th to 9th October 1947	—	Surface Chemistry (joint meeting with Societé de Chimie Physique)	Bordeaux, France	'Research' Spec. Supp. 1949. 1–334	
19th December 1947	—	Colloidal Electrolytes and Solutions	London	43	811–830
31st March to 2nd April 1948	3	The Interaction of Water and Porous Materials	Southampton	—	1–294
23rd to 25th September 1948	4	The Physical Chemistry of Process Metallurgy	Ashorne Hill, Leamington Spa	—	1–344
12th to 14th April 1949	5	Crystal Growth	Bristol	—	1–366
29th to 31st August 1949	6	Lipo-Proteins	Birmingham	—	1–168
22nd to 24th September 1949	7	Chromatographic Analysis	Reading	—	1–336
12th to 14th April 1950	8	Heterogeneous Catalysis	Liverpool	—	1–366
25th to 28th September 1950	9	Spectroscopy and Molecular Structure and Optical Methods of Investigating Cell Structure	Cambridge	—	1–504
11th to 13th April 1951	10	Hydrocarbons	Oxford	—	1–339
18th to 20th July 1951	11	The Size and Shape Factor in Colloidal Systems	Ashorne Hill, Leamington Spa	—	1–252
8th to 10th April 1952	12	Radiation Chemistry	Leeds	—	1–319
6th to 8th August 1952	13	The Physical Chemistry of Proteins	Cambridge	—	1–288
8th to 9th September 1952	14	The Reactivity of Free Radicals	Toronto, Canada	—	1–256
16th to 18th April 1953	15	The Equilibrium Properties of Solutions of Non-Electrolytes	Royal Inst., London	—	1–292
8th to 10th September 1953	16	The Physical Chemistry of Dyeing and Tanning	Leeds	—	1–256
7th to 9th April 1954	17	The Study of Fast Reactions	Birmingham	—	1–236
15th to 17th September 1954	18	Coagulation and Flocculation	Sheffield	—	1–372
4th to 6th April 1955	19	Microwave and Radio-Frequency Spectroscopy	Cambridge	—	1–282
10th to 12th August 1955	20	Physical Chemistry of Enzymes	Oxford	—	1–317
10th to 12th April 1956	21	Membrane Phenomena	Nottingham	—	1–288
20th to 21st September 1956	22	Physical Chemistry of Processes at High Pressures	Glasgow	—	1–226
15th to 18th April 1957	23	Molecular Mechanism of Rate Processes in Solids	Amsterdam, Holland	—	1–241
17th to 19th September 1957	24	Interaction in Ionic Solutions	Oxford	—	1–239
15th to 17th April 1958	25	Configurations and Interactions of Macromolecules and Liquid Crystals	Leeds	—	1–235
9th to 11th September 1958	26	Ions of the Transition Elements	Dublin, Irish Republic	—	1–192

Date	No.	Subject	Place	Vol.	no. and pp.
14th to 16th April 1959	29	Energy Transfer with Special Reference to Biological Systems	Nottingham	—	1–273
2nd to 4th September 1959	28	Crystal Imperfections and the Chemical Reactivity of Solids	Queen's Univ., Kingston, Ont., Canada	—	1–252
11th to 13th April 1960	29	Oxidation–Reduction Reactions in Ionizing Solvents	Durham	—	1–260
13th to 15th September 1960	30	The Physical Chemistry of Aerosols	Bristol	—	1–229
11th to 12th April 1961	31	Radiation Effects in Inorganic Solids	Centre d'Etudes Nucléaires de Seclay, France		1–275
5th to 7th September 1961	32	The Structure and Properties of Ionic Melts	Liverpool	—	1–268
10th to 12th April 1962	33	Inelastic Collisions of Atoms and Simple Molecules	Cambridge	—	1–300
17th to 19th September 1962	34	High Resolution Nuclear Magnetic Resonance	Oxford	—	1–202
2nd and 3rd April 1963	35	The Structure of Electronically Excited Species in the Gas-Phase	Dundee	—	1–240
2nd to 4th September 1963	36	Fundamental Processes in Radiation Chemistry	Univ. of Notre Dame, Ind.	—	
2nd and 3rd April 1964	37	Chemical Reactions in the Atmosphere	Edinburgh	—	1–225
15th to 17th September 1964	38	Dislocations in Solids	Göttingen, W. Germany	—	1–320
12th to 14th April 1965	39	The Kinetics of Proton Transfer Processes	Newcastle-upon-Tyne	—	1–277
14th to 16th September 1965	40	Intermolecular Forces	Bristol		1291
4th to 6th April 1966	41	The Role of the Adsorbed State in Heterogeneous Catalysis	Bristol	—	1–291
27th to 28th September 1966	42	Colloid Stability in Aqueous and Non-Aqueous Media	Nottingham	—	1–322
11th to 13th April 1967	43	The Structure and Properties of Liquids	Exeter	—	1–248
5th to 7th September 1967	44	Molecular Dynamics of the Chemical Reactions of Gases	Toronto, Canada	—	1–308
2nd to 4th April 1968	45	Electrode Reactions of Organic Compounds	Newcastle-upon-Tyne	—	1–282
17th to 19th September 1968	46	Homogeneous Catalysis with Special Reference to Hydrogenation and Oxidation	Liverpool	—	1–228
25th to 27th March 1969	47	Bonding in Metallo-Organic Compounds	Cambridge	—	1–205
16ht to 18th September 1969	48	Motions in Molecular Crystals	Oxford	—	1–221
15th to 17th April 1970	49	Polymer Solutions	Manchester	—	1–287
22nd to 24th September 1970	50	The Vitreous State	Bristol	—	1–241
14th to 16th April 1971	51	Electrical Conduction in Organic Solids	Nottingham	—	1–229
13th to 15th September 1971	52	Surface Chemistry of Oxides	Brunel Univ.	—	1–381

Appendix D

ANALYSIS OF MEMBERSHIP

The primary membership data are given in Table D.1. Though they are not used in the analysis the numbers of students are included in the Summary for completeness.

Table D.2 first gives the group numbers in the six lists, both for total and for home, personal members only: then it given the net gains, losses, and thence the recruitments during the five periods. Obviously $D = B - C$. The proportional rates of change of the last three quantities are expressed as percentage changes *per annum*, relative to the numbers at the beginning of each period and assuming a linear change during that period, which is a regrettable but inevitable simplification. Obviously, again, $D' = B' - C'$.

TABLE D.1 Analysis of Membership

Primary Data

Sub-Group Estimated Membership:	Date 1 Apr. 1911	1 Apr. 1913	31 Jan. 1917	1 Sep. 1919	1 Oct. 1926	1 Jan 1938
(a) Academic Group						
Personal,						
Home	42	41	51	55	80	202
Foreign	5	8	13	16	39	137
Libraries,						
Home	3	4	5	5	10	13
Foreign	0	0	0	0	8	32
Sub-Totals	50	53	69	76	136	384
Estimated Losses:						
Personal,						
Home	4	8	3	18	21	
Foreign	1	0	5	6	15	
Libraries,						
Home	0	0	0	0	2	
Foreign	0	0	0	0	0	
Sub-Totals	5	8	8	24	38	
(b) Research Institutions Group						
Estimated Membership:						
Personal,						
	8	8	9	16	36	41
	6	5	4	7	8	25
Libraries,						
Home	1	1	0	0	0	2
Foreign	0	0	0	0	0	5
Sub-Totals	15	14	13	23	44	73
Estimated Losses:						
Personal,						
Home	1	1	1	8	15	
Foreign	0	1	0	2	3	
Libraries,						
Home	0	1	0	0	0	
Foreign	0	0	0	0	0	
Sub-Totals	1	3	1	10	18	

Appendix D

Sub-Group Estimated Membership:	Date 1 Apr. 1911	1 Apr. 1913	31 Jan. 1917	1 Sep. 1919	1 Oct. 1926	1 Jan. 1938
(c) Industrial Group						
Estimated Membership:						
Personal and Corporate:						
116	108	122	204	231	192	
Foreign	20	21	17	16	28	121
Libraries,						
Home	0	1	1	1	3	1
Foreign	0	0	0	0	0	6
Sub-Totals	136	130	140	221	262	320
Estimates Losses:						
Personal and Corporate,						
Home	20	29	15	84	158	
Foreign	4	1	9	16		
Libraries,						
Home	0	0	0	0	2	
Foreign	0	0	0	0	0	
Sub-Totals	24	36	16	93	176	

Summary

	1911	1913	1917	1919	1926	1938
Full Members:						
Academic	50	53	69	136	384	
Research Institutions	15	14	13	23	44	73
Industrial	137	130	140	221	262	320
Uncertain Class.	—	—	—	—	—	6
Totals	201	197	222	320	442	783
Students	4	4	5	4	5	37
Grand Totals	205	201	227	324	447	820

TABLE D.2 Analysis of Changes of Membership

Column Designations: a = Number, B = Net gain, C = Loss, D = Recruitment;
B', C', and D' = Annual rates of Net Gain, Loss and Recruitment

Year	Interval (Month)	Basic Data								Annual percentage Rates of Change					
		Total membership				Home, personal				Total			Home, personal		
		A	B	C	D	A	B	C	D	B'	C'	D'	B'	C'	D'
(a) Academic Group															
1911	24	50	3	−5	8	42	−1	−4	3	3.0	−5.0	8.0	−1.2	−4.8	3.6
1913	46	53	16	−8	24	41	10	−8	18	7.9	−3.9	11.8	6.4	−5.1	11.5
1917	31	69	7	−8	15	51	4	−3	7	3.9	−4.5	8.4	3.0	−2.3	5.3
1919	85	76	60	−24	84	55	25	−18	43	11.1	−4.5	15.6	6.4	−4.6	11.0
1926	135	136	248	−38	286	80	122	−21	143	16.2	−2.5	18.7	13.6	−2.6	15.9
1938		384				202									
(b) Research Institutions Group															
1911	24	15	−1	−1	0	8	0	−1	1	−3.3	−3.3	0	0	−6.2	6.2
1913	46	14	−1	−3	2	8	1	−1	2	−1.9	−5.6	3.7	3.3	−3.3	6.5
1917	31	13	10	−1	11	9	7	−1	8	29.8	−3.0	32.8	30.1	−4.3	34.4

	1	2	3	4	5	6	7	8	9	10	11	12	13	14	15
1919	85	23	21	−10	31		20	−8	28	12.9	−6.1	19.0	17.6	−7.1	24.7
1926	135	44	29	−18	47	16	5	−15	20	5.9	−3.6	9.5	1.2	−3.7	4.9
1938		73				36									
						41									

(c) Industrial Group

	1	2	3	4	5	6	7	8	9	10	11	12	13	14	15
1911	24	136	−6	−24	18	116	−8	−20	12	−2.2	−8.8	6.6	−3.4	−8.6	5.2
1913	46	130	10	−36	46	108	14	−29	43	2.0	−7.2	9.2	3.4	−7.0	10.4
1917	31	140	81	−16	97	122	82	−15	97	22.4	−4.4	26.8	26.0	−4.8	30.8
1919	85	221	41	−93	134	204	27	−84	111	2.6	−5.9	8.6	1.9	−5.8	7.7
1926	135	262	58	−176	234	231	−39	−158	119	2.0	−6.0	7.9	−1.5	−6.1	4.6
1938		320				192									

Index of Names

Abadie, P., 286
Acree, S. F., 58
Adam, N. K., 72, 242, 254, 348, 388
Adamson, A. W., 305
Adzic, R., 324
Agar, J. N., 76, 271, 278, 287, 311, 324, 334, 340
Albright, G. S., 54, 55
Alexander, A. E., 233, 255, 331, 337
Allen, A. O., 295
Allen, G., 315, 319, 335, 336, 342
Allen, P., 50
Allmand, A. J., 55, 56, 57, 69, 101, 105, 111, 120, 151, 190, 193, 227, 228, 229, 232, 241, 268, 271, 289, 290, 330, 331, 341, 387, plate
Allsopp, C. B., 294
Amat, G., 308
Amdur, I., 376
Anderson, J. S., 332, 337
Andrade, E. N. da C., 68, 69, 83, 195
Andreev, N. S., 320
Angus, R., 128
Armstrong, H. E., 14, 22, 28, 34, 92
Arndt, F., 100
Arrhenius, S., 19, 20, 34, 37, 58, 63, 115, 134, 302
Ashmore, P. G., 334, 340

Astbury, W. T., 79, 84, 86, 96, 128, 203, 206, 242, 304, 355
Aston, J. G., 99, 294
Atherton, N. M., 354
Awberry, J. H., 115

Badash, L., 60
Badger, R. M., 80
Badoz, J., 323
Baekeland, L. H., 5
Bak, K., 319
Baker, C. W., 233
Baker, H. B., 101
Baldwin, R. L., 300
Baly, E. C. C., 93
Bamford, C. H., 333, 334, 340
Bancroft, W. D., 25, 64, 188, 190, 191, 249
Bard, A. J., 317
Barker, C., 179
Barker, E. F., 80
Barker, J. A., 312, 315
Barrer, R. M., 132, 275, 300, 332, 337
Bartlett, P. D., 288
Bauer, E., 113
Bauer, S. H., 298
Baur, E., 196, 197, 286
Bawn, C. E. H., 77, 100, 110, 225, 242, 271, 279, 294, 313, 315, 316, 317, 331, 332, 333, 334, 335, 337, 364, 367, 368, 369, 370, 388, plate

Index of Names

Baxendale, J. H., 305
Baxter, J. B., 287
Beacall, T., 133, 331, 345
Beadle, C., 26
Beeck, O., 292
Beilby, G. T., 11, 15, 34, 52, 107, 108
Belfield, R., 28
Bell, R. P., 67, 75, 119, 120, 130, 203, 242, 246, 259, 267, 279, 294, 298, 300, 301, 302, 311, 331, 332, 333, 334, 335, 337, 340, 343, 350, 361, 369, 387, plate
Bennett, G. M., 291, 331, 332, 337
Benson, S. W., 294
Bergman, F., 300
Berkeley, Earl of, 22, 23
Berkner, L. V., 310
Bernal, J. D., 52, 79, 84, 86, 97, 98, 113, 114 143, 203, 204, 287, 297, 303
Bernstein, H. J., 351
Berthoud, A., 93
Bessemer, H., 65
Bevan, E. J., 26
Biilman, E. C. S., 270
Bingen, E., 368, 369
Bircumshaw, L. L., 96
Birge, R. T., 80
Birkeland, Kr., 15, 21, 28, 37
Bishop, D. M., 323
Bishop, E., 4
Bjerrum, N., 75, 137, 268, 303, 347, 388
Blackman, M., 107
Bloom, A., 307
Bloom, H., 355
Blount, B., 11, 18, 19, 30, 147
Bockris, J. O'M., 307
Bodenstein, M., 69, 93, 187

Bohr, N., 53, 86, 126
Bolam, T. R., 254
Bonet-Maury, P., 295
Bonhöffer, K. F., 94, 120, 300, 325
Bonino, G. B., 80
Booij, H. L., 291
Booth, C., 319
Boreskov, G. K., 313
Born, M., 82
Borns, H., 27, 28, 39, 43, 55, 57, 60, 213, 214
Bottcher, C. J. F., 229, 286, 301, 311, 333, 340
Bourke, J. J., 104, 166, 205, 208, 209, 210, 375, 390
Bousfield, E. G. P, 8, 12, 28, 29
Bousfield, W. R., 27, 28, 37, 46, 55, 175, 194
Bowden, F. P., 301, 331, 332, 337
Bowen, E. J., 67, 69, 70, 93, 102, 122, 123, 124, 183, 332, 337
Boys, C. V., 6, 7
Brackman, W., 317
Bradfield, J. R. G., 293
Bragg, W. H., 59, 79, 84, 97, 132, 203, 206, 211, 213, 354
Bragg, W. L. B., 71, 84, 114, 144, 203
Brdicka, R., 376
Bresler, S. E., 303
Bretscher, E., 103
Brill, R., 84
Brønsted, J. N., 75, 77, 312
Brot, C., 318
Brown, C., 10, 19, 27, 116, 269
Brown, R. D., 294
Brown, T. E., 317
Buckingham, A. D., 334, 340, 383
Buff, F. P., 306
Burton, M., 295, 296
Bury, C. R., 126, 198

Butler, J. A. V., 83, 132, 254, 257, 259, 260, 271, 275, 304, 306, 332, 333, 337
Byers Brown, W., 312

Cabannes, J., 80
Caglioti, V., 305
Cahn, R. S., 158, 169
Caillat, M. M., 306
Caldin, E. F., 311, 335, 342
Campbell, N., 115
Caress, A., 136, 331, 345
Carothers, W. H., 109
Carrington, A., 335, 342, 354
Catchpole, H. B., 293
Chandresekhar, S., 324
Chapman, D. L., 68, 69
Chargaff, E., 291
Chelmsford, Lord, 209
Christiansen, J. A., 69, 75, 77, 101
Christie, J., 53
Churchill, W. S., 50
Cimino, A., 313
Claesson, S., 291, 304
Claudet, A. C., 26
Clausius, R. J. E., 102
Clusius, K., 122
Clyne, M. A. A., 354
Coates, J. E., 70, 126
Cohen, E., 37, 200, 213, 270
Cohn, E., 291, 295
Cole, K. S., 117
Cole, R. H., 308
Coles, I., 226
Collie, C. H., 286
Commoner, B., 293
Conway, B, E., 312, 317
Cook, M. A., 301
Cooper, W. R., 3, 7, 11, 18, 19, 22, 32, 45, 46, 48, 55, 57, 71, 147, 148, 161, 162, 183, 184
Cornforth, J., 129
Correns, C. W., 290, 291

Cotton, F. A., 317
Cottrell, A. H., 76, 271, 311, 345
Cottrell, T. L., 310, 333, 340
Coulson, C. A., 16, 242, 288, 313, 327, 331, 332, 335, 337, 352
Coumoulos, G. D., 132, 285
Cowper-Coles, S., 3, 5, 6, 7, 8, 12, 13, 15, 17, 18, 19, 147, 148, 178, 206, 210, 269
Cox, E. G., 97, 204, 271, 290
Crawford, B., 294
Cripwell, F. J., 286
Crommelin, C. A., 63
Crookes, W., 19
Cross, C. F., 26
Crowfoot Hodgkin, D., 52, 81, 97, 98, 204
Cruickshank, D. W. J., 317, 318
Csanji, L. J., 305

Dahl, L. F., 317
Dailey, B, P., 308
Dainton, F. S., 58, 122, 123, 125, 134, 244, 246, 256, 264, 278, 279, 295, 304, 305, 306, 310, 311, 312, 313, 314, 325, 332, 334, 335, 337, 343, 345, 347, 350, 370, 388, plate
Dale, H., 132
Dampier Whetham, W. C., 14, 15, 23
Daniels, F., 62
Dannatt, C. W., 289
Danon, F., 312
Danon, J., 323
Darling, C. R., 70, 212
Davies, C. N., 276
Davies, C. W., 76, 242, 271, 302, 303, 331, 338
Davies, M. M., 126, 318, 333, 335, 336, 340
Dawson, H. M., 77, 130
Day, A., 24
de Boer, E., 305

Index of Names

de Boer, J. H., 121, 227, 285, 292, 302, 321
de Ferranti, S. Z., 26
de Gennes, P. G., 273, 274, 324
de Hemptinne, X., 317
Debye, P. J. W., 58, 75, 95, 102, 103, 325, 346
Dennison, D. M., 137
Derjaguin, D. V., 298, 306, 314
Dervichian, D. G., 229, 287, 294, 298, 303
Desch, C. H., 30, 41, 48, 56, 57, 72, 74, 76, 105, 106, 108, 121, 187, 206, 268, 331, 344, 345, 347, 387, 388, plate
Dewar, M. J. S., 288
Digby, W. P., 6, 18
Dirac, P. A. M., 328
Dixon, M., 299, 300
Djeransk, A., 324
Djerassi, C., 82
Dobbie, J., 127
Dodson, R. W., 305
Domb, C., 319
Donnan, F. G., 3, 5, 6, 7, 8, 17, 19, 29, 41, 45, 49, 50, 55, 56, 68, 69, 72, 74, 78, 79, 81, 84, 103, 105, 106, 107, 108, 111, 114, 118, 121, 127, 134, 143, 144, 189, 193, 200, 214, 215, 216, 217, 232, 266, 267, 268, 269, 271, 285, 301, 325, 338, 345, 347, 365, 387, 388, plate
Dorner, B., 315
Doty, P. M., 287, 295
Douglas, A. E., 309
Douglas, H. W., 320
Dowden, D. A., 291
Dows, D. A., 309
Drickamer, H. G., 301
Dryden, J. S., 318
Duchesne, J., 299
Dufour, B., 323
Duncan, J. F., 323

Dworkin, A. S., 307

Edsall, J. T., 291, 295, 296, 355
Edwards, S. F., 319
Egelstaff, P. A., 318
Egerton, A. C. G., 80, 92, 94, 182, 332, 348, 388
Ehrlich, G., 313
Eigen, M., 298, 302, 303, 313, 341, 348, 352
Eirich, F. E., 287
Eisenberg, H., 319
Eley, D. D., 49, 51, 132, 201, 258, 260, 277, 279, 292, 297, 298, 299, 300, 304, 305, 313, 314, 320, 321, 332, 333, 334, 338, 350, 366
Ellingham, H. J. T., 101, 107, 289
Ellis, J. W., 80
Elton, G. A. H., 306
Emeleus, H. J., 114, 242, 256, 294, 331, 338, 366
Engstrom, A., 293
Errera, J., 103
Eucken, A., 123, 292
Evans, A. G., 332, 338
Evans, M. G., 130, 133, 134, 203, 227, 243, 267, 274, 285, 288, 294, 296, 331, 332, 338, 364
Evans, U. R., 110
Everett, D. H., 275, 302, 309, 311, 312, 314, 315, 320, 332, 333, 334, 335, 336, 338
Ewald, A. H., 301
Ewald, P. P., 79, 83
Ewing, A., 55
Eyring, H., 113, 118, 119, 133, 134, 292, 308

Fajans, K., 75, 127, 213, 270
Falkenhagen, H., 302
Faraday, M., 210, 362, 365, 384
Farkas, A., 100, 296

Farkas, L., 100, 285
Farren, W., 53
Faure, F., 291
Fawcett, E. W., 110
Fay, R. C., 323
Ferguson, A., 114, 115, 347, 388
Ferguson, G. W., 376
Ferguson, J., 331
Fieser, L. F., 288
Finch, G. I., 107, 203
Findlay, A., 134, 135, 230, 330
Firth, E. M., 135
Fite, W. L., 310
Fixman, M., 315
Fleck, A., 52
Flood, H., 291
Flory, P. J., 299, 319, 355
Forland, T., 307
Fornander, S., 289
Forster, Th., 304, 355
Fowler, R. H., 75, 114, 119, 122, 123, 241
Fox, J. J., 126, 128, 137
Franck, E. U., 312, 315, 335, 342, 352
Franck, J., 69, 70, 100
Frank, C., 384
Frank, F. C., 290, 291, 311
Frankevich, E. L., 320
Frankland, P. F., 76
Franklin, R., 257, 339
Fraser, M. J., 300
Frazer, A. C., 291
Freese-Green, C., 70
Freeth, F. A., 5, 50, 56, 68, 69, 73, 81, 215
Freundlich, H., 34, 60, 94, 100, 104
Frey, H. M., 335, 342
Frey-Wyssling, A., 291
Frohlich, H., 286
Frumkin, A., 287

Fuoss, R. M., 294
Furukawa, K., 307

Gabor, D., 53
Gallaher, P. K., 323
Gallais, F., 197
Garner, W. E., 72, 76, 80, 120, 121, 193, 203, 226, 227, 229, 230, 231, 233, 234, 243, 268, 271, 286, 287, 288, 290, 291, 305, 330, 347, 387, 388, plate
Garrison, W. M., 295
Gaskell, P. N., 320
Gatty, O., 200
Gee, G., 132, 135, 234, 243, 317, 318, 319, 320, 331, 332, 333, 335, 338, 345, 370, 388, plate
George, P., 300
Gerischer, H., 352
Ghosh, J. C., 58
Gibbs, W. E., 78
Gibson, J., 379
Glasser, L., 320
Glasstone, S., 110
Glazebrook, R. T., 27, 28, 46, 51, 213, 268, 269, 344, 387, plate
Glazman, Yu., M., 314
Gledrye, F. O. W., 285
Glocker, G., 294
Gold, V., 311, 334, 340
Goldfinger, P., 376
Goldschmidt, V. M., 79
Gomer, R., 351
Goodeve, C. F., 80, 110, 111, 112, 143, 233, 266, 267, 268, 271, 287, 289, 294, 295, 330, 331, 332, 342, 344, 345, 348, 360, 387, 388, plate
Goodings, E. P., 320
Gordy, W., 299
Gorter, E. N., 72, 285
Grabowski, Z. R., 304, 317

Index of Names

Gray, P., 250, 260, 335, 336, 342, 354
Gray, T. J., 292
Greenwood, H. C., 48, 49, 50, 52
Greenwood, N. N., 317
Griffith, J. S., 354
Griffith, R. O., 80, 111, 330, 338
Gucker, F. T., 306
Guggenheim, E. A., 119, 123, 233, 241, 242, 274, 297, 325, 331, 332, 338, 341, 348, 376, 388
Guinier, M., 306
Gustavson, K. H., 298
Gutfreund, H., 299
Gutowsky, H. S., 308
Guye, P., 24
Gwinn, W. D., 299

Haanel, E., 30
Haber, F., 137
Hadfield, R. A., 25, 28, 35, 37, 44, 45, 46, 47, 48, 51, 55, 57, 59, 60, 73, 91, 106, 143, 156, 158, 160, 173, 174, 177, 207, 213, 268, 269, 344, 362, 387, plate
Haissinsky, M., 287, 295, 306
Haldane Gee, W. W., 41, 67
Haldane, J. B. S., 88
Hall, A. D., 61
Halpern, J., 317
Hamann, S. D., 301
Hampson, G. C., 103
Hansen, K. H., 376
Harbord, F. W., 26
Harcourt, A. G. V., 67, 68
Hardy, W. B., 83, 88, 103, 344
Harker, J. A., 27, 46, 48, 49, 56, 66
Harned, H. S., 75, 302, 348, 355, 388
Harteck, P., 100
Hartley, E. G. J., 24

Hartley, G. S., 255, 289, 331, 338
Hartley, H., 50, 53, 65, 67, 75, 78, 81, 92, 94, 101, 123, 126, 201, 226, 267, 268, 347, 388
Hassel, O., 103
Hasted, J. B., 286
Hatfield, W. H., 61, 65
Hatschek, E., 29, 34, 55, 57, 59, 73, 84, 106, 195, 196
Haworth, W. N., 95
Helfferich, F., 300
Hemmings, F. C., 48
Henderson, D., 315
Hengstenberg, J., 84
Henri, V., 80
Henry, D. C., 88, 271, 275, 298
Hering, H., 309
Herington, E. F. G., 135
Hermann, C., 79
Hermans, J. J., 287, 351
Herschbach, D. R., 308, 352
Herzberg, G., 80, 292, 293, 309, 328, 355
Herzfeld, K. F., 308
Herzog, R. O., 76, 84, 96
Hevesy, G., 75
Hey, D. H., 131
Heyrovsky, J., 287
Hickling, A., 287
Hildebrand, J. H., 113, 114, 267, 268, 297, 347, 348, 355, 388
Hill, A. V., 85
Hill, L., 362
Hill, T. M., 295
Hills, G. J., 279, 302, 335, 336, 342
Hinshelwood, C. N., 65, 67, 77, 78, 79, 101, 110, 119, 122, 123, 124, 130, 133, 137, 144, 192, 203, 226, 243, 264, 267, 285, 288, 294, 299, 306, 307, 308, 331, 332, 333, 334, 348, 366, 388, plate

Hirst, E. L., 95, 128
Hissink, D. J., 66
Hittorf, W., 22, 26
Hocart, R., 290
Hochstrasser, R. M., 323
Hoijtink, G. J., 305, (see Hoytink)
Hoogland, J. G., 287
Hopkins, F. G., 85
Hopkins, G., 344
Hopkinson, B., 27, 52, 53
Hornig, D. H., 292, 293, 299, 308, 351
Howe, H. M., 59
Hoytink, G. J., 335, 343, (see Hoijtink)
Huckel, E., 58, 75
Huggins, M. L., 319
Huggins, W., 21, 95
Hughes, E. D., 129, 130, 312
Hume, J., 97
Huntington, A. K., 11, 18, 19, 22, 29, 32, 48, 52, 55, 72, 148, 160
Hutton, R. S., 22, 25, 27, 41, 64, 66, 67

Ibers, J. A., 318
Ing, R. H., 133
Ingold, C. K., 87, 129, 130, 137, 242, 312, 328, 332, 338, 348, 366, 376, 383
Ingram, D. J. E., 305
Inokuchi, H., 320
Ivin, K. J., 335, 343

Jackson, H., 22, 52
Jackson, W., 286
James, B. R., 317
James, R. O., 321
Janik, J., 318, 324
Japp, F. R., 136
Jeans, J., 11
Jellinek, H. H. G., 135

Joffé, A. F., 76
Johns, C., 30, 35, 62, 63
Johnson, F. H., 304
Johnson, G. R., 320
Johnson, R. C., 80
Johnson, S., 115
Johnson, R. C., 80
Johnston, E. H., 62
Johnston, H. S., 298, 351
Jones, R. V., 53
Jordan Lloyd, D., 343, 344
Jordan, J., 317
Joseph, A. J., 52, 80
Jost, W. J., 121, 298, 308, 311, 318, 333, 334, 340, 351, 379
Juliard, A., 289

Kahlenberg, L., 23, 24
Kammerlingh Onnes, H., 63
Karasz, F. E., 319
Kaziz, C., 306
Keen, B. A., 65, 66
Keightley, E., 33
Kelvin, Lord, 10, 19, 21, 25, 116, 268, 269, 344, 387, plate
Kemball, C., 115, 279, 313, 333, 335, 341
Kendall, J., 113, 114, 115, 117, 348
Kennedy, C., 211
Kerker, M., 306
Kershaw, J. B. C., 15, 43
Kestner, N. R., 312
Ketelaar, J., 361
Kever, K. O., 376
Keys, A., 117
Kharasch, S., 131, 296
Kirke Rose, T., 55, 57
Kirkwood, J. G., 286, 297, 300
Kiselev, A. V., 321
Kistiakowsky, G. B., 294, 296, 298, 299
Kitchener, J. A., 297, 306, 314

Index of Names

Klemim, A., 307
Klemperer, W., 352
Knor, Z., 313
Knowles, J., 87
Knox, J., 48
Knox, R., 211
Koefoed, J., 297, 319
Koehler, J. S., 301
Kohler, F., 315
Kolthoff, I. M., 298
Komissarov, V. D., 317
Kondratjew, V., 80
Kornfeld, G., 100
Kornitzer, B. E., 180, 181, 182, 186, 216, 217, 218, 219, 226, 230, 232, 233, 271, 285, 295, 331, 333, 334, 367, 379
Kozyrev, D. M., 299
Kratky, O., 294
Kraus, C., 75
Krebs, H., 86, 320
Kreevoy, M. M., 312
Kresge, A. H., 312
Krogh, A., 117
Kronig, A., 327
Kropatshev, K. V. A., 303
Kruyt, H. R., 285
Kryszewski, M., 303
Kuenen, J. P., 63
Kuhn, W., 82, 83
Kuoyman, E. C., 294
Kupperman, A., 309

La Mer, V. K., 296, 314
Labes, M. M., 320
Laidler, K. J., 292, 300
Lamb, J., 319, 335, 343
Lambert, B., 34
Langevin, P., 102
Langmuir, I., 261, 270
Lapworth, A., 55, 129
Larin, I. K., 308

Lawrence, A. S. C., 242, 299
Le Chatelier, H., 35
Le Roy, D. J., 316
Leach, S., 292, 308
Leadbetter, A. J., 320
Lecomte, J., 80, 292
LeFevre, R. J. W., 286
Lehfeldt, R. A., 12, 21
Lennard-Jones, J. E., 16, 76, 80; 90, 91, 143, 203, 232, 238, 239, 241, 242, 268, 290, 291, 292, 294, 315, 327, 328, 331, 332, 339, 341, 378, 387, plate
Leslie, R. B., 320
Lessing, R., 63, 66, 105, 123, 133, 267, 275, 306, 331, 332, 333, 343, 345, 347, 352, 388
Lewis, G. N., 14, 64, 65, 348
Lewis, J., 317
Lewis, T. J., 321
Lewis, W. C. McC., 56, 57, 62, 76, 84, 106, 110, 111, 126, 270, 338
Lind, S. C., 190, 191, 192
Lindemann, F. A., 62, 81, 101, 270 (Lord Cherwell)
Lingane, J. J., 287
Linnett, J. W., 276, 304, 308, 320, 331, 333, 334, 335, 336, 341, 345, 388, plate
Lipscomb, W. N., 299
Liquori, A. M., 352
Lloyd George, D., 47
Lodge, O., 11, 19, 25, 56, 206, 212, 268, 269, 271, 345, 354, 387, plate
London, F., 113, 119, 297
Long, F. A., 312
Longuet-Higgins H. C., 244, 276, 277, 299, 303, 308, 312, 328, 332, 333, 339, 350, 355, 383
Lonsdale, K., 79

Lowry, T. M., 22, 23, 24, 28, 29, 43, 48, 52, 55, 56, 57, 60, 63, 64, 67, 77, 78, 79, 82, 83, 84, 85, 96, 99, 105, 106, 107, 108, 114, 165, 187, 213, 268, 271, 287, 328, 345, 387, plate
Luckhurst, G. R., 354
Lundegardh, H., 289
Lunt, R. W., 133, 134, 331, 345
Lyklema, J., 314, 321

Magat, M., 113, 229, 233, 285, 286, 287, 288, 292, 294, 306, 373
Maier, W., 303
Manecke, G., 300
Marconi, G., 213, 270
Marcus, R. A., 305, 317
Mark, H., 79, 84, 94, 95, 96, 109, 110, 125
Marlow, G. S. W., 73, 74, 75, 83, 84, 93, 103, 175, 179, 180, 181, 185, 186, 192, 193, 199, 200, 203, 207, 208, 213, 216, 217, 218, 219, 223, 226, 227, 228, 229, 230, 231, 232, 233, 234, 253, 262, 268, 271, 285, 331, 344, 346, 363, 375, plate
Marrack, J. M., 298
Marsden, R. J. B., 103
Marseille, A., 324
Marta, F., 317
Martin, A. E., 128
Mason, R. E., 317
Mathews, D. B., 312
Mathieu, M., 285
Matijevic, E., 314
Maugin, Ch., 79, 213, 270
Maxted, E. B., 49
Mayo, F. R., 288, 351
Mazurin, O. V., 320
McBain, J. W., 63, 104, 287, 347, 348, 388

McBain, L., 104
McCall, D., 318, 370
McCance, A., 289, 290, 347
McConell, H. M., 352
McGlashan, M. L., 315, 334, 341
McInnes, D. A., 75
McKeown, A., 81, 110, 111, 200
McLellan, J. C. M., 80
Mecke, R., 80, 92
Mees, C. K., 64
Meier, G., 324
Melville, H. W., 109, 125, 133, 134, 135, 136, 225, 242, 247, 248, 271, 285, 288, 291, 294, 298, 303, 305, 306, 331, 332, 333, 334, 335, 345, 361, 364, 366, 368, 369, 382, 387, 388, plate
Merton, T. R., 52, 53
Messel, R., 48
Meyer, K. H., 84, 95, 109, 118
Michels, A., 376
Miles, F. D., 123, 124
Milner, S. R., 58
Mitchell, J. W., 305
Mitrofanov, K. P., 323
Moelwyn-Hughes, E. A., 130, 252, 332, 333, 339, 345
Moltkenhausen, K., 213, 271
Mond, A., 215
Mond, E. S., 90, 91, 123, 175, 214, 215, 270
Mond, L., 11, 19, 90, 91, 269
Mond, R. L., 27, 30, 46, 54, 78, 90, 91, 93, 96, 106, 123, 175, 180, 193, 194, 206, 208, 211, 212, 213, 268, 354, 387, plate
Monypenny, J. H. G., 76, 77
Moore, H., 54, 57, 64
Moore, T. S., 365
Moritz, E. R., 29
Morrison, J. A., 292, 305, 318

Index of Names

Morrison, M. W., 21, 48
Morton, R. K., 300
Moseley, R., 140, 141
Mott, N. F., 121, 203
Moullin, J. B., 286
Moulton, Lord, 50
Muller, A., 79
Mulliken, R. S., 80, 299, 328
Murrell, J. N., 316, 335, 336, 343
Musgrave, F. F., 225, 247, 248, 264, 334, 335, 343, 345, 357
Mysels, K. J., 314

Nagasawa, M., 300
Nagel, D. H., 101, 126
Nancollas, G. S., 4
Nathan, F., 91
Nelson, G. F., 320
Nernst, H., 55, 65, 81, 91
Neurath, H., 300
Nicholson, J. W., 53
Nicolet, M., 310
Nielsen, H. H., 292, 293
Nikitin, E. E., 308
Norrish, R. G. W., 96, 125, 245, 258, 259, 261, 263, 264, 276, 277, 294, 298, 299, 300, 301, 310, 311, 326, 332, 334, 341, 345, 348, 361, 387, 388, plate
North, A. M., 316, 354
Noyes, W. A. Jr., 69, 294
Nurnberg, H. W., 312
Nuttall, T. Y., 27
Nyholm, R. H., 317

Oden, S., 61
Ogata, L., 317
Ogston, A. G., 299, 300
Olby, R., 86, 88, 95
Olin, J., 304
Olofsson, B., 298
Onsager, L., 58, 75, 302

Oosterhoff, L. J., 305, 323
Orgel, L. E., 304
Orton, K. J. P., 129
Ostwald, F. W., 6, 34, 60, 77, 87, 104, 136, 315, 365
Ottewill, R. H., 312, 314
Overbeek, J. Th. G., 287, 299, 314, 351

Pankhurst, K. G., 297
Parfitt, G. D., 314, 321
Parsons, C. L., 190, 191, 192, 217, 218
Partington, J. R., 49, 55, 57, 83, 106, 110, 164, 190, 191, 214, 331
Pasternak, M., 323
Paterson, C. C., 56, 126
Patterson, D., 319
Pauli, W., 34, 60, 104
Pauling, L., 84, 86, 103, 295, 296, 304, 328, 348
Pedersen, K. O., 295
Penney, W. G., 103, 203, 331, 339
Penrose, R. P., 286
Peretti, E. A., 289
Perkin, M. F., 28, 30, 32, 33, 46, 48, 147, 148
Perkin, W. H., 25, 173, 268, 269, 345, 365, 387, plate
Perrin, J., 62, 270
Perrin, M. W., 125
Pershan, P. S., 323
Perutz, M. F., 287
Peterlin, A., 351
Pethica, B, A., 335, 336, 343, 345, 364, 369
Philip, J. C., 90, 91, 92, 101, 365
Picard, H. F. K., 26
Pigeon, L. M., 289
Pings, C. J., 315
Pirie, N. W., 88
Pitzer, K. S., 294

Plyler, E. K., 292, 293
Pohl, H. A., 121, 320
Polanyi, J. C., 134, 316, 308, 354
Polanyi, M., 76, 77, 79, 84, 94, 99, 100, 110, 111, 118, 119, 120, 130, 134, 203, 227, 232, 316, 326, 330, 339, 348
Pollack, M., 320, 339
Polonovski, J., 300
Ponec, V., 313
Pople, J. A., 277, 308, 323, 327, 333, 341, 354
Porter, A. W., 29, 46, 54, 56, 57, 63, 71, 73, 83, 84, 105, 106, 161, 162, 163, 164, 193, 212, 268, 279, 345, 366, 387, plate
Porter, G., 293, 298, 304, 306, 310, 316, 333, 334, 335, 341, 348, 368, 369
Poshkus, D. P., 312
Pourbaix, M. J. N., 289
Pousfield, W. R., 64
Powell, H. M., 204
Powles, J. G., 286
Prebble, W. C., 4, 5, 6
Preston, R. D., 260, 289
Price, W. C., 292, 293
Prigogine, I., 297
Pring, J. N., 62, 73
Prokhorov, A. M., 299, 306
Prokhorov, P. S., 299
Prue, J. E., 302, 335, 336, 343
Pryce, M., 304
Ptitsyn, O. B., 319
Pullman, A., 288, 292
Pullman, B., 292
Purnell, J. H., 335, 336, 343
Putzeys, P., 287

Quirk, J. P., 321

Rabinowitch, E., 120, 304

Rajan, V. S. V., 324
Raksani, K., 324
Raman, C. V., 80
Ramsay, D. A., 309
Ramsay, W., 5, 7, 11, 19, 21, 29, 32, 68, 78, 80, 81, 97, 136, 269
Randall, J. T., 242, 245, 257, 260, 292, 293, 331, 332, 339
Randall, M., 75
Randic, M., 323
Rankine, A. O., 56, 57, 123
Rao, C. N. R., 354
Rawlins, F. I. G., 80, 93, 97, 107, 137, 143, 232, 268, 271, 331, 345, 376
Rayleigh, Lord, 11, 19, 26, 80, 268, 269
Rayner, E. H., 29, 46, 47, 55, 69
Reader, W. J., 50, 136, 215
Remington, W. R., 298
Remy, R., 75
Rice, F. O., 100, 124
Rice, O. K., 296, 312
Rice, S. A., 303, 315, 320, 352, 354
Richards, J. W., 44
Richards, R. E., 299, 308, 333, 334, 341
Richardson, J. C., 211
Richardson, O. W., 80
Richmond, G., 211
Rideal, E. K., 49, 51, 62, 72, 83, 84, 94, 97, 105, 110, 116, 122, 123, 124, 125, 128, 131, 132, 134, 135, 136, 137, 143, 144, 197, 199, 200, 201, 224, 227, 233, 242, 266, 268, 271, 274, 285, 289, 290, 291, 294, 326, 327, 330, 331, 332, 345, 347, 366, 378, 387, 388, plate
Rideal, S., 33, 110
Rinne, F., 98

Index of Names

Rintoul, W., 90, 105, 107, 108, 112, 114, 124, 173, 268, 344, 387, plate
Ritson, R. M., 286
Robb, J. C., 316, 334, 341
Roberts, D., 309
Robertson, J. M., 128, 204
Robertson, R., 56, 63, 66, 67, 74, 77, 80, 91, 105, 106, 112, 120, 127, 151, 183, 193, 211, 216, 268, 331, 344, 387, plate
Robinson, R., 65, 87, 129, 130, 195
Robinson, R. A., 302
Robson, S., 289
Rodenbush, W. H., 296
Romo, L., 314
Rosen, B., 309
Rosenhain, W., 55, 57
Roughton, F. J. W., 258, 260, 298, 299, 300
Rowlinson, J. S., 279, 312, 315, 319, 334, 335, 336, 341, 354, 370, 375
Ruck-Keene, J., 362, 378, 381
Rushbrooke, G. S., 315, 355
Russell, E. J., 61, 65
Russell, J., 81
Rutgers, A. J., 351
Rutherford, E., 100, 108, 383

Sabatier, P., 37
Sachs, G., 76
Sack, H., 103
Sadron, C., 352
Salamon, A. G., 30, 54
Samuel, Lord, 267
Scatchard, G., 75, 300
Schaefer, C., 80
Schenck, R., 99
Schiebold, E., 79
Schiff, H. I., 310
Schnepp, O., 318
Schofield, R. K., 66, 72, 271, 289, 298
Schulman, J. H., 132, 242, 254, 255, 258, 289, 291, 294, 295, 297, 300, 331, 332, 339, 347
Schuster, A., 25, 27, 60
Schwab, G. M., 291, 314
Scott, R. L., 297, 319
Scott-Hansen, A., 37
Seitz, F., 290, 301
Seligman, R., 28, 29
Semenov, N., 77
Senter, G., 27, 28, 46, 56, 106, 160
Sgamellotti, A., 323
Sheppard, N. S., 335, 336, 343
Sheppard, S. E., 64
Sheraga, H. A., 312
Shimaji, M., 320
Shimanouchi, T., 319
Shimoji, M., 307
Shoolery, J. N., 309
Shoppee, C. W., 130
Shutt, W. J., 81, 127, 133
Sidgwick, N. V., 65, 68, 73, 75, 76, 94, 96, 100, 103, 106, 143, 166, 192, 268, 325, 345, 347, 387, 388, plate
Siebrand, W., 320
Siemens, A., 10, 11, 19, 27, 29, 60, 147, 269
Siemens, W., 11, 12
Simpson, D., 226
Simpson, H. V., 3, 7
Skinner, H. A., 276, 335, 343
Skpinel, V. S., 323
Slade, R. E., 73, 123, 144, 195, 214, 215, 216, 223, 227, 229, 233, 237, 239, 331, 345
Smit, W. M., 291
Smith, A., 115
Smyth, C. P., 103, 286

Snoek, J. L., 103
Snow, C. P., 80
Soddy, F. S., 136
Soloman, M. M., 28, 29, 41
Solvay, E., 26
Speakman, J. B., 297
Spence, R., 333, 341
Spencer, J. F., 80
Spiers, F. S., 3, 6, 7, 8, 9, 12, 13, 18, 19, 28, 32, 39, 40, 46, 48, 66, 73, 74, 77, 147, 148, 150, 153, 156, 178, 187, 188, 206, 207, 216, 217, 268, 269, 375, plate
Squire, W. S., 12
Stassano, E., 30
Staudinger, H., 86, 94, 95, 96, 109, 110, 125, 131, 135
Staverman, M. J., 300
Steacie, E. W. R., 124, 288, 296, 304, 305, 333, 341, 344, 361, 387, plate
Stecki, J., 315
Stein, B. R., 306
Steinhart, O. J., 12, 18, 19, 26, 147
Stephens, P. J., 323
Stewart, G. W., 98, 113
Stockbarger, D. C., 290
Stockmayer, W. H., 351
Stokes, R. H., 351
Stranski, I. N., 290, 291
Strathdee, R. B., 136
Style, D. W. G., 93, 120, 123, 129, 242, 264, 271, 286, 309, 333, 341
Sugden, S., 120, 121, 133, 143, 331
Sugden, T. M., 299, 309, 333, 335, 341
Sulman, H. L., 26
Sundheim, B, R., 307
Sutherland, G. B. B. M., 103, 128, 137, 203, 260, 286, 292, 293, 331, 339

Sutton, J., 305
Sutton, L. E., 120, 271, 317, 328, 333, 342, 367
Svedberg, T., 59, 86, 88
Swan, J. W., 4, 5, 7, 8, 9, 10, 14, 18, 19, 20, 21, 24, 25, 27, 30, 35, 173, 260, 267, 269, 344, 387, plate
Swinburne, J., 3, 4, 6, 7, 8, 9, 10, 11, 15, 19, 22, 26, 27, 28, 39, 45, 173, 178, 266, 268, 269, 344, 387, plate
Sykes, P. H., 27
Symons, M. C. R., 334, 342
Szabo, Z. G., 321
Szent-Gyorgyi, A., 294, 304

Tait, W. H., 19
Tatlock, R. R., 90
Taylor, A. M., 80, 93, 137
Taylor, H. S., 49, 51, 94, 114, 116, 133, 266, 267, 268, 270, 271, 292, 296, 298, 332, 339, 347, 348, 350, 355, 361, 387, 388, plate
Taylor, J., 368
Taylor, T. W. J., 62
Temperley, H. N. V., 315
Teorell, E. T., 300, 355
Tezak, B., 299, 314
Theorell, H., 285, 300
Thomas, J. M., 320
Thompson, F. C., 177
Thompson, G. P., 107
Thompson, H. W., 102, 136, 137, 203, 286, 292, 293, 331, 342
Thompson, J. J., 14, 19, 22, 25, 26, 64, 70
Thorell, B., 293
Thorpe, T. E., 20, 23, 151
Timmermans, I., 297
Tiselius, H., 88, 291, 295

Index of Names 413

Tizard, H. T., 53, 65, 72, 73, 81, 83, 268, 347, 388
Tobolsky, A. V., 288
Todd, A., 264
Todd, Lord., 129
Tompkins, F. C., 223, 230, 233, 234, 235, 262, 268, 271, 277, 278, 279, 295, 301, 331, 333, 334, 335, 336, 363, 367, 370, 373, 377, 378, 379, 381, 382, 383, plate
Tooth, E., 180, 181
Topley, B., 264, 333, 342, 345
Townend, D. T. A., 63
Townes, C. H., 299
Toy, F. C., 68
Traill, D., 247, 301
Trapnell, B. W. M., 291
Travers, M. W., 78, 105, 114, 192, 209, 212, 239, 240, 268, 325, 331, 345, 347, 387, 388, plate
Tronstad, L., 200
Tryhorn, F. G., 78, 79
Tsutsumi, S., 317
Tuckett, R. F., 132, 135
Turkevich, J., 292, 294, 314
Turnbull, D., 320
Twigg, G. H., 248, 332, 333, 339, 345, 357, 376

Ubbelohde, A. R. J. P., 201, 224, 243, 247, 248, 264, 277, 301, 306, 307, 308, 309, 310, 311, 316, 332, 333, 334, 335, 339, 364, 369, 370, 388, plate
Ugo, R., 317
Ulich, H., 75
Urbain, G., 324
Uytterhoeven, J. B., 321

van Arkel, A. E., 103, 285, 286

van den Honert, J. H., 289
van der Waals, J. H., 309, 323, 324, 335, 343, 352, 383
van 't Hoff, J. H., 6, 365
van Reijen, L. L., 314
Van Vleck, J. H., 304
Vernon, W. H. J., 64, 75
Verwey, E. J. W., 285
Vetter, K., 324
Vichnievsky, H., 285
Vineyard, G. H., 306
Vleck, A. A., 305
Vodar, B., 297, 301
Voevodsky, V. V., 307, 351
Vogel, J. L. F., 26
von Hippel, A., 286
Vouk, V. B., 306

Wagner, C., 121
Wagner, H. G., 379
Wakeley, G., 181, 218, 271
Walden, P., 24
Walker, J., 27, 51, 115, 116
Walker, O. J., 127
Walker, S., 320
Walsh, A. D., 309, 334, 343
Wansborough-Jones, O. H., 136, 137, 223, 239, 243, 244, 247, 263, 264, 331, 333, 340, 345, 348, 360, 388
Wappling, R., 323
Wassermann, A., 120
Waters, W. A., 131, 243, 288
Watson, H. B., 129
Waugh, D. F., 303
Waugh, J. S., 309
Weaver, W., 84
Weissberger, A., 100
Weissenberg, K., 79, 84
Werner, A., 37, 86
Westgren, A. F., 79
Whalley, E., 320

Whiffen, D. H., 278, 279, 286, 318, 334, 342, 368, 369, 375
White, H. J., 298
White, J. W., 318, 354
Whytlaw-Gray, R., 96, 97, 112, 123
Widom, B., 308
Wigner, E., 120, 134
Wild, W., 310
Wilkins, M., 257, 339
Williams, E. C., 70, 78
Williams, J. W., 300
Williams, R. J. P., 320
Willis, M. R., 316, 320
Willows, R. S., 70
Wilse Robinson, G., 352
Wilsmore, N. T. M., 26, 28, 39, 42, 43, 214

Wilson, E. B., 21, 292, 293
Wolf, H., 352
Wood, R. W., 80
Wyart, J., 290
Wynne Jones, W. F. K., 4, 195, 243, 287, 305, 311, 332, 333, 334, 340

Yeddanapalli, L. M., 376
Young, D. A., 336, 373
Young, J. Z., 117

Zachariasen, W., 79
Zarzycki, J., 307, 320
Zecchina, A., 321
Zeil, W., 318
Zerfors, S., 290